EUROPEAN WOODS AND FORESTS
Studies in Cultural History

European Woods and Forests

Studies in Cultural History

Edited by

Charles Watkins

Department of Geography
University of Nottingham
UK

CAB INTERNATIONAL

CAB INTERNATIONAL
Wallingford
Oxon OX10 8DE
UK

CAB INTERNATIONAL
198 Madison Avenue
New York, NY 10016-4314
USA

Tel: +44 (0)1491 832111
Fax: +44 (0)1491 833508
E-mail: cabi@cabi.org

Tel: +1 212 726 6490
Fax: +1 212 686 7993
E-mail: cabi-nao@cabi.org

© CAB INTERNATIONAL 1998. All rights reserved. No part of this publication may be reproduced in any form or by any means, electronically, mechanically, by photocopying, recording or otherwise, without the prior permission of the copyright owners.

A catalogue record for this book is available from the British Library, London, UK.

Library of Congress Cataloging-in-Publication Data
European woods and forests: studies in cultural history/edited by C. Watkins.
    p.   cm.
   Selected papers presented at an international conference held at the University of Nottingham in Sept. 1996.
    Includes bibliographical references (p.   ) and index.
    ISBN 0–85199–257–9 (alk. paper)
    1. Forests and forestry--Europe--History--Congresses. 2. Forest management--Europe--History--Congresses. 3. Forest ecology--Europe--History--Congresses. I. Watkins, C.

SD177.E87   1998
333.75'094--dc21

98–14969
CIP

ISBN 0 85199 257 9

Typeset by Solidus (Bristol) Limited
Printed and bound in the UK at the University Press, Cambridge

CONTENTS

PREFACE vii

INTRODUCTION
Themes in the history of European woods and forests 1
C. Watkins

CHAPTER 1
The word 'Forst/forest' as an indicator of fiscal property and possible consequences for the history of Western European forests 11
R. Kiess

CHAPTER 2
Medieval forests and parks in southern and central England 19
D. Hooke

CHAPTER 3
Royal Forests in England and their income in the budget of the feudal monarchy from the mid twelfth to the early thirteenth centuries 33
Y.J. Serovayskaya

CHAPTER 4
English cathedrals as sources of forest and woodland history 39
G. Simpson

CHAPTER 5
The rise, decline and extinction of spring wood management in south-west Yorkshire 55
M. Jones

CHAPTER 6
The continuous conflict between sustainable management regulations and over-utilization of woodland caused by local demands in Austria from the thirteenth century onwards 73
E. Johann

CHAPTER 7
An ecological revolution? The 'Schlagwaldwirtschaft' in western Germany in the eighteenth and nineteenth centuries 83
C. Ernst

Chapter 8
'A solemn and gloomy umbrage': changing interpretations of the ancient oaks of Sherwood Forest — 93
C. Watkins

Chapter 9
Landed estates, the 'spirit of planting' and woodland management in later Georgian Britain: a case study from the Dukeries, Nottinghamshire — 115
S. Seymour

Chapter 10
Need versus greed? Attitudes to woodland management on a central Scottish Highland estate, 1630–1740 — 135
F. Watson

Chapter 11
Woodland management and timber supply for ship masts in eighteenth century western Liguria (Italy) — 157
G. Paola and F. Ciciliot

Chapter 12
Afforestation policy of the Zionist Movement in Palestine 1895–1948 — 165
N. Liphschitz and G. Biger

Chapter 13
The expansion of the forest and the defence of nature: the work of forest engineers in Spain 1900–1936. — 181
E. Rico Boquete

Chapter 14
The promotion of participation in planning for soil and water conservation through reforestation: a case study of Guadalajara (Spain) — 191
J.D. García Pérez

Chapter 15
Making the invisible visible: ancient woodlands, British forest policy and the social construction of reality — 215
J. Tsouvalis-Gerber

References — 231

Index — 243

Preface

Most of the chapters in the book are based on presentations given at the International Conference on Advances in Forest and Woodland History which was held at the University of Nottingham in September 1996. I would like to thank the Forest Ecology Group of the British Ecological Society and the International Union of Forest Research Organisations (IUFRO) for supporting and publicizing this meeting. I would also like to thank the Department of Geography, University of Nottingham, for providing secretarial and organizational assistance, especially Pat Bridges and Ben Cowell; the staff of Hugh Stewart Hall at the University of Nottingham where the conference was based; Nottingham City Council, Graham Whalley and the staff of Wollaton Hall for providing the venue for a reception; George Peterken for presenting the post-conference tour to the Wye Valley, and Tim Hardwick of CAB International Publishing Division and his staff for their assistance with the publication of this book. A second book drawing on papers from the conference, *The Ecological History of European Forests*, edited by Keith Kirby and Charles Watkins, is also published by CABI PUBLISHING.

Charles Watkins
Nottingham, March 1998

INTRODUCTION
Themes in the history of European woods and forests

Charles Watkins
Department of Geography, University of Nottingham, Nottingham NG7 2RD, UK

Introduction

The history of European woods and forests has long remained somewhat on the edge of academic study; it has never been central to any particular discipline or sub-discipline. Several books have been written which purport to be histories of forestry, but these have usually been based on narrow definitions of forestry. Some have considered, for example, the history of medieval forests, others have concentrated on the rise of scientific forestry. Few have considered the role of woods and forests in a broader social context. Geographers and historians have tended to view woods and forests, rather heroically, as areas from which agricultural land can be won, rather than as land uses which might deserve study in themselves. Indeed, it can be argued that the complexity of forest history has been disguised by its treatment as a relatively simple natural category. Less interest has been shown in the way trees and woodland were managed, interpreted and valued than in the way in which they have been extirpated. For many the clearance of woodland has been merely noticed as a symptom of increasing agricultural production and population: the broader cultural meanings of woods and forests have been largely ignored. Increasingly, however, researchers and scholars from a variety of disciplines are studying the interactions and connections between humans and trees and woodland from a broader perspective. They are interested in how people have made use of different types of trees and woodland and how this use relates to economic and social power structures (see, for example, Rackham, 1980; Wickham, 1994; Williams, 1989).

Perhaps one reason for a lack of historical research into woods and forests is an inherent ambiguity in their definition. What is meant by these terms varies through time and from country to country. To take England as an example, it is well known that the medieval Royal Forests were areas where the monarchy retained special hunting rights (Young, 1979; Rackham, 1980, 1989; James, 1981). Some forests, such as Exmoor, had very few trees; most were made up of tracts of land which could contain villages, heaths, arable land, pasture and woodland. A survey of Sherwood Forest made by Richard Bankes in 1609 shows that by far the greater part of the forest was settled agricultural land and heath and that it included the whole town of Nottingham (Mastoris and Groves, 1997). There was no direct connection between the idea of forest and the concept of woodland: medieval forests were administrative units more akin to a modern national park than a plantation of trees. With the decline in Crown interest, especially from the eighteenth century onwards, the term forest became increasingly associated with those wooded areas, such as the New Forest and the Forest of Dean, which survived as Royal Forests. It was not however until the establishment of professional forestry in the nineteenth century, and the intermixture of traditional estate woodland management with ideas of scientific forest management introduced from the Continent, that the terms 'forestry' and 'forester' began to be the normal terms used to describe woodland management and managers. It was the state Forestry Commission, established in 1919, which introduced the term forest to describe their administrative units and then went on to use the term National Forest Park for recreational areas. Their use of the term forest was ambiguous: it linked the coniferous afforestation of massive areas of both upland and lowland with ideas of ancient deciduous woods associated with the remnant Royal Forests. As such, the ambiguity of the term may have assisted the Forestry Commission to beguile those concerned with the revolutionary effect of large-scale plantations on the landscape (Revill and Watkins, 1996).

If the term forest needs care in interpretation, we are on no firmer ground with the terms wood and woodland. Recent research in historical ecology is increasingly showing that the boundaries between woodland, wood pasture, pasture with trees and arable land are difficult to define with precision (Kirby and Watkins, 1998). In general terms, distinct boundaries between dense woodland and more open woodland have become more pronounced through time. These changes are related to the more intensive management of land and the greater control over grazing, and are very closely tied to changes in the pattern of land ownership and control. Even today, however, there remain many areas of semi-natural vegetation where it is difficult to map distinct boundaries between woodland and non-woodland.

These definitional ambiguities emphasize the great care needed in interpreting the history of woods and forests. The changing uses of the terms can obscure; the changing meanings themselves can, however, be used to uncover changing attitudes and practices. One way around the difficulty is to draw on a wide range of sources and methods. Historical ecology has brought great benefits in developing an understanding of woodland history through the combination of ecological survey and historical analysis (Rackham, 1980; Peterken, 1981, 1996; Russell, 1997; Kirby and Watkins, 1998). Ecological surveys are not, however, used in the studies detailed in this book. Rather the authors have drawn on a wide range of documents, images, maps and oral histories to explore the ambiguities of woodland history in a series of case studies.

Approaches

The chapters in this book can be grouped into four broad sections which draw on different types of evidence and have differing approaches to woodland history. The first group, Chapters 1–3, deals with medieval woodland. Rudolf Kiess, through a careful analysis of place names, wood names and field names in south-west Germany, recovers a number of small forests which may have been attached to estates and castles and tries to establish their origin. Della Hooke combines information on boundaries from Anglo-Saxon charters with place-name evidence in southern and central England. She then compares this evidence with later forest boundaries and with knowledge about geology and soils to reconstruct the notion of Anglo-Saxon forests. By way of contrast, Yu Serovayskaya analyses the economic value of the English Royal Forests to the feudal monarchy using the Pipe Rolls.

The second group of chapters illustrates the benefits of using a wide range of sources to explore particular issues over a long time period. Gavin Simpson shows how English cathedrals can themselves be used as evidence. He demonstrates the value of applying dendrochronology to cathedral timbers, especially when this evidence can be combined with cathedral accounts and maps to reconstruct timber supply sources. Melvyn Jones uses the Domesday Book, estate records and accounts and maps to provide a long view of the development of the woodland pattern and changes in woodland use in south-west Yorkshire. Elisabeth Johann uses a wide range of documents over a similar time span to show the interplay between different woodland interests, and the development of ideas of sustainable forestry in Carinthia, Austria, from the thirteenth century onwards.

The third group of chapters makes extensive use of estate and forest records of the seventeenth and eighteenth centuries. Christoph Ernst studies the Schlagwaldwirtschaft system of growing and felling trees section by section in the forests of Hunsrueck and Eifel in the eighteenth and nineteenth centuries. I use official reports, estate records and literary descriptions to uncover changing ways of interpreting the ancient oaks of Sherwood Forest. Susanne Seymour uses a wide range of estate documents for three neighbouring estates in north Nottinghamshire to examine the enthusiasm for making plantations in the eighteenth century. Scottish Barony Court records are used by Fiona Watson to explore attitudes to woodland management on the Menzies' estates in central Scotland 1630–1740. Finally in this group, Gaudenzio Paola and Furio Ciciliot's examination of early eighteenth century documents on the supply of ship masts allows them to consider the type and management of woods in western Liguria, Italy.

A different range of source material comes into play in the final group of chapters. This deals largely with the twentieth century and there is an increasing dependence on official published documentation and reports. Nili Liphschitz and Gideon Biger make extensive use of official reports, statistics and scientific papers in addition to archive documents and correspondence in their examination of the afforestation policy of the Zionist movement in Palestine 1895–1948. A similar range of documents is used by Eduardo Rico Boquete in his study of the work of forest engineers in Spain 1900–1936. The final two chapters also draw on official documents and scientific papers, but they are able, in addition, to make use of oral history and interview material because they are dealing with more recent events. José García Pérez has carried out extensive geographical surveys and combines this with an examination of written history and the collection of oral history through anthropological methods of fieldwork in his study of soil and water conservation measures through reafforestation in Guadalajara (Spain). Judith Gerber combines textual analysis of official documentation with interviews to explore the definition of ancient woodland, the social construction of nature and recent British forest policy.

The range of approaches used by the authors is wide, and the stories they tell have widely differing contexts: the definition and use of forests in early medieval southern Germany and Anglo-Saxon and Norman England; long-term changes in woodland uses in England and Austria; changing ideas of woodland management, aesthetics, control and expansion in eighteenth century Germany, England, Scotland and Italy; the ideas behind twentieth century afforestation in Israel and Spain; and the development of the notion of ancient woodland in England. Together they enlarge our understanding of woodland and environmental history and several broad themes emerge.

Power

Princes, bishops, kings, aristocrats and state authorities of all kinds have a particular interest in controlling their forests. Rudolf Kiess argues that at least from the seventh century onwards in Germany, forests were defined to impose order over large areas. Strict control over hunting and other rights was a means of policing border zones. Indeed there is increasing evidence that many of the special laws linked with forests and woodland were associated with their status as frontier zones. This is emphasized in Della Hooke's chapter in which she outlines the location of forests within broad geographical regions. She reclaims hunting as a major use of land and status enhancer in England in the pre-Norman period. The spectacle of hunting, its association with military prowess and horsemanship, and the control of hunting grounds were key elements in the maintenance of princely, royal and aristocratic power and prestige. Chris Wickham (1994, p. 160) has argued that in the early medieval period the 'symbolism of hunting ever more clearly came to match that of royal charisma' and that the dispersal of forests and hunting rights to churches and aristocrats matched 'the dispersal of that charisma and of royal power in general to the aristocracy'. Yu Serovayskaya and Gavin Simpson's work show the crucial role that Royal Forests played in enhancing the power and status of the monarchy through income generation and as sources of timber for major ecclesiastical buildings.

The form of woodland management has been strongly influenced by the balance between communal rights over land and private ownership. Communal rights to grazing and pasture began to be seen by many landowners as restricting their woodland management. Christoph Ernst argues that it is only with the removal of conflicts with game and agricultural use that modern forestry techniques such as the Schlagwaldwirtschaft could be established. When individual landowners had gained full control over grazing, other agricultural practices and hunting, they had the confidence to develop new forms of woodland management. The obverse of new forms of woodland management was the removal and extinction of various common rights.

It was in the long-term interest of established landowners to maintain control over the management of timber on their estates. Frequently the management of woodland and trees became the domain of the landowner as opposed to the farmer, although this varies from place to place depending on the exact form of landownership and the nature of rights over the use of trees. In England there is a legal distinction between 'timber' trees, which were usually owned and managed by the landowner, and other trees which were managed by their tenants, and with the rise of the large landed estate from the seventeenth century, tree and woodland management became increasingly disassociated from other agricultural practices. Even hedgerow trees on farms were maintained and controlled by the landlord (Daniels and Watkins, 1991). The wholesale domination of woodlands by the landed interest was not, however, established without frequent social protest and theft of wood (Mooser, 1986; Neeson, 1993).

Fiona Watson demonstrates the links between grazing and woodland management in seventeenth century Scotland and notes that 'for those who worked the land, the responsibilities of farming were more likely to override any concern for woodlands'. Other chapters show how aristocrats and other private landowners have used woodlands to display and confirm their power over land. I show, for example, how landowners in eighteenth century Nottinghamshire were able to appropriate the remnants of the Royal Forest of Sherwood, while Susanne Seymour details how private landowners' management of their woods and new plantations asserted their dominance of the local landscape and, moreover, how woodland products were used to help enclose their farmland. Nili Liphschitz and Gideon Biger show how in early twentieth century Palestine plantations were made with the clear intent to claim land and stop traditional grazing rights. That forestry programmes continue to both demonstrate and symbolize power relations is emphasized by recent work on forest mapping and sustainable forestry in Indonesia and the Dominican Republic (Peluso, 1995; Rocheleau and Ross, 1995).

Industrialization

Greater emphasis is now being placed on the importance of woodlands and forests as sources of raw materials for buildings, agriculture, transport and industry. Research is showing the crucial importance of the charcoal industries through history (Metaille, 1992). Chris Wickham (1994, p. 185), while emphasizing regional variation in the uses of woodland in early medieval Europe, argues that 'wood was the only material that was used, or (in most of Western Europe) even available for heating, and thus no one could do without it altogether; it was the material *par excellence* for tools and building'. Several specific links between woodland management and industrialization are brought out in this book. Melvyn Jones points to the direct link between rising profits from coppicing for charcoal manufacture and the decline in woodland grazing on the 7th Earl of Shrewsbury's estate at Wentworth in Yorkshire. By 1723 the annual profits from the coppice were seen as sustainable and relatively risk free. Charcoal production was of vital importance in the development of the Industrial Revolution (Hammersley, 1973).

Christoph Ernst, in his documentation of the rise of the system of controlled felling, section by section, in Germany known as Schlagwaldwirtschaft, brings out the move away from an agricultural forest of sustenance to an industrial and productive forest and relates this to the perceived scarcity of woodland. Gaudenzio Paola and Furio Ciciliot show how the value of timber for ship masts resulted in the careful management of woodlands in western Liguria for this specific market. Industrialization led, of course, to many new markets for timber and wood products. Ted Collins (1992) has shown, for example, how many new markets for coppice products developed in the late eighteenth and nineteenth centuries. The stage is set for further analyses of regional and national variations in the links between woodland management and industrialization and agricultural change.

Knowledges and Professionalization

An important theme brought out by several chapters in this book is the way in which traditional knowledges about woodland management have become replaced, especially from the eighteenth century onwards, with the ideas of scientific forestry. This transformation has been associated with the development of the profession of forestry and forestry education. It is often quite difficult to discover the nature of many of the traditional practices of woodland management, such as coppicing, pollarding, woodland grazing, the making of temporary arable fields and the use of fire. This is partly because their very prevalence and normality meant that authors felt they did not need to comment on them. For example, the English agricultural improver and land agent Nathaniel Kent noted in 1774, when discussing coppicing on the Foxley estate in Herefordshire, that 'the management of it is so well known that it is needless for me to give any particular direction about it' (Daniels and Watkins, 1991, p. 154). Moreover, when such practices were described, they tend to be decried as old-fashioned, damaging, uneconomic: most authors and commentators wished to see these practices stopped and replaced with their modern approaches and methods. Subsequently these traditional modes of management became rarer and overshadowed by modern forestry techniques. Many became forgotten and it has taken the assiduous scholarship of woodland historians such as Oliver Rackham (1980) and Diego Moreno (1990) to uncover them in recent years.

While the slow decline of traditional woodland management techniques went largely unremarked by contemporary commentators, the rise of scientific forestry is well documented. Foresters both practical and academic have been prolix in forging the new discipline and approach to land management. James (1996, p. 44), in his review of the forestry literature of Germany, France and the UK, concludes that 'Early forestry in Germany engendered the scientific and intellectual basis for advancements and knowledge in many other countries'. The development and spread of scientific forestry is a key element of several of the chapters in this book. Christoph Ernst emphasizes that the control of traditional agricultural and hunting practices was a prerequisite for the development of the new, sustainable, forestry which concentrated on timber production. The rise of scientific forestry went hand in hand with the removal of traditional common rights in many countries. In Sherwood Forest, the rise in plantation forestry was predicated by the collapse of traditional common grazing rights in conjunction with the increasing domination of private landed estates.

Forest science developed apace in the nineteenth and twentieth centuries. The chapters by Eduardo Rico Boquete and José García-Pérez explore the promulgation and spread of modern forestry in Spain. They establish the strong influence of German and American forest science on the development of afforestation in that country and show the inappropriateness and counter-productive effects of such 'technological fixes' as widespread terracing. The search for more productive tree species has been a key theme over the last two centuries (see, for example, Doughty, 1996). Plant collecting, acclimatization and experimental plantations and the establishment of special forestry nurseries are all important aspects of this. One result of such experiments has been the rise of monocultures of trees, many introduced from other countries or continents, formed from those species which are seen as the most productive and profitable. Nili Liphschitz and Gideon Biger demonstrate the complicated history of experimentation, scientific and political debate associated with the acclimatization of tree species in Palestine and show how *Pinus halapensis* came to dominate the new plantations.

The Culture of Trees, Woods and Forests

The development of a strong interest in the cultural associations of trees and woodland has been a productive and exciting area of research and scholarship in recent years. Keith Thomas's (1984) work on changing attitudes to trees and woodland 1500–1800 in England has been influential. Other important studies include those by Stephen Daniels (1988) on the political iconography of trees in eighteenth century England and Simon Schama's (1995) broad review of the culture of wood in Europe and the USA. These studies have parallels with the long tradition of studying the folklore and myths associated with different kinds of trees by historians and social anthropologists (Davies, 1988). Much recent work on the relationship between society and nature has been informed by debates in environmental history (Demeritt, 1994; Williams, 1994). Radkau (1997), for example, has examined the development of nature worship in Germany and the way in which it interlinks with the development of German forestry in the nineteenth century.

The way in which trees are catalogued and categorized plays an important part in their protection and survival. Individual trees can be so long-lived that they develop several layers of meaning that can be documented through time, as I have shown with the ancient oaks of Sherwood Forest. Although the reasons for their appreciation have varied, these ancient oaks have survived because they have been allotted some kind of value, whether as constructional timber, props of picturesque aesthetics, harbourers of romantic legends or habitats for rare beetles. It is only fairly recently that they have been given some form of legal protection. Eduardo Rico Boquete shows in his study of Spanish trees and forestry in the early part of the twentieth century the importance of registers and catalogues as a means of protecting individual trees and conserving woodland.

In the final chapter, the importance of inventing particular categories which are believable within different cultural contexts is emphasized by Judith Tsouvalis-Gerber. She shows how the recent re-invention of the category 'ancient woodland' in Britain has given value to a type of woodland much of which had remained unmanaged and unprofitable for most of the twentieth century. The decline of traditional markets for coppice products and the rise of modern forestry had resulted in large areas of old coppice and coppice-with-standards woodland in Britain being officially categorized as 'scrub' and hence ripe for clearance or conversion to plantations. When, in the 1960s and 1970s, ecologists and conservationists began to ascribe this woodland with nature conservation and historical values, 'scrub' was reborn as 'ancient woodland'; it was only with the invention of this new vocabulary that the forestry establishment was convinced that this type of woodland was valued by society.

Conclusion

The authors in this book all in their different ways contribute to an increasing uncertainty about woodland and forest history. Research is drawing out the enormous variation in European woodland use at different scales and through time. It is becoming increasingly difficult to accept woodland as a simple category from which a settled landscape has been wrought. Rather it must be seen as a complex type of land use which has varied dramatically in the density, age, species and form of trees and shrubs of which it consists. The utility of woodland and the cultural values ascribed to it are also diverse: whether it is tilled or grazed by domestic stock, whether it is a provider of status and symbol of power; whether it is a site of traditional woodmanship or scientific experimentation. To explore this complex bundle of uses and values is no simple matter. This book demonstrates the breadth and vitality of current research and shows that scholars are taking a wide variety of approaches. As woodland and forests have frequently been found throughout history at the borders of regions, so much of the most interesting research is being undertaken on the edges of disciplines, and it is interdisciplinary research, combining varied sources and approaches, which is proving most fruitful.

References

Collins, E.J.T. (1992) Woodlands and woodland industries in Great Britain during and after the charcoal iron era. In: Metaille, J.P. (ed.) *Protoindustries et Histoire des Forêts, (Les Cahiers de l'ISARD* 3). University of Toulouse, Toulouse, pp. 109–120.

Daniels, S. (1988) The political iconography of woodland in later Georgian Britain. In: Cosgrove, D. and Daniels S. (eds) *The Iconography of Landscape.* Cambridge University Press, Cambridge, pp. 43–82.

Daniels, S. and Watkins, C. (1991) Picturesque landscaping and estate management: Uvedale Price at Foxley, 1770–1829. *Rural History* 2, 141–170.

Davies, D. (1988) The evocative symbolism of trees. In: Cosgrove, D. and Daniels S. (eds) *The Iconography of Landscape.* Cambridge University Press, Cambridge, pp. 32–42.

Demeritt, D. (1994) Ecology, objectivity and critique in writings on nature and human societies. *Journal of Historical Geography* 20, 22–37.

Doughty, R. (1996) Not a koala in sight: promotion and spread of eucalyptus. *Ecumene* 3, 200–214.

Hammersley, G. (1973) The charcoal iron industry and its fuel. *Economic History Review. Second Series* 26, 593–613.

James, N.D.G. (1981) *A History of English Forestry.* Basil Blackwell, Oxford.

James, N.D.G (1996) A history of forestry and monographic forestry literature in Germany, France, and the United Kingdom. In: McDonald, P. and Lassoie, J. (eds) *The Literature of Forestry and Agroforestry.* Cornell University Press, Ithaca, pp. 15–44.

Mastoris, S.N. and Groves, S.M. (1997) Sherwood Forest in 1609. A Crown Survey by Richard Bankes. *Thoroton Society Record Series* 40, Thoroton Society, Nottingham.

Metaille, J.P. (ed.) (1992) *Protoindustries et Histoire des Forêts* (*Les Cahiers de l'ISARD* 3), University of Toulouse, Toulouse.

Mooser, J. (1986) Property and wood theft: agrarian capitalism and social conflict in rural society, 1800–50. A Westphalian case study. In Moeller, R. (ed.) *Peasants and Lords in Modern Germany.* Allen & Unwin, Boston, pp. 52–80.

Moreno, D. (1990) *Dal Documento al Terreno. Storia, e Archeologia dei Sistemi Agro-silvo-pastorali.* Il Mulino, Bologna.

Neeson, J.M. (1993) *Commoners: Common Right, Enclosure and Social Change in England, 1700–1820.* Cambridge University Press, Cambridge.

Peluso, N.L. (1995) Whose woods are these? Counter-mapping forest territories in Kalimantan, Indonesia. *Antipode* 27, 383–406.

Peterken, G.F. (1981) *Woodland Conservation and Management.* Chapman & Hall, London.

Peterken, G.F. (1996) *Natural Woodland. Ecology and Conservation in Northern Temperate Regions.* Cambridge University Press, Cambridge.

Rackham, O. (1980) *Ancient Woodland: Its History, Vegetation and Uses in England.* Edward Arnold, London.

Rackham, O. (1989) *The Last Forest. The Story of Hatfield Forest.* Dent, London.

Radkau, J. (1997) The wordy worship of nature and the tacit feeling for nature in the history of German forestry. In: Teich, M., Porter, R. and Gustafsson, B. (eds) *Nature and Society in Historical Context.* Cambridge University Press, Cambridge.

Revill, G. and Watkins, C. (1996) Educated access: interpreting Forestry Commission Forest Park Guides. In: Watkins, C. (ed.) *Rights of Way: Policy, Culture, and Management.* Pinter, London, pp. 100–128.

Rocheleau, D. and Ross, L. (1995) Trees as tools, trees as text: struggles over resources in Zambrana-Chaucey, Dominican Republic. *Antipode* 27, 407–428.

Russell, E.W.B. (1997) *People and Land Through Time. Linking Ecology and History.* Yale University Press, New Haven and London.

Schama, S. (1995) *Landscape and Memory.* Harper Collins, London.

Thomas, K. (1984) *Man and the Natural World. Changing Attitudes in England 1500–1800.* Allen Lane, London.

Wickham, C. (1994) European forests in the early Middle Ages: landscape and land clearance. In: *Land and Power. Studies in Italian and European Social History, 400–1200.* British School at Rome, London, pp. 155–199.

Williams, M. (1989) *Americans and their Forests. A Historical Geography.* Cambridge University Press, Cambridge.

Williams, M. (1994) The relations of environmental history and historical geography. *Journal of Historical Geography* 20, 3–21.

Young, C.R. (1979) *The Royal Forests of Medieval England.* University of Pennsylvania Press, Pennsylvania/Leicester University Press, Leicester.

CHAPTER 1
The word 'Forst/forest' as an indicator of fiscal property and possible consequences for the history of Western European forests

Rudolf Kiess
Ludwigstraße 54, 70176 Stuttgart, Germany

This chapter investigates the institution of the forest in Western Europe and in particular in Germany from the seventh century onwards. Forests as opposed to ordinary woods, normally called *silvae*, had more or less well defined boundaries and were usually described and understood as royal hunting-grounds. In contrast to these normally large forests (or in German 'Wildbann' districts) there are traces of obviously much smaller forests which have survived in wood-names, field-names or place-names. I have tried to collect them systematically in south-west Germany and sporadically in other parts of Germany. My thesis is that at least some of these names originated in the seventh or eighth centuries, not so much as traces of ancient forest districts, but of smaller forest units attached to estates or castles. Their distribution suggests that they indicate Frankish settlements in Merovingian and Carolingian times. These names also serve as indicators of what was once Crown property. This is otherwise difficult to find in Germany where the central power of the kings suffered a steady decline during the Middle Ages.

The emphasis in this chapter is on the 'forest' as an important political and social institution in Western Europe and Germany from the first appearance of the term in the middle of the seventh century until the beginning of the nineteenth century. A great number of forests are documented, beginning with the famous charter for the monasteries of Stablo and Malmédy in the Ardennes in 648 (Halkin and Roland, 1909 no. 2 and 4). Afforestation was a royal prerogative both in Britain and on the Continent. When in the Holy Roman Empire a large number of such forests were passed on to, or newly created for, regional princes, rulers and sovereigns, both lay and ecclesiastical (the lords we call in German *Landesherren*), hunting was, according to the texts, still the principal attribute.

There were, of course, other elements of the forests which increased the income of the holder. These included clearances providing additional agricultural land, the provision of grazing in the woods and the sale of timber and firewood. In addition to these economic aspects, forests became more and more important for these *Landesherren* as a means of extending their territory or at least of imposing rights or privileges over parts of a neighbouring territory. One has to bear in mind the growing diversification of feudal rights in Central Europe and the decline of a central power in Germany during the Middle Ages. Forest boundaries also served as the first territorial boundaries.

Forest districts could form an independent system of administrative units such as in the duchy of Wuerttemberg, a state in south-west Germany, where I come from and where I started my research into forest history. A map from the sixteenth century (Fig. 1.1) shows the forest boundaries of Wuerttemberg which did not coincide with the boundaries of ordinary administrative districts called *Aemter* (Kiess, 1958, p. 22). It is obvious that the whole country was not covered with woods. By no means all existing woods belonged to the sovereign; a considerable part belonged to towns, villages, other noble families and private persons. Irrespective of ownership of the ground, forest authority was enforced all over the country guaranteeing the hunting prerogative of the sovereign and, from the end of the fifteenth century onwards, ensuring that woods as natural resources were not devastated.

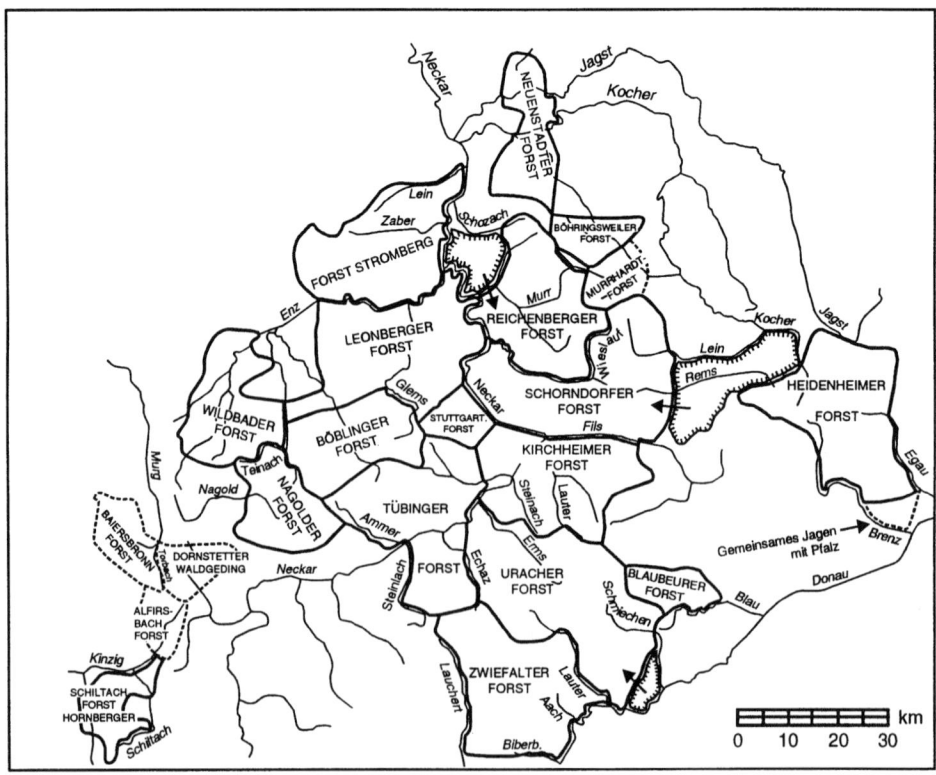

Fig. 1.1. The forest districts of the Duchy of Wuerttemberg (south-west Germany), mid sixteenth century.

Some of these forest districts can be traced back to older forests which may have been granted to church institutions by the King. In Germany such a district is often called a *Wildbann*, a term in which the syllable 'Wild' does not refer to 'Wild' meaning game, but to 'wild' in the sense of desolate, waste land which the King originally regarded as his personal property and which was often called 'eremus', land, in this case woodland, not yet disposed of. Later on the terms Wildbann and Forst were applied to areas with and without woods (Kaspers, 1957, pp. 50ff., 232f.).

The description of this situation in Germany forms only the background of what I have in mind. When studying forest districts 40 years ago I occasionally came across the word 'Forst' in a clearly different context. Smaller woods, but also stretches of open country were called 'Forst'. This led me to collect such wood-names and field-names systematically, in the first place from detailed nineteenth century local maps, and then also from other sources, either printed or from manuscripts in archives.

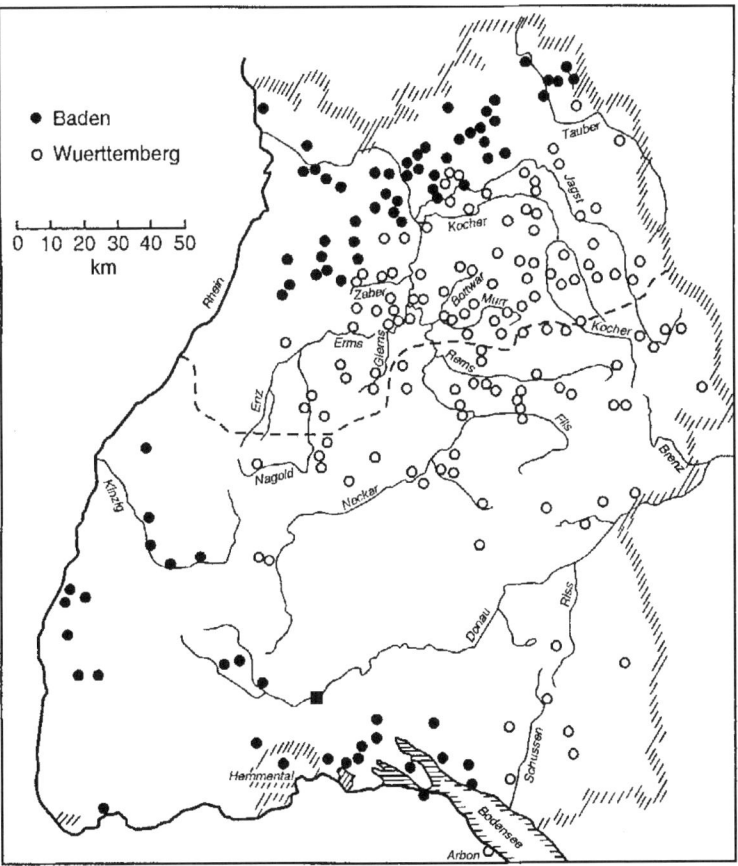

Fig. 1.2. Forest-names in Baden-Wuerttemberg (south-west Germany).

The map of the forest-names found in south-west Germany shows no equal distribution (Fig. 1.2; Kiess, 1996, p. 48). There are, for example, hardly any such names in the Black Forest which was colonized much later than other parts of south-west Germany. There is, on the other hand, a concentration of references in the northern part and around Lake Constance, both regions occupied by the Franks in the eighth century when they defeated the Alemanic tribe. This distribution of forest-names gave rise to the claim that at least some of these names can be traced back to the eighth or even seventh centuries when the term 'Forst' first appeared in the charter of 648 for the Monasteries of Stablo and Malmédy in the Ardennes. This hypothesis was supported by the fact that these forest-names were often found in positions which are interesting from an historical point of view.

Going back to the Black Forest, it is remarkable that in that large region otherwise devoid of forest-names a small number can be found along a valley where a former Roman road leading from Strasbourg to Rottweil, also used in Frankish times, and the old abbey of Gengenbach, an Iro-Scottish foundation, bear witness to Frankish activities (Fig. 1.3). Further examples can be found elsewhere. The term 'old abbey' is relevant because such forest-names are indeed conspicuous in places where abbeys were founded or possessed land. Many of their possessions had been passed on to them by the King or

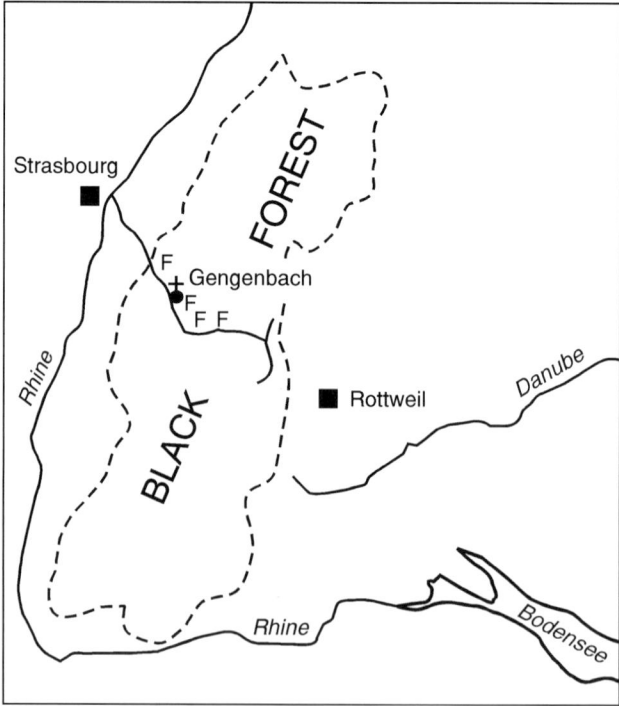

Fig. 1.3. Forest-names in the Black Forest along an originally Roman road from Strasbourg to Rottweil.

indirectly via donations of noble families formerly endowed by the King.

The list of possessions of the Abbey of Weissenburg in northern Alsace in France shows the difference between ordinary woods called *silva* and the forest called *forastis*, that is a wood under special right (Zeuss, 1842 no. 45). The size of such a forest is, as in parts of England, expressed by the number of pigs which can be fed in the particular forest. Therefore agricultural purposes seem to have been at least as important as hunting. The proximity to Roman remains – roads, castles – can be seen as a continuity of state or later royal property. This remark reveals the special interest of the historian in searching for these ancient forest-names. As there was no Domesday Book on the Continent, it is very difficult to find traces of royal property or what is called the royal demesne.

Another similar example is the Abbey of Fulda founded around 744 on the advice of St Boniface, the famous English missionary, in a place which was not a wilderness, as described in the foundation document, but close to a former royal estate. In later centuries Fulda had a system of large forest districts or *Wildbaenne* (Bramforst, Zundernhart), but there is also a source which mentions in passing as part of a parish boundary an obviously small, perhaps already obsolete forest called Forastis Mastur (Fig. 1.4). This could be

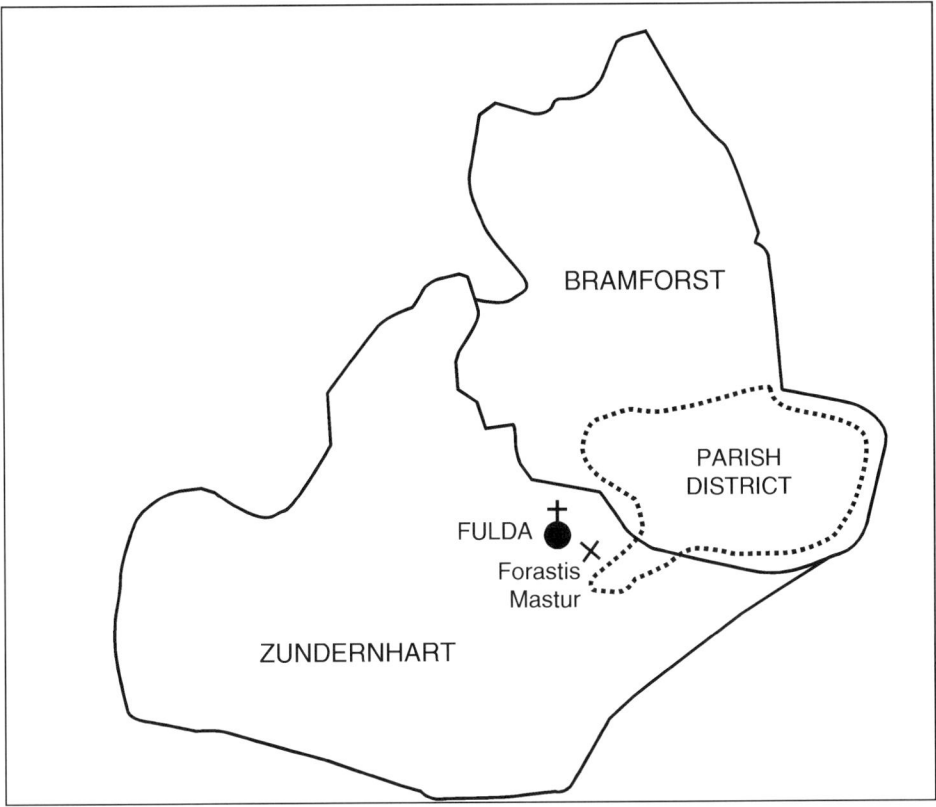

Fig. 1.4. The forests and Wildbaenne of the Abbey of Fulda.

the trace of a royal wood which had belonged to the royal estate preceding the monastery of Fulda (Kiess, 1996, 108f.).

Another feature of the forest-names in question is their position on hills, often plateaus, table-mountains or slopes running down from such elevations. The position on hills is to be seen in connection with the fact that forest-names, that is small forests, are also found behind castles on the edge of hills. Perhaps one could call them hill-forts. In Germany castles on hills were usually built in the eleventh century. Recently, however, quite a few hill-forts of earlier times have been investigated, and therefore archaeologists should be

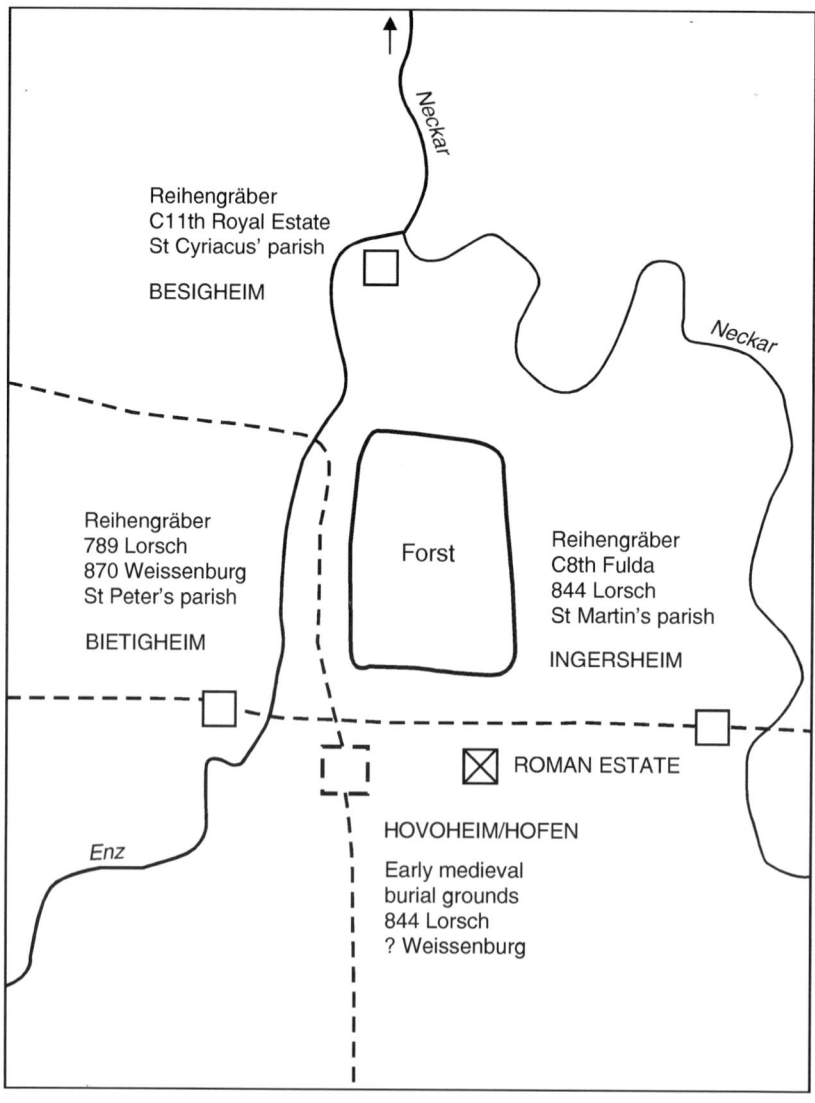

Fig. 1.5. The Bietigheim Forest.

encouraged to get interested in forest-names as an additional indication of feudal property.

The example of Bietigheim (Fig. 1.5) shows a combination of features which support the argument that such a forest can indicate a place of historical interest (Kiess, 1995). This forest, which has probably survived from Frankish times right to the present day, is located near Stuttgart and offers, apart from the still existing name 'Forst', references to old roads, *Reihengräber* (aligned graves), Frankish patron saints of churches, possessions of the old monasteries of Fulda, Lorsch and Weissenburg, Roman remains and a written testimony in a collection of records of the monastery of Weissenburg which mentions an estate with a 'forastis' (Zeuss, 1842, no. 26).

In the course of my studies I have come to the conclusion that in contrast with large forest districts these small forests have not attracted adequate attention in research and have perhaps not even been noticed sufficiently. Admittedly such names could originate from a more recent forest organization. This has to be checked in every case. In order to give these small forests their right place in the history of forests other versions of woods and forests should be studied. The difference between, and the existence of forests on the one hand and of woods not called forests, but just *silva/silvae* on the other hand has been documented since Carolingian times. In this context it is important to notice that even large wooded areas in royal possession were not necessarily called forests, but just *silvae*, although foresters (*forestarii*) held office in these woods. Some of these woods were also called by the Latin word

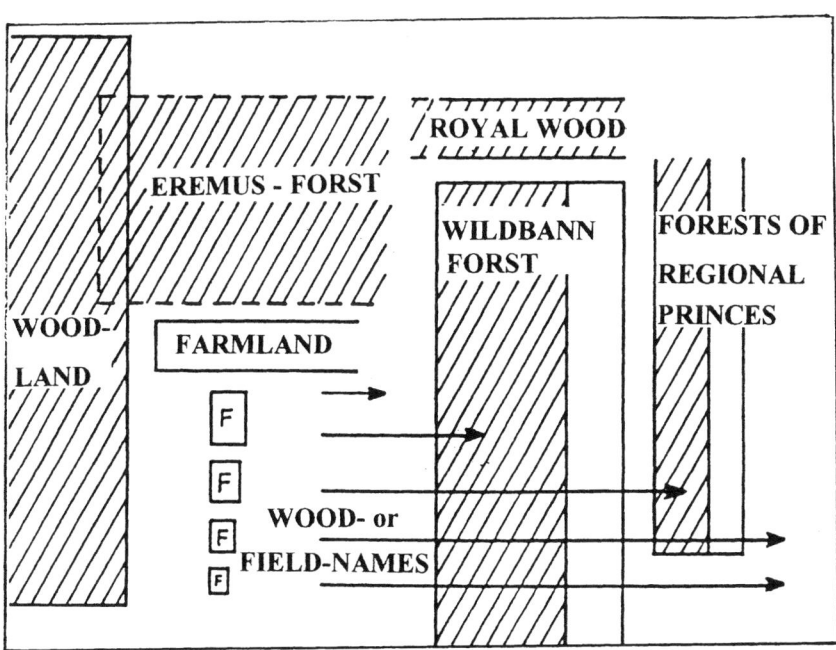

Fig. 1.6. Distribution and development of the terms applied to ancient forests.

nemus, in plural *nemora*, although this word is not generally accepted as a term indicating royal property. One example from my native region of a wood not called forest before 1348 is a former royal wood, in German 'Reichswald', called Schönbuch with an elaborate system of regulations as far as hunting and the use of timber and grazing-land is concerned (Grees, 1969).

My conclusion is that the term 'Forst, forest' was only applied when it was necessary to secure the property in question against the interests of other claimants. I have tried to summarize my conclusions and, above all, the position of the small forests in a diagram which shows the possible distribution and development of the various terms applied to ancient woods (Fig. 1.6). The original woodland including the eremus could develop in three ways. It could first develop into royal woods without being named as forest, second, it could become part of large forest districts, later in the hands of regional princes (*Landesherren*) and third it could become small forests attached to royal estates whose names may have survived as names of woods and fields. The main purpose of this chapter, however, is to present the existence of small forests in contrast to the larger districts which have always attracted the attention of forest historians and of historians in general.

References

Grees, H. (ed.) (1969) *Der Schönbuch. Beiträge zu seiner Landeskundlichen Erforschung.* Konkordia, Bühl/Baden.

Halkin, J. and Roland, C.G. (eds) (1909) *Recueil des Chartes de l'abbaye de Stablo-Malmédy.* Kiessling, Bruxelles.

Kaspers, H. (1957) *Comitatus Nemoris. Die Waldgrafschaft zwischen Maas und Rhein.* Dürener Geschichtsverein, Düren and Aachen.

Kiess, R. (1958) *Die Rolle der Forsten im Aufbau des württembergischen Territoriums bis ins 16.* Jahrhundert (Veröffentlichungen der Kommission für geschichtliche Landeskunde in Baden-Württemberg Reihe B Forschungen 2. Band). Kohlhammer, Stuttgart.

Kiess, R. (1992) Forst-Namen als Spuren frühmittelalterlicher Geschichte in Württemberg. *Zeitschrift für Württembergische Landesgeschichte* 51, 11–116.

Kiess, R. (1995) Der Bietigheimer Forst im Rahmen der allgemeinen Geschichte der Forsten. *Blätter zur Stadtgeschichte* 12, 7–28. Hg. vom Archiv der Stadt Bietigheim-Bissingen.

Kiess, R. (1996) Forst-Namen als Spuren frühmittelalterlicher Geschichte II: Beispiele aus Baden und angrenzenden Territorien. Folgerungen für die Forstgeschichte. *Zeitschrift für die Geschichte des Oberrheins* 144, 47–123.

Zeuss, C. (ed.) (1842) *Traditiones possessionesque Wizenburgenses.* Kranzbühler, Speyer.

CHAPTER 2
Medieval forests and parks in southern and central England

Della Hooke
91 Oakfield Road, Selly Park, Birmingham B29 7HL, UK

Introduction

It is generally assumed that forests in southern and central England were made by Norman kings in areas which were relatively densely wooded (apart from some moorland areas such as Exmoor). But the location of former royal forests is considerably more complex. Land had already been designated for hunting in late Anglo-Saxon England, with areas of woodland or heath set aside for this purpose and with enclosures for the retention and capture of deer. It is essential to understand late Anglo-Saxon land use, especially the distribution of woodland, together with systems of territorial organization within the wider geographical framework, if the location of hunting areas is to be understood. This is important because studies show that such areas frequently influenced the location of the new post-Conquest forests. The regeneration of woodland may itself have been affected by the setting aside of land. The evidence is reviewed on a county scale for parts of southern and central England but more detailed studies are necessary to correlate the documentary, place-name and field evidence for selected areas if the interrelationship between Anglo-Saxon hunting areas and medieval forests and parks is to be understood. Problem areas are briefly touched upon, such as those in which extensive woodland occurs but is not formally used for either early medieval hunting or later forest or those in which late Anglo-Saxon hunting is clearly attested but which failed to become designated as forest.

The Location of Medieval Forests

Medieval forests, areas subject to the forest law which was introduced into England after the Norman Conquest, are usually thought to have coincided with the most heavily wooded regions of England (Young, 1979, p. 2; Birrell, 1980, p. 78; Cantor, 1982, pp. 56–70), although some – such as Exmoor – were obviously made in moorland country (Rackham, 1986, p. 131),[1] and even in wooded areas are likely to have comprised much open land. Given the difficulty of identifying areas of Anglo-Saxon woodland (see, for instance, the contrasting views of the late Sir Frank Stenton and the Ordnance Survey (Hill,

1981, p. 16)), there are many questions about why some areas of woodland should have become royal forest while others obviously did not. There was, for instance, no forest in the Weald of south-eastern England although this was one of the most heavily wooded regions of early medieval England, 'the great wood which we call Andred' said to extend for 120 miles from east to west and 30 miles from north to south (Swanton, 1996, pp. 84–85).

This chapter will propose that it is necessary to understand the nature of the landscape in the *early* medieval period if one is to investigate the distribution of the later, medieval, forests. Hunting was not just a preoccupation of Norman kings; it was important to royal and aristocratic landowners in late Anglo-Saxon England. Several charters grant the right to hunt and others attempt to claim immunity from entertaining the king and huntsmen, dogs and hawks in Mercia in the ninth century (Sawyer, 1968, S 1281, S 198, S 207).[2] The most reliable of these exempted the minster at Blockley (formerly in Worcestershire but now in Gloucestershire) from dues which included 'the feeding and maintenance of all hawks and falcons in the land of the Mercians, and of all huntsmen of the king or ealdorman except only those who are in the province of the Hwicce' (Sawyer, 1968, S 207; Birch, 1885–99, B 489). The Anglo-Saxon kingdom of the Hwicce, which was based upon the Severn basin and extended over most of what was later to be Gloucestershire and Worcestershire, together with the south-western part of Warwickshire, was gradually incorporated into Mercia after the seventh century but its rulers obviously attempted to protect their rights after its assimilation.

Although hawks and falcons and hunting dogs are discussed in Anglo-Saxon England and depicted in early manuscripts, it is unlikely that deer-hunting had acquired the ritual with which the sport was associated in medieval times: huntsmen frequently used dogs to drive the deer towards gaps in a hedge or fence across which nets had been placed to ensnare the fleeing animals, the barrier often forming a sort of enclosure (Thorpe, 1846, p. 21; Hooke, 1998). This remained one of the most effective ways of hunting and killing a quantity of deer throughout the Middle Ages, the nets referred to as 'hayes', but permanent enclosures were constructed to impede the movement of game by the ninth century (Cummins, 1988, p. 57). Cummins suggests that the 'haye', a term used in other contexts to mean both 'hedge' and 'long-net', may have been somewhat flimsier in nature than the boundary of a medieval park, perhaps 'a series of palisades or hurdles designed to guide driven game towards waiting archers or nets in what was otherwise an open wood, or to act as a pen for the capture of wild deer which might be transferred to a permanent park' (Cummins, 1988, pp. 57–58). In Scotland, too, the same method was in use in medieval times, the drive occasionally guiding the deer towards a narrow gap in the pale of a deer-park which led into an extension where the huntsmen would be waiting (Cummins, 1988, p. 54). Domesday Book carries this practice further back in time and *haiae* are referred to on several occasions in, for instance, the Midland folios: a *haia* in Kington, Worcestershire, was described as a place in *qua capiebant ferae* 'in which wild animals used to be captured' (Thorn and Thorn, 1982, 18,4) and at Lingen, now in Herefordshire but earlier in Shropshire, three hays were described as specifically for *capreolis capiendis*, 'taking roe-deer' (Thorn and Thorn, 1986, 6,14); others were described as *haiae firmae*, 'fixed hays' (Darby, 1977, p. 204). It is argued that the Anglo-Saxon *haga* described the same type of feature. More extensive areas of land, however, had also been designated as forests within Frankish territories by the seventh century, areas which were intended to act as game reserves for the king and nobility, and the concept was obviously well known in England in early medieval times (Hooke, 1989, pp. 125–126).

Haga Features

In the early medieval period, *haga* features noted in pre-Conquest charters and occasional place-names refer to enclosures in woodland country that seem to have been associated with the taking of deer. The first documented deer-park, occurring in an eleventh century will bequeathing a park at Ongar in Essex, specifically refers to this as a *derhage* (Thorpe, 1865, p. 574), but the term was also used for a type of boundary fence. The name *haga*, 'haw', suggests that the hawthorn formed part of the enclosure but the feature occurs in quite different localities to the normal 'hedgerow' and a dead hedge may be indicated (Hooke, 1981, pp. 234–247). Essentially, 'enclosure' is indicated, probably in a protective capacity, and the term has survived in Scandinavia for a fenced pasture. In its early usage in England an argument can be presented for interpreting *haga* as an enclosure associated with the retention and taking of deer, as noted above (Hooke, 1989). This supposition is strengthened by the use of the term along known forest boundaries in Germany in the tenth century where such features coincided with the boundaries of the forests of Bramforst and Zunderhart (Metz, 1954, pp. 41–42). There the term came to mean 'enclosure of a wood containing wild animals' or 'defensive enclosure', the way I believe it was used in England in late Anglo-Saxon times. The usage of the term is discussed in numerous German dictionaries but a meaning for *hagen* as *die umzäuning eines waldes, worin wild gehegt wird*, 'the enclosure of a wood in which game are preserved' (Grimm and Grimm, 1877, p. 152), or *einen wildzaun machen*, 'to make a game enclosure' (Lexer, 1992, p. 79) is always prominent and the hunting terminology *jagen und hagen* is current from medieval times (Kiess, pers. comm.). The use of the term for an enclosure within the walls of a defended *burh* appears to be a secondary usage and in this context the urban *haga* eventually came to mean little more than a messuage.

The Distribution of Royal Medieval Forests

While forests remained sources of pasture, timber and wood, even reserves of potential arable, their primary purpose as hunting reserves must not be underestimated (Birrell, 1996, pp. 437–438). The reason for examining the location of the term *haga* is that it appears to be very closely related to the distribution of later, medieval forests. Figure 2.1a illustrates the correlation between the

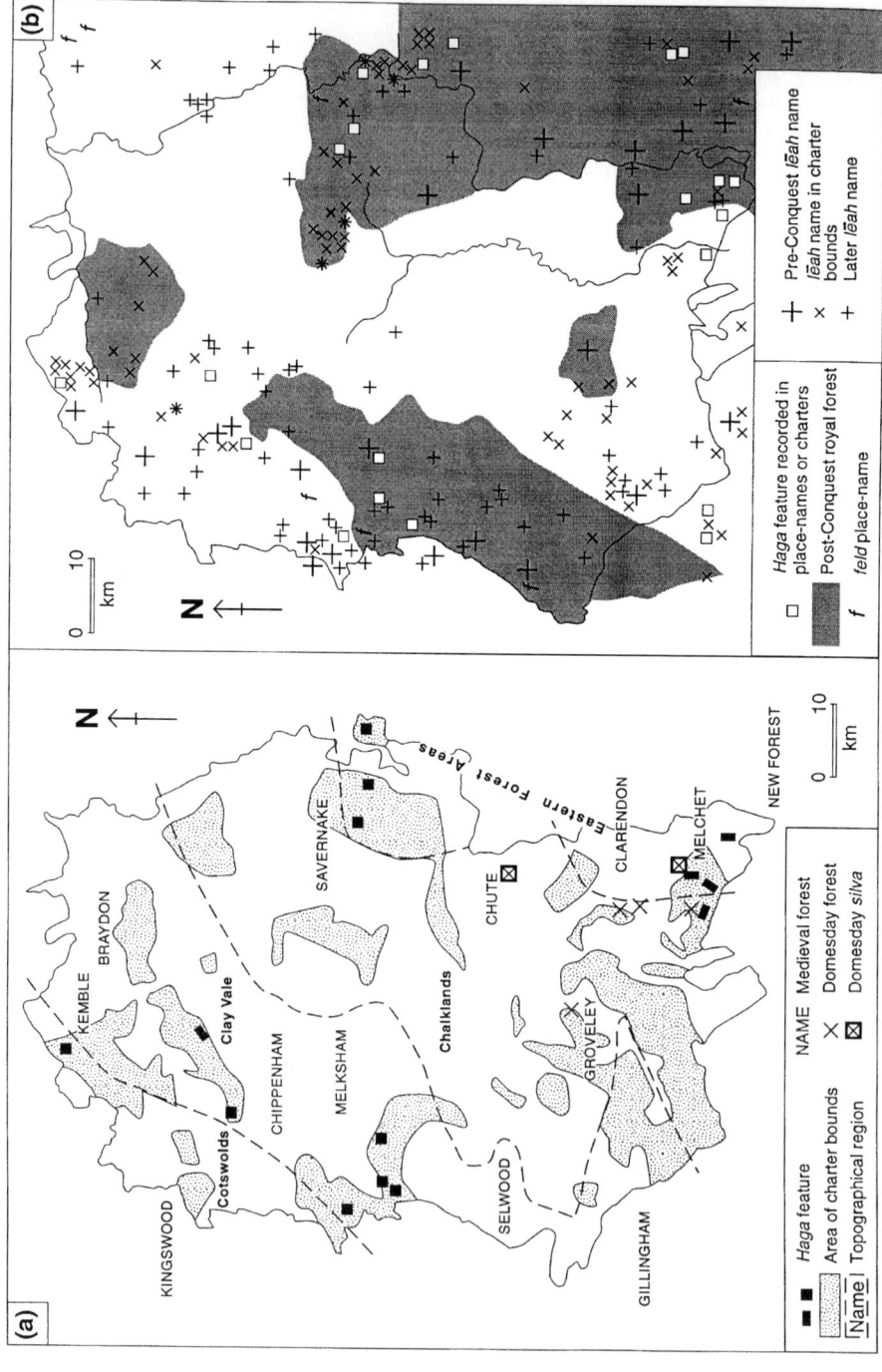

Fig. 2.1. (a) The incidence of the *haga* term in Wiltshire shown against the distribution of medieval forests (source: charters and place-names). (b) Wiltshire woodland indicated by the Old English *lēah* term (source: charters and place-names).

incidence of the term in charter boundary clauses with later recorded forest in the county of Wiltshire. The term was obviously only recorded where there was charter coverage (shown dotted); it cannot show the full distribution of such features, as royal lands that were never granted away do not appear: most charters record grants made to the church. This may explain why so often *haga* features appear on the *fringes* of royal forests; however, most of the later Wiltshire forest areas have *haga* features recorded along estate boundaries in pre-Conquest charters, many of them associated with royal land. In Wiltshire, the medieval forests of Kemble, Chippenham, Melksham and Selwood lay in the west of the county, largely coinciding with the clay lowlands which lie between the limestone escarpment of the Cotswolds and the chalk. To the east, where the chalk gives way to Eocene sands and gravels, *haga* features are again found in the areas of the later Savernake and Clarendon–Melchet Forests. Land in the districts of Chute and Melchet had clearly been put into royal forest by 1086, the entries for Milford and Laverstock in the south-eastern sector of the county noting that part of the land was *in foresta regis*, 'within the king's forest', the forests themselves recorded as the *silvae* of *Milcheti* and *Cetum* (Darby, 1977, pp. 195–198; Thorn and Thorn, 1979, 1310; 18; 119). The great woodlands of Chippenham and Melksham are noted but the presence of the later forests of Braydon, Savernake and Selwood 'is suggested only by the sparsity of vills in their neighbourhood' (Welldon Finn, 1979, p. 39). However, many areas of known woodland which are omitted in the Domesday folios may already have been taken into forest (Hooke, 1994; Darby, 1977, pp. 195–196). Apart from the Domesday evidence relating forests to woodland, it is also possible to plot the charter and forest evidence against early place-names which confirm the presence of woodland. The distribution of the *leah* term clearly shows that the same regions were wooded in late Anglo-Saxon times (Fig. 2.1b).

In Dorset, *haga* features are similarly concentrated in the low-lying clayland area of the Vale of Blackmoor, the vale later to lie within the royal forest of Blackmoor with the forest of Gillingham to the north; outlying references also occur on the margin of Cranborne Chase and Holt Forest (Fig. 2.2a). The latter forest was earlier known as the *foresta de Winburne* in 1086 (Darby, 1979, p. 103). The claylands were particularly well wooded in early medieval times, as was the greensand escarpment which bounded them on the east. *Haga* features are recorded along the boundaries of Sturminster Newton and Stalbridge and Weston (Fig. 2.2b), the former estate granted by King Edgar to Glastonbury Abbey in 968, the latter allegedly granted by Athelstan to the church of Sherborne in 933 (Birch, 1885–1899, B 1214, B 696; Sawyer, 1968, S 764, S 423). Further south, Buckland Newton, with no fewer than four *haga* features recorded along its boundary, was still in royal hands in the mid tenth century; a deer-park was officially recognized here at Duntish in the mid thirteenth century. A further 'old' *haga* lay along the boundary of Woolland in 833 a few kilometres to the east (Birch, 1885–1899, B 410; Sawyer, 1968, S 277), the boundary, marked by a substantial bank, coinciding with that of the park later made at Melcombe (Cantor and Wilson, 1964, pp. 154–158). In all cases except the latter, woodland indicated by *leah* place-names lay in close

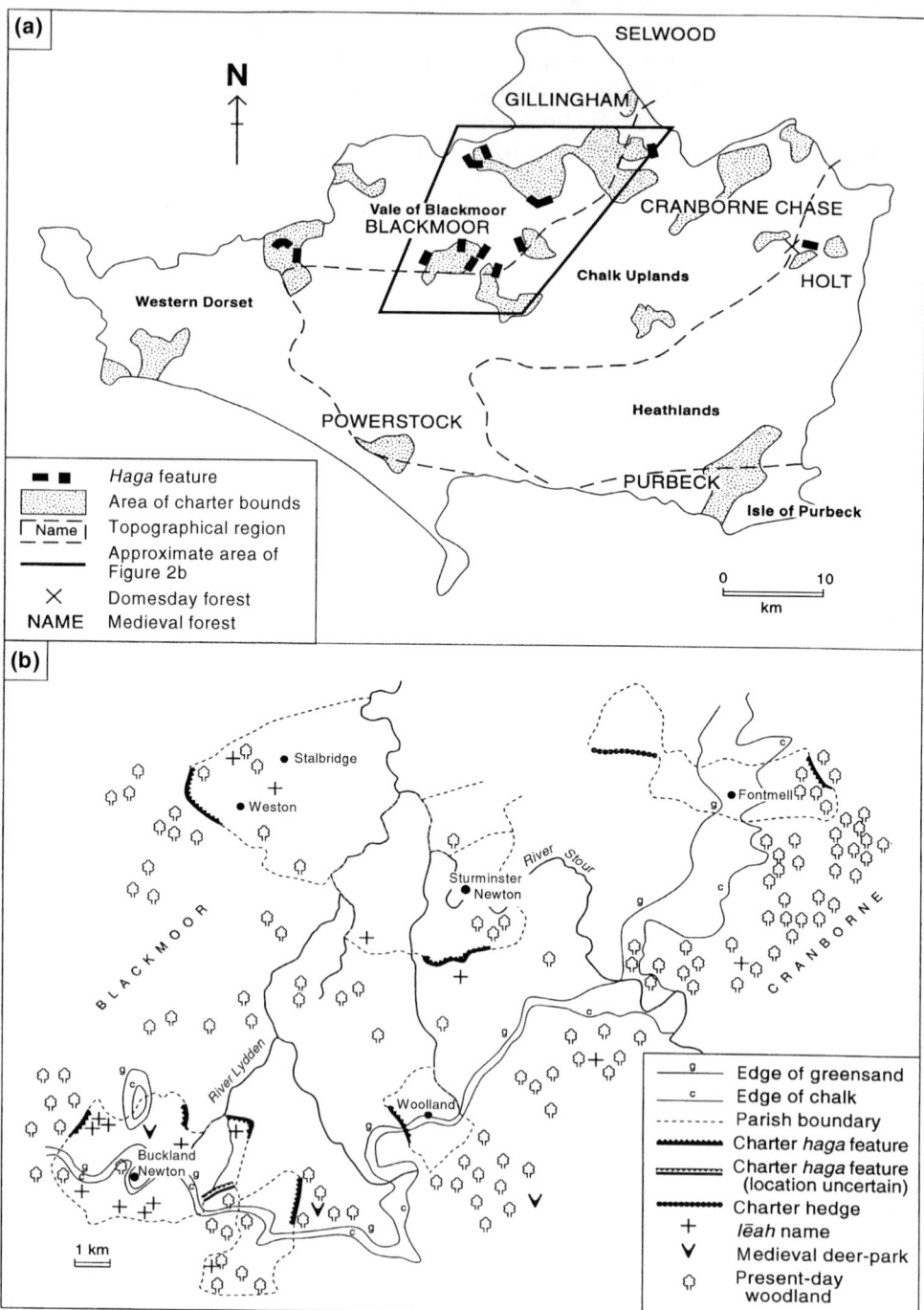

Fig. 2.2. (a) The incidence of the *haga* term in Dorset shown against the distribution of medieval forests (source: charters and place-names). (b) *Haga* features in the Vale of Blackmoor.

proximity to the *haga* boundaries and hunting was to be a major land use in the region.

The evidence suggests that this connection with hunting, represented by the *haga* enclosures, laid the foundation for the later forests when Norman kings seized the right to hunt on anybody's land as well as their own, a licence also granted to the marcher lords in the Borderland (Linnard, 1982, p. 32). One can observe three main kinds of region in which hunting was a dominant land use in early medieval England. Firstly, one may note frontier regions; some of these had been heavily wooded for a long time – Selwood, along the boundary between Wiltshire and Somerset, had been known to the British as *Coit Maur*, 'the great wood', and had held back the Anglo-Saxons until the mid seventh century (Stenton, 1971, pp. 63–64).[3] The kingdom of the Hwicce in midland England was similarly surrounded by woods (Hooke, 1985, p. 79). Second, one may recognize 'areas of opportunity'. Where Iron Age and Roman agriculture and settlement had been abandoned, such as on the higher chalklands of northern Hampshire (Fig. 2.3), land had become available for pasture or other uses. One such area is described in greater detail later in this chapter. The Welsh Border was used for hunting at the time of the Norman Conquest, according to the Domesday Book, because of the amount of land which had been wasted, presumably by conflicts between the Welsh and the English. The Herefordshire folios note how Osbern fitz Richard hunted on the devastated lands of 11 vills straddling the border to the east of Radnor which were overgrown with wood: 'On these waste lands have grown woods in which Osbern goes hunting and he has from them what he can catch' (Thorn and Thorn, 1983, 24,3). Again, in Shropshire, near Montgomery, the wasted lands of 13 vills had been used for hunting in 1066 (Darby, 1977, p. 202; Thorn and Thorn, 1986, 4,1,35). This was not yet true legal forest but hays for the capture of deer were plentiful (Darby, 1977, pp. 204–207) and were probably the equivalent of the Anglo-Saxon *hagan*. Third, hunting was prevalent on some royal land: concentrations of both *haga* features and later forests are found around the capital of Winchester in Hampshire. The same association is found between many later forests and royal palaces (Rackham, 1986, p. 133) but one might ask what factors affected the choice of palace site in the first instance for a substantial number may have been established as hunting lodges.

Areas set aside for hunting were not necessarily densely wooded. Much of this woodland was probably relatively open in character and the occurrence of the *leah* term, denoting woodland with open glades, suggests that much of it had been used as wood-pasture. There were also areas of heathland, and open country on the margins of woodland is often indicated by *feld* names. It may have been the use of land in this way, restricting clearance, that actually led to woodland regeneration. Where environmental evidence has been available, tree growth seems to increase in later Anglo-Saxon England rather than earlier (Day, 1989). Rackham has rightly drawn attention to the fact that the medieval forests did not consist entirely of woodland. Nevertheless, studies like that carried out for Wiltshire (above, Fig. 2.1b) confirm that areas which were deliberately set aside for hunting in southern and midland England usually did contain abundant woodland while the Domesday folios show that hunting was

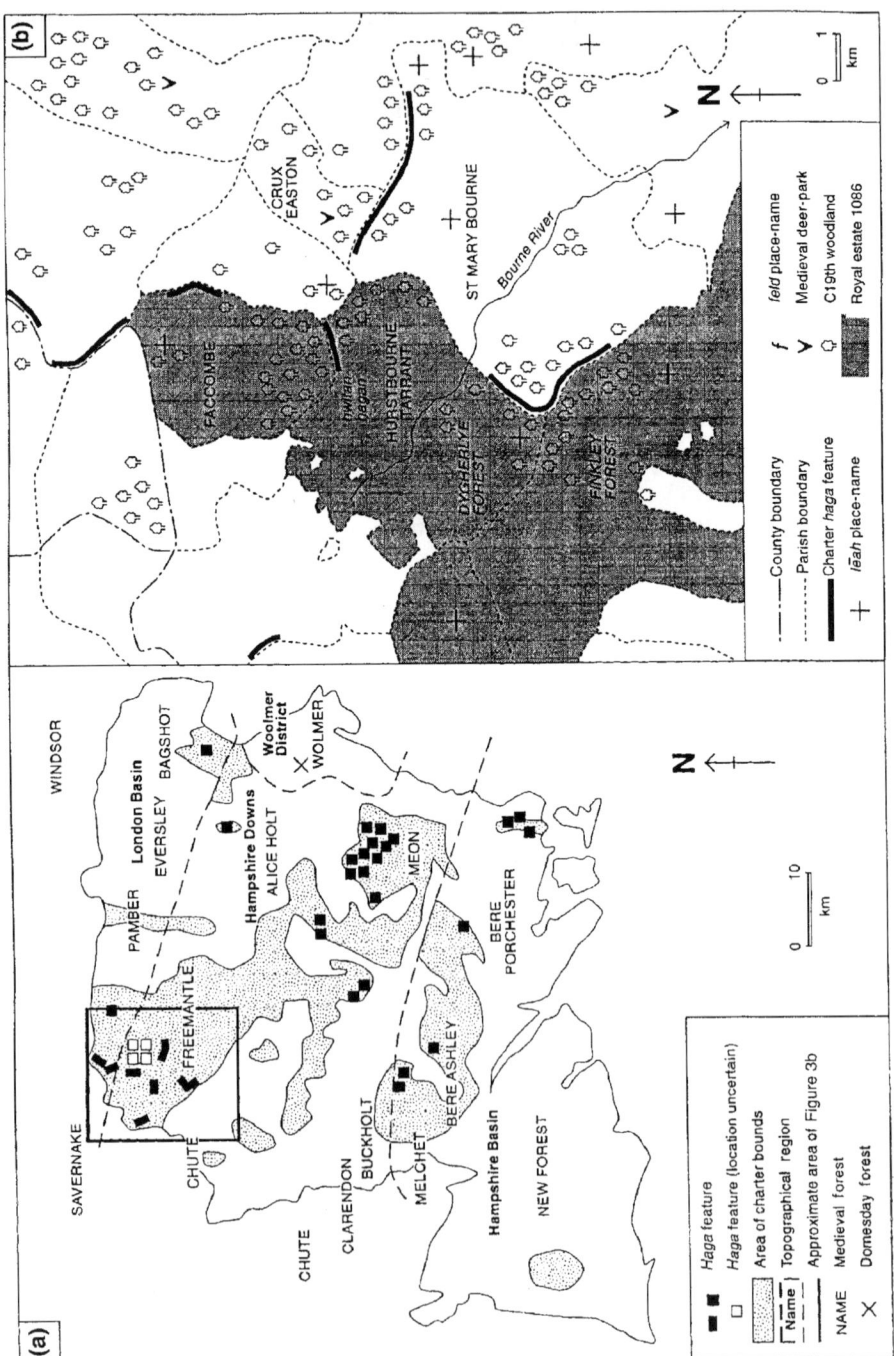

Fig. 2.3. (a) The incidence of the *haga* term in Hampshire shown against the distribution of medieval forests (source: charters and place-names). (b) North Hampshire: *haga* and forest features to the north-east of Andover.

sometimes the only economic use recorded in areas that had become overgrown after raiding.

To study the relationship of natural resources, land use and land ownership in depth requires detailed studies which relate charter boundaries, place-name evidence and the later forest boundaries to geology and soils: in north Hampshire the *hagan* of the chalk downlands occur where there was some coverage of clay-with-flints but also spread over abandoned chalkland arable (Hooke, 1989, pp. 127–128). Relict Iron Age and Roman field systems, together with related settlements, lay within the walks of Finkley and Doiley in Chute woodlands but *leah* place-names clearly indicate the presence of woodland by later Anglo-Saxon times, woods which have persisted to the present day. The name Doiley (*Digerlega* in 1155) is *Digerleah* 'thick wood' (Coates, 1989, p. 66). These woods go unnoticed in the Domesday folios, either because the peasants had no right to pasture their pigs in the lord's woods (Hampshire is one of the counties in which woodland is entered only according to its pig-rent) or because the woods had already become part of the Forest of Chute (Hooke, 1994, p. 43). Although the New Forest is recorded in some detail, other Hampshire forests receive only incidental mention (Darby, 1977, p. 198).

Although close to the royal centre of Andover, this seems to have represented a marginal region in the Anglo-Saxon period and several estates here were alienated to the church in the later eighth, ninth and early tenth centuries. Faccombe, however, remained a royal estate and was the morning gift of Wynflæd, the grandmother of kings Edwy and Edgar, in the tenth century; it was still a royal manor in 1086. There is archaeological evidence here of a range of timbered buildings having been erected in the mid ninth century and extended in the tenth, with a new manorial hall added in the mid tenth century to give a range of palace buildings akin to those at Cheddar (Rahtz, 1979). This was apparently a hunting lodge and the bone evidence shows that the hunting of red and roe deer was mainly replaced by the taking of fallow deer after the mid twelfth century (Fairbrother, 1984). *Haga* features were plentiful along boundaries in this region (Fig. 2.3), one recorded in 900 (Birch, 1885–1899, B 594; Sawyer, 1968, S 359) coinciding with the boundary of Finkley in 1298 (then referred to as a *hayam*) (Shore, 1888, pp. 40–60). Such boundaries apparently extended for several kilometres, as along the southern boundary of Crux Easton, a vill held by a huntsman in 1066, and along the southern boundary of Faccombe. The latter is referred to as the 'white *haga*' in 961 (Birch, 1885–1899, B 1080; Sawyer, 1968, S 689) and is represented today by a woodbank full of white-coloured flints. In the field, several *haga* boundaries appear to be marked by abraded banks often today carrying shrubs or trees.

West Worcestershire was also a frontier region, in this case of the Anglo-Saxon kingdom of the Hwicce (Hooke, 1985). Much of the land to the north of the river Teme was given away by the king to the church of Worcester although the crown retained the most western estates of Martley and part of Wichenford. Charters of Grimley refer to the *ealdan kyninges hagan*, 'the king's old fence/enclosure', along part of the boundary separating royal and church lands (Birch, 1885–1899, B 462; Sawyer, 1968, S 201; Hooke, 1990, pp. 69–78, 115–118, 287). Large areas to the south of the Teme were also granted to the churches of Worcester and Pershore but Hanley Castle was retained by the crown and was to become the administrative centre of the Forest of Malvern, which extended across all the land west of the Severn by 1086 (Hooke, 1989, pp. 126–127). In this area *haga* features are again abundant and one such boundary, recorded in a charter of Upton-on-Severn in 962 (Hooke, 1990, pp. 244–247), coincided with that of one of the woods of the medieval deer-park attached to Hanley Castle and, probably, to the *haia* recorded here in the Domesday Book (Moore, 1982, 1,34). In this same region, on the Worcestershire Plain at the foot of the Malvern Hills, there are charter references to 'wolf' and 'swine' *hagan* but in what is now cultivated countryside no field remains have been found to give any indication of the nature of these features. By 1086, a considerable part of the county had been put under forest law, including the Malvern region shown in Fig. 2.4b, and by the thirteenth century only the south-eastern part of the county, the fertile Vale of Evesham, lay outside the forest. By the thirteenth century royal forest covered a quarter of the land area of England (Bazeley, 1921, p. 146), and was by them somewhat reduced from its greatest extent.

Finally, one must pose the question as to why, if the presence of woodland was such an important factor in forest location, some heavily wooded regions did not become forest. The Weald never became royal forest and neither did it have any *haga* features recorded; all those that do occur in Kentish charters lay close to the river Thames. Was it the long-established tradition of the use of forest dens[4] or swine-pastures by surrounding communities (Witney, 1976) that influenced this? Were ownership rights, first by communities, later by manors (Hooke, 1993), too strong? Conversely, there are other areas in which *haga* features did exist but where these did not lead to royal forest: Meon in Hampshire (Fig. 2.3) may be quoted as an example. Only detailed study might cast further light upon this problem but is beyond the scope of the present chapter.

Conclusion

In conclusion, it is suggested that medievalists interested in royal forests should be aware of an earlier history in the use of woodland, and particularly in its use for hunting, for the core regions appear to have been already recognized in late Anglo-Saxon times as hunting areas. The distribution of Anglo-Saxon hunting grounds, where they can be identified, appears to have influenced the location of the royal forests established after the Norman Conquest.

Fig. 2.4. (Opposite). (a) The incidence of the *haga* term in Worcestershire shown against the distribution of medieval forests (source: charters and place-names). (b) West Worcestershire: *haga* and forest features to the south of Worcester.

Parks and forests in medieval England

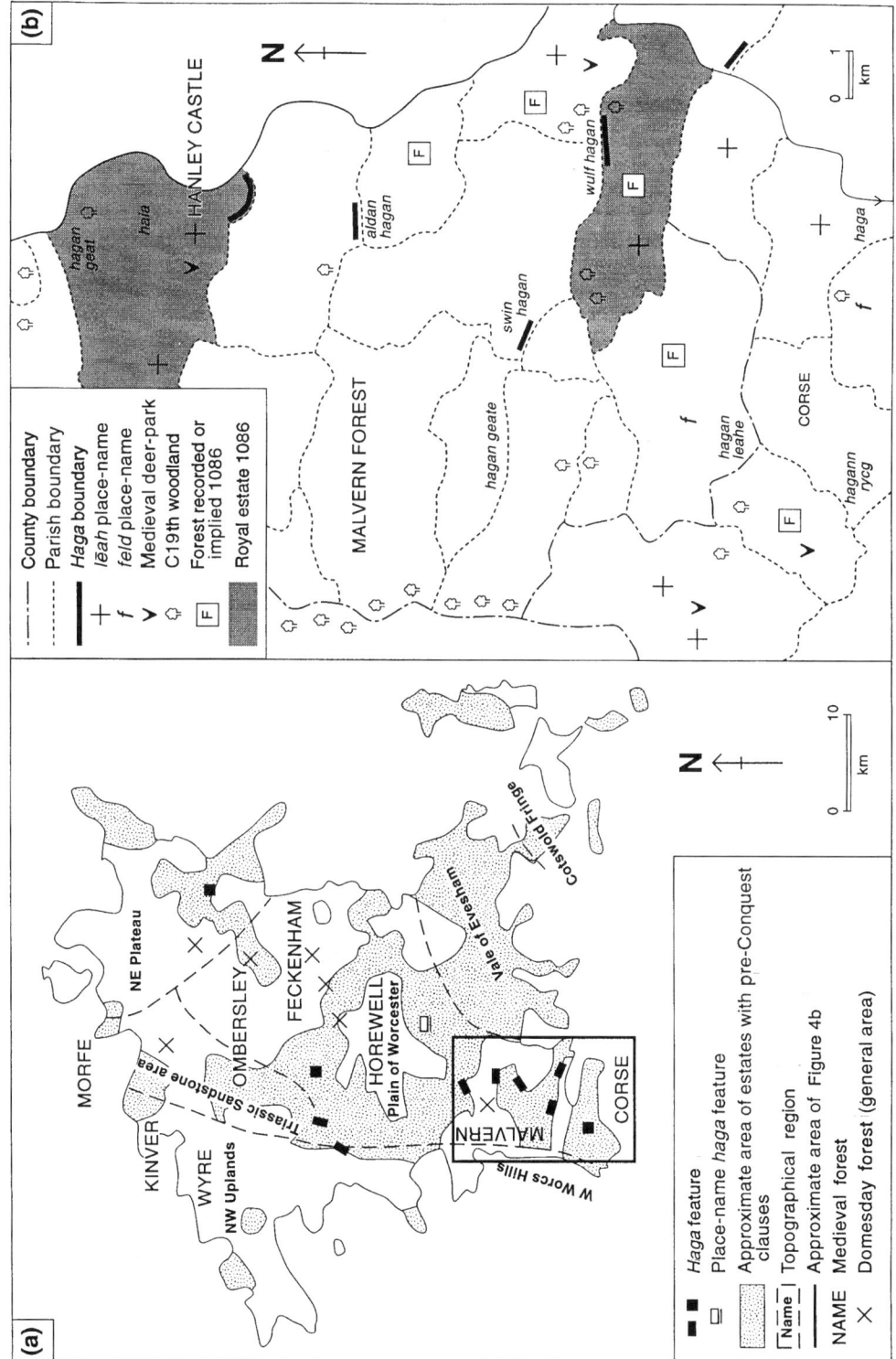

Acknowledgement

I would like to thank Kathryn Sharp for her cartographic assistance.

References

Bazeley, M.L. (1921) The extent of the English forests in the thirteenth century. *Transactions of the Royal Historical Society*, 4th series 4, 140–172.
Birch, W. de Gray (1885–1899) *Cartularium Saxonicum*. Whiting and Co., London; reprint 1964, Johnson Reprint Co., New York and London.
Birrell, J.R. (1980) The English medieval forest. *Journal of Forest History* 24, 78–85.
Birrell, J.R. (1996) Hunting and the royal forest. *L'Uomo e la Foresta Secc. 12–18*, Atti delle Settimane di Studi, Prato 1995, 27 (Prato), 437–457.
Cantor, L. (1982) Forests, chases, parks and warrens. In: Cantor, L. (ed.) *The English Medieval Landscape*. Croom Helm, London, pp. 56–85.
Cantor, L.M. and Wilson, J.D. (1964) The medieval deer-parks of Dorset: 3. *Proceedings of the Dorset Natural History and Archaeological Society for 1963* 86, 141–152.
Coates, R. (1989) *The Place-Names of Hampshire*. Batsford, London.
Cummins, J. (1988) *The Hound and the Hawk, the Art of Medieval Hunting*. St Martin's Press, New York.
Darby, H.C. (1977) *Domesday England*. Cambridge University Press, Cambridge.
Darby, H.C. (1979) Dorset. In: Darby, H.C. and Welldon Finn, R. (eds) *The Domesday Geography of South-West England*. Cambridge University Press, Cambridge, pp. 67–131.
Day, S.P. (1989) Reconstructing the environment of Shotover Forest, Oxfordshire. *Medieval Settlement Research Group Annual Report* 4, 6.
Fairbrother, J.R. (1984) Faccombe Netherton. *Archaeological and Historical Research* 1, City of London Archaeological Society.
Grimm, J. and Grimm, W. (1877) *Deutsches Wörterbuch von Jacob Grimm und Wilhelm Grimm, 4, 2*. Leipzig.
Hill, D. (1981) *An Atlas of Anglo-Saxon England*. Blackwell, Oxford.
Hooke, D. (1981) *Anglo-Saxon Landscapes of the West Midlands: The Charter Evidence*. British Archaeological Reports, British Series 95. Oxford.
Hooke, D. (1985) *The Anglo-Saxon Landscape, The Kingdom of the Hwicce*. Manchester University Press, Manchester.
Hooke, D. (1988) The Warwickshire Arden: the evolution and future of an historic landscape. *Landscape History* 10, 51–59.
Hooke, D. (1989) Pre-Conquest woodland: its distribution and usage. *Agricultural History Review* 37, 113–129.
Hooke, D. (1990) *Worcestershire Anglo-Saxon Charter-Bounds*. Boydell Press, Woodbridge.
Hooke, D. (1993) Woodland in the peasant economy of England. In: Brandl, H. (ed.) *Geschichte der Kleinprivatwaldwirtschaft, Geschichte des Bauernwaldes, Mitteilungen der Forstlichen Versuchs – und Forschungsanstalt Baden-Württemberg*, Heft 175. Freiburg, 202–210.
Hooke, D. (1994) The woodlands of England in Domesday Book. In: Billen, C. and Vanrie, A. (eds) *Les Sources de l'Histoire Forestière de la Belgique*. Archives et Bibliothèques de Belgique, Numéro spécial 45. Brussels, 35–51.
Hooke, D. (1998) *The Landscapes of Anglo-Saxon England*. Leicester University Press London and Washington (in press).

Lexer, M. (1992) *Mittelhochdeutschen Taschenwörtebuch* 38. Stuttgart.
Linnard, W. (1982) *Welsh Woods and Forests: History and Utilization*. National Museum of Wales, Cardiff.
Metz, W. (1954) Das 'Gahagio regis' der Langobardan und die deutschen Hagen-Ortsnamen. In: *Beitrage zur Namenforschung in Verbindung mit Ernst Dickenmann, herausgegeben von Hans Krahe*, Band 5. Winter, Heidelburg, pp. 39–51.
Moore, J.S. (ed. and trans.) (1982) *Domesday Book, 15, Gloucestershire*. Phillimore, Chichester.
Neilson, N. (1940) The forests. In: Willard, J.F. and Morris, W.A (eds) *The English Government at Work, 1327–1336*, Vol 1. Medieval Academy of America, Cambridge, Massachusetts.
Rackham, O. (1986) *The History of the Countryside*. Dent, London.
Rahtz, P. (1979) *The Saxon and Medieval Palaces at Cheddar*. British Archaeological Reports, British Series 65. Oxford.
Sawyer, P.H. (1968) *Anglo-Saxon Charters: An Annotated List and Bibliography*. Royal Historical Society, London.
Shore, T.W. (1988) Ancient Hampshire forests. *Hampshire Field Club and Archaeological Society* 1, 40–60.
Smith, A.H. (1970) *English Place-Name Elements, Part 1*. English Place-Name Society, Vol. 25, 2nd Edn. Cambridge University Press, Cambridge.
Stenton, F.M. (1971) *Anglo-Saxon England*, 3rd Edn. Clarendon Press, Oxford.
Swanton, M.J. (ed. and trans.) (1996) *The Anglo-Saxon Chronicle*. Dent, London.
Thorn, C. and Thorn, F. (ed. and trans.) (1979) *Domesday Book, 6, Wiltshire*. Phillimore, Chichester.
Thorn, C. and Thorn, F. (ed. and trans.) (1982) *Domesday Book, 16, Worcestershire*. Phillimore, Chichester.
Thorn, C. and Thorn, F. (ed. and trans.) (1983) *Domesday Book, 17, Herefordshire*. Phillimore, Chichester.
Thorn, C. and Thorn, F. (ed. and trans.) (1986) *Domesday Book, 35, Shropshire*. Phillimore, Chichester.
Thorpe, B. (ed.) (1846) *Colloquium ad Pueros Lingua Latinae Locutione Exercendis ab Ælfrico Compilatum, Analecta – Satonica*. London.
Thorpe, B. (1865) *Diplomatarium Anglicum Ævi Saxonici*. London.
Welldon Finn, R. (1979) Wiltshire. In: Darby, H.C. and Welldon Finn, R. (eds) *The Domesday Geography of South-West England*. Cambridge University Press, Cambridge, pp. 1–66.
West, J. (1964) The forest offenders of medieval Worcestershire. *Folk Life* 2, 80–115.
Willis-Bund, J.W. and Page, W. (eds) (1906) *The Victoria History of the County of Worcester*, Vol. 2. Constable, London
Witney, K.P. (1976) *The Jutish Forest: a Study of the Weald of Kent from 450 to 1380 AD*. Athlone Press, London.
Young, C.R. (1979) *The Royal Forests of Medieval England*. University of Pennsylvania Press, Pittsburgh, Pennsylvania/Leicester University Press, Leicester.

Notes

1. Rackham (1986, p. 131) rightly notes that forests were not all wooded. Doubts must be cast on Rackham's estimates of forest coverage and of the nature of the land involved, however, when he includes within 'The big wooded areas' which had 'few

forests' Worcestershire and north-west Warwickshire (p. 133). Virtually the whole of Worcestershire (apart from the Vale of Evesham) was forest at some time in the medieval period (*VCH Worcestershire II* 1906, pp. 315–321; West, 1964) and there are indications that the Arden woodland may also have been forest for a short time (see Hooke, 1988). Darby, 1977, fig. 65, p. 197, records forests recorded and implied in 1086 while the distribution of medieval royal forests is shown in Bazeley 1921, p. 165, and Neilson 1940, map V. Neither of the last two maps, however, is entirely accurate.

2. Sawyer's, *Anglo-Saxon Charters: an Annotated List and Bibliography* is currently being brought up to date by S. Kelly.
3. A battle fought *aet Peonnum* (believed to be near Penselwood on the edge of Somerset and Wiltshire) in AD 658 is traditionally thought to have opened up east Somerset to English settlement, see Stenton, 1971, pp. 63–64.
4. *den-baer**, 'swine-pasture', *denn*, 'woodland pasture, esp. for swine' (Smith, 1970, p. 128); probably related to MDu *dann* 'a forest, a haunt of wild beasts'; occurs in the woodlands of Kent and Sussex with a few examples in Surrey and Essex; probably obsolete by the end of the OE period.

* A postulated form.

CHAPTER 3
Royal Forests in England and their income in the budget of the feudal monarchy from the mid twelfth to the early thirteenth centuries

Yu J. Serovayskaya
University of Karaganda, University St 28, 470012, Karaganda 12, Kazakhstan

The institution of the Royal Forest played a very important role in English medieval feudal society from the eleventh to the fourteenth centuries. The main aim of my study is to reveal the income of the Royal Forests and show what proportion of the State Exchequer they formed. The incomes of the Royal Forests have yet to become a subject of systematic study either in England or the historiography of post-Soviet Russia and other states of the Commonwealth of Independent States. The statement of the problem and attempts to solve it open up new opportunities for improving our knowledge of the structure of the budget of the feudal monarchy, the Crown's fiscal and forest policies, the function of the Royal Forests and their place in English feudal society. It also helps us to understand the role the Forests played in sharpening the social and political contradictions in England in the mid twelfth and early thirteenth centuries.

Although I am aware of recent developments in historical ecology (Rackham, 1980; Kirby and Watkins, 1998) and of the benefits of combining documentary evidence with field survey, my chapter is based entirely on written historical sources and I agree with Hilton (1966) that historians of the English economy of the Middle Ages are fortunate in the abundant survival of historical sources of all kinds.

The principal source for elucidating the problem under study is the many-volumed Pipe Rolls. The overall investigation and statistical analysis of the figures in the Pipe Rolls for the fiscal years 1158–1188 (Henry II) and for the fiscal years 1204–1205 and 1208–1209 (John) show us the early revenue of the Exchequer under various heads.[1]

Even such authentic sources as the Pipe Rolls do not reflect completely the true size of the income from the Forests. Much income in the form of special royal dispositions by-passed the Exchequer. Moreover the Pipe Rolls do not account for the value of timber, game, wildfowl, fuel and so forth which was used for the maintenance of the Royal Court, nor for that distributed to representatives of the feudal elite.² In addition, part of the Forest incomes, in particular fines paid by those who had broken forest laws, went not to the State Exchequer but direct to the Chief Forester of the special Exchequer of the Forests based at Nottingham.³ Moreover, it is impossible to account for those sums plundered by the legal and administrative staff of the Forests such as foresters, forest officers and bailiffs. All these factors suggest that the true income from the Forests was much greater than that represented by the statistics in the Pipe Rolls.

The annual records in the Pipe Rolls normally consist of two parts. First, ready money and second, debts to the Exchequer (*debita pro foresta*). In this study I have combined the two parts and consider the total income for different financial years. The data of the Pipe Rolls allowed us to determine only the size of the income from the Forests. To determine the relative importance of the Forest income, it was necessary to compare these data with others which provide the total income for the Crown. To do this we made use of Ramsay's (1925) research on the gross income of the Crown.⁴

When considering the income from the Forests, we must take account of the area and distribution of the Forests. The Forests were not simply a local phenomenon in medieval England (Young, 1979; Rackham, 1980). The whole county of Surrey, for example, and parts of Bedfordshire were afforested by Henry II and remained under Forest law until 1204. The county of Rutland was in the bounds of the Forest until 1223. The limits of the Royal Forests of Huntingdonshire coincided with the borders of the shire from 1224 until 1286. Essex, which was afforested in the reign of William I, was reputed as Forest until 1301 (Pearson, 1887; Malden, 1905; Bazeley, 1921). The New Forest in Hampshire covered around 95,000 acres; the Forest of Dean about 100,000 acres and the forests of Yorkshire about 90,000 acres (Baring, 1909; Page, 1910; Darby and Campbell, 1962).

The income from the Royal Forests was accounted under several different headings. The first, the *cost of timber sold*, was the regular item of Exchequer income. It consisted of receipts from two categories: the returns of sales of wood by the Royal administration of the Forests and payments for licences on the sale of wood. Rubner (1965) argues that 43% of the gain from such licences entered the Crown's treasury. Our study shows that during the 29 years of the reign of Henry II and the two selected years of King John, the Exchequer received under this item £4093 4s. 4d. which forms 6.6% of the total Forest income (see Tables 3.1 and 3.2).

The second heading is *census and firma of the foresters* which consisted of fees paid into the Exchequer for entries into post by forest stewards and foresters. This item received £2418 12s. 8d., 4.2% of the total income. The third heading related to all income associated with the *right of chase*. It includes income from game, game licences and permission to make private parks, free warrens and free chases. The income under this category in this period is rather small (£398 1s. 2d.) and forms only 0.6% of the total income. The revenues from *pasture and fattening cattle in the forests*, including grazing of the forest floor and the use of leaf fodder, were also small, forming only 1.1% of the total income of the Forest.⁵

Table 3.1. Incomes of the Crown from the Forests 1158–1179.

Item of receipts in the Exchequer from the Forests	1158–1168				1168–1179			
	Receipts for 10 years	Average for year	Maximum yearly receipt		Receipts for 10 years	Average for year	Maximum yearly receipt	
			Year	Sum			Year	Sum
1. Cost of the timber sold	£1358 9s. 2d.	£135 16s.	£1165 66s.	£188 6s.	£1537 15s.	£153 14s.	£1169 70s.	£367 16s.
2. Census and firma of the foresters	£227 15s.	£22 14s.	£1166 67s.	£70 18s.	£97 1s. 8d.	£9 12s.	£1173 74s.	£30 0s. 2d.
3. Right of chase	£86 16s.	£8 12s.	£1159 60s.	£49 16s.	£46 13s.	£4 12s.	£1169 70s.	£25 13s.
4. Revenues from pasture and fattening cattle in the forests	£16 10s.	£1 12s.	£1167 68s.	£16 10s.	£289 6s.	£28 9s.	£1168 69s.	£136 16s.
5. Fines on the sentence of travelling judges	£481 15s. 8d.	£48 2s.	£1167 68s.	£170 11s.	£6024 0s. 11d.	£602 8s.	£1169 70s.	£968 1s.
6. Arbitrary fines	£31 3s. 7d.	£3 2s.	£1167 68s.	£8 1s. 11d.	£13,703 4s. 10d.	£1370 4s.	£1176 77s.	£6292 5s. 2d.
7. Redemption money for deafforestation	–	–	–	–	£333 6s.	£33 6s.	£1177 78s.	£333 6s.
Total	£2202 9s. 5d.	£220 10s.			£22,031 7s. 5d.	£2203 2s.		–
% of total sum	3.7	(0.37)			36.6	(3.66)		

Table 3.2. Incomes of the Crown from the Forests.

Item of receipts in the Exchequer from the Forests	1179–1188		Maximum yearly receipt		1204–05	1208–09	Total			
	Receipts for 9 years	Average for year	Year	Sum			Act to the Exchequer	Average for year	Total	% of total sum
1. Cost of timber sold	£1006 6s. 7d.	£112	1180–81	£174 7s.	£132 9s.	£58 4s.	£4093 4s. 4d.	£132		6.6
2. Census and firma of the foresters	£65 14s.	£7 4s.	1186–87	£81 15s.	£1373 2s.	£653	£2418 12s. 8d.	£78		4.2
3. Right of chase	£3 9s.	7s. 8d.	1181–82	£3	£221 3s. 2d.	£40	£398 1s. 2d.	£12 14s.		0.6
4. Revenues from pasture and fattening cattle in the forests	£337 16s. 7d.	£37 12s.	1186–87	£93 13s.	–	–	£643 12s. 7d.	£20 14s.		1.1
5. Fines on the sentence of travelling judges	£5676 4s.	£630 12s.	1184–85	£1997 18s.	£59 14s. 6d.	–	£12,242 5s. 1d.	£395		20.3
6. Arbitrary fines	£8598 14s. 5d.	£970 12s.	1180–81	£1344 18s. 5d.	£116 16s. 4d.	–	£28,222 15s. 6d.	£910 12s.		46.8
7. Redemption money for deafforestation	£885	£98 6s.	1184–85	£443	£10,146 17s.	£918 17s.	£12,286	£396 6s.		20.4
Total	£16,571 4s. 7d.	£1841	–	–	£12,050 2s.	£7444 7s. 4d.	£60,307 13s. 4d.	£1945 6s.		100
% from total sum	27.5	(3.05)			19.9	12.3				

Payment of fines formed by far the largest category of income. *Arbitrary fines* provided 46.8% of all income and *fines on the sentence of travelling judges* provided 20.3%. In the single year 1175 the records indicate that 1500 offenders were fined. The final income category was *redemption money for deafforestation*. In 1204–05 the treasury received 100 marks under this heading from the forests of Staffordshire, 100 marks from Yorkshire, 500 marks from Essex, 2000 marks from Cornwall and 5000 marks from Devon.[6] This category provided just over a fifth (20.4%) of all income to the Exchequer from the Forests.

This analysis of income shows that the Royal Forests provided a flexible and universal instrument of fiscal exploitation. Incomes from the Royal Forests provided, along with other feudal payments such as scutum money, receipts from towns, receipts from judicial curia on lawsuits from the common right and so forth, a substantial component of the Crown's money resources. In our period the Royal Forests provided just under a tenth (9.8%) of the total income received by the Exchequer, and, moreover, the Royal Forest became increasingly profitable through the twelfth and fourteenth centuries.

This growth in income was partially due to improving economic conditions and the consequent growth in demand for the various products of the forest economy. However, in order to increase Forest income, the Crown abused the system of arbitrary fines and redemption money and plundered certain strata of English society, first and foremost the peasants and artisans, but also feudal proprietors of all ranks. Therefore, the Crown's attempt to retain and increase the area of the Royal Forests, keep intact its rights and privileges as regards the forests and increase the income received met with ever-growing resistance from most strata of English society.

The Royal Forest should be seen in this period as a perfect instrument for the monopolization of natural resources in the interests of the rulers of feudal England. It was simultaneously the source of power and weakness and was an historical phenomenon of great importance. The objective result of the Forests' existence also had a great effect on the interrelations between society and forest ecosystems. Further work is required to explore the correlation between the legal and social form of the Royal Forests and their ecological characteristics.

References

Baring, F.H. (1909) *Domesday Tables for the Counties of Surrey, Berkshire, Middlesex, Hertford, Buckingham and Bedford, and for the New Forest.* London.

Bazeley, M.L. (1921) The extent of the English forests in the thirteenth century. *Transactions of the Royal Historical Society*, 4th series 4, 146–148.

Darby, H.C. and Campbell, E.M.J. (ed.) (1962) *The Domesday Geography of South-East England.* Cambridge University Press, Cambridge.

Hilton, R.H. (1966) *A Medieval Society.* Weidenfeld & Nicholson, London.

Kirby, K.J. and Watkins, C. (eds) (1998) *The Ecological History of European Forests.* CAB International, Wallingford.

Malden, H.E. (ed.) (1905) *The Victoria History of the County of Surrey.* Constable, London.
Page, W. (ed.) (1910) *The Victoria History of the County of Yorkshire.* Constable, London.
Pearson, C.H. (1887) *Historical Maps of England During the Thirteenth Century.* London, 10.
Rackham, O. (1980) *Ancient Woodland: its History, Vegetation and Uses in England.* Edward Arnold, London.
Ramsay, J.H. (1925) *History of the Revenues of the Kings of England.* Oxford.
Rubner, H. (1965) *Untersuchungen zur forstverfassung des mittelalterlichen Frankreichs.* Wiesbaden.
Young, C.R. (1979) *The Royal Forests of Medieval England.* Leicester University Press, Leicester.

Notes

1. The Publications of the Pipe Rolls Society, London 1884–1925. Vol. I–XXXVIII; Hunter, J. (ed.) (1930) *The Great Rolls of the Pipe for the Second, Third and Fourth Years of the Reign of King Henry the Second.* Record Commission, London; (1940) *The Great Roll of the Pipe for the Sixth Year of the Reign of King John 1204–1205.* London, Vol. 18; (1968) *Calendar of Memoranda Rolls, Exchequer, Preserved in the Public Record Office.* London, pp. 23–66.
2. Some natural receipts from the Forests were recorded in other rolls. See Hardy, T.D. (ed.) (1844) *Rotuli de Liberate ac de Misis et Praestitis Regnante Johanne.* London; (1916–1964) *Calendar of the Liberate Rolls Henry III 1226–1272.* London, Vols 1–6; Hardy, T.D. (ed.) (1833–1844) *Rotuli Litterarum Clausarum.* London, Vols 1–2.
3. For example in the rolls of the Chief Forester Hugo Neville in 1207 debts for different delinquencies in the forests amounted to £5500 and in the previous 6.5 years when Gugo Neville held this post the total debts were £9400. See Holt, J.C. (1961) *The Northerners: a Study in the Reign of King John.* Oxford, p. 157.
4. Ramsay, J.H. (1925) *History of the Revenues of the Kings of England.* Oxford. Vol. 1, pp. 74, 90, 100–101, 125–126, 134, 143, 161, 165, 195, 232.
5. *The Great Roll of the Pipe for the Twenty-second Year of the Reign of King Henry the Second AD 1175–1176* (1904). London. Vol. 25, p. 203.
6. *The Great Roll of the Pipe for the Sixth Year of the Reign of King John 1204–1205* (1940). London, Vol. 18, pp. 40, 189.

CHAPTER 4
English cathedrals as sources of forest and woodland history

Gavin Simpson
Historic Buildings Research Unit, Department of Archaeology, University of Nottingham, Nottingham NG7 2RD, UK

Cathedrals and monasteries were the centres of learning in the Middle Ages. Their prelates were the guardians of great wealth in land and property and did business with the government and merchant classes. They kept detailed records and accounts of their everyday activities. The survival of these records of transactions of church, state and commerce has usually been fortuitous. However, those that have been preserved can give very detailed information about the use of natural resources from woodlands and forests, their changing role in the national economy and how they were managed, as shown for example by Rackham's (1975) study of Hayley Wood, Cambridgeshire, part of the estates of the Bishops of Ely. I give examples in this chapter which I have encountered as an archaeologist researching the architectural history of cathedrals. English cathedrals, medieval monastic buildings and parish churches contain (or sadly, too often, contained) some of the earliest and most elaborate timber-framed structures in the country. The oak timber is often of the best quality that could be obtained from medieval forests. Its origin may be documented and its tree-rings can be analysed. Analysis will demonstrate the growing conditions or woodland management practised in the forest and dendrochronology will determine when it was felled.

Twelfth and thirteenth century cathedral roofs were based on tie-beams which prevented the timber structure from pushing the wall-tops apart. The width of the cathedral was as much dependent upon the length of the tie-beams available as upon the ingenuity of the architect in the use of materials. At Lincoln cathedral – built between 1190 and 1280 – the tie-beams are 46 ft long (Foot *et al.*, 1986). At Salisbury – built between 1220 and 1266 – they are only 40 ft long. The rafters of the earliest roofs at Lincoln are also single lengths of over 40 ft with a scantling of 8×9 in. A single tree could make four rafters (Fig. 4.1). However, it must be assumed that the demand for such timber soon outstripped the ability of woodlands to supply it, for about half the rafters used to build roofs c. 1210 were made up of two lengths scarfed together. Up to this time most of the timbers were derived from trees 100–250 years old when felled. However, by the 1270s, in the last decade of the building programme, the carpenters were mostly using trees of between 50 and 100 years of age (Fig. 4.2). At least 300 of the trees used at Lincoln came from the

Fig. 4.1. A cross-section of a collar beam from the roof of St Hugh's Choir, Lincoln cathedral. It came from a tree which was at least 250 years old when it was felled in the twelfth century.

Royal Forest of Sherwood. Others came from the estates of the bishop but, as no fabric accounts have survived, details are lacking.

Henry III throughout his long reign (1216–1272) was a great sponsor of ecclesiastical building and his generous gifts of timber from the royal forests are recorded in the Close Rolls.[1] He was especially generous to the new cathedral at Salisbury, which was only 2½ miles from Clarendon, one of the largest of his palaces. He gave a total of 418 trees towards the building of the cathedral from 16 Royal Parks or Forests, plus another 160 trees, coppice and underwood for non-structural purposes such as fuel for limekilns. The two largest gifts of 100 trees each came from the Forests of Dean and Trivelle, which are also the furthest distant (Simpson, 1996b). The cost of their transport, some of which was no doubt by water, would have had to be met by the fabric fund (Edwards and Hindle, 1991). The trees that the king gave were less than a third of the total of more than 1550 required to build the cathedral. The majority of the timber must have come from elsewhere but again the absence of fabric accounts means that details are lacking. This total may be compared with the master carpenter's estimate of the oaks required for the old St Paul's cathedral, London, after the total destruction of its roofs and leaded timber spire by fire in 1561 (Kitching, 1986, p. 125). The largest, 25 in number, were 50 ft long × 16 in. square and even the 2440 smallest trees were 40 ft

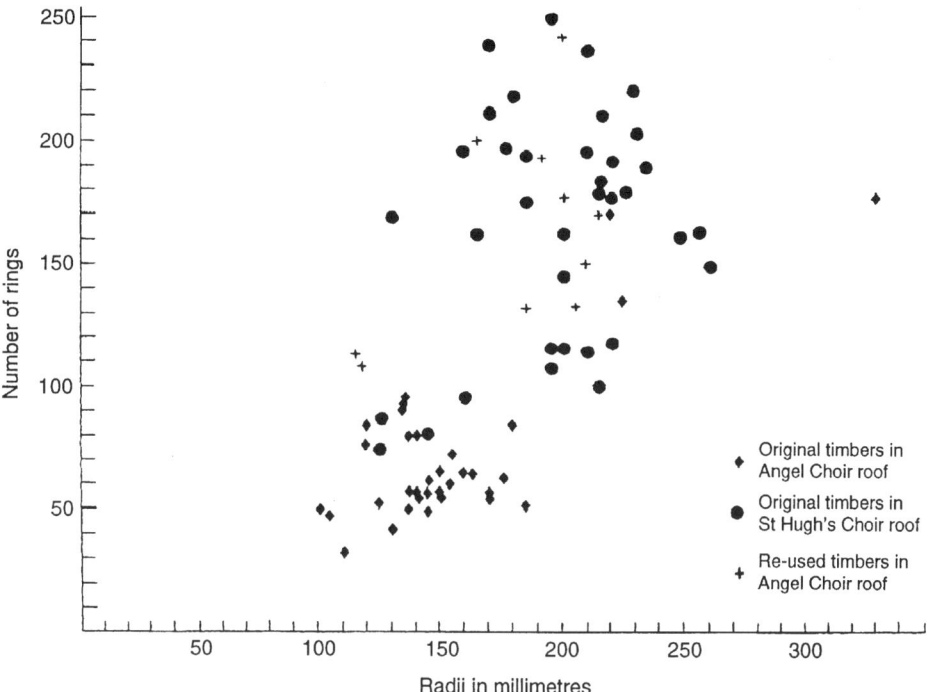

Fig. 4.2. An analysis of the size : age at felling of timbers used to build the earliest (*c.* 1200 – dots and crosses) and the latest (*c.* 1275 – diamonds) medieval roofs of Lincoln cathedral. Numbers and measurements of the earlier timbers are minimal since the centre of the tree and its sapwood has usually been removed by the carpenters.

long × 12 in. square. In total 2740 trees were required with, in addition 800 ft of 3 in. oak planking and 88,000 ft. of 1½ in. for making two floors in the tower and for covering the rafters. The total cost was just over £3250. The majority of the timber came from the Welbeck estate, Nottinghamshire (SK 5674), where the frames for the eastern (choir) roof were prepared, and also from Yorkshire, particularly near Guisborough (NZ 6014), where the frames for the western (nave) roof were prepared.

Both cathedrals of Ely and Exeter have extensive fabric accounts surviving from the late thirteenth to the mid fourteenth centuries (Chapman, 1907; Erskine, 1981, 1983). These give a much fuller picture of the sources and variety of woodland products that were required during long building campaigns and provide an interesting contrast between a monastic and a secular cathedral as well as between two different ecological and geographical regions.

The Bishop of Ely received modest grants of timber from the king for building construction in the thirteenth century and mature timber was no doubt available from his estates to provide the major portion (for locations see Heal, 1973, p. 201; for timber resources see Rackham, 1980, p. 18).[1] However, the account rolls of the sacrist which include payments for those building works for which the prior was responsible at Ely in the first half of the fourteenth century indicate that timber, indeed most types of building material, were purchased from other landowners or on the open market at local fairs (Litton and Simpson, 1996, p. 195).[2] They do not, like the contemporary Exeter fabric accounts, show costs of felling timber on the cathedral estates and bringing it into the city. When the abbey church became the seat of a bishop in 1109 the monastic lands were divided between him and the prior. Those allotted to the monks provided primarily their food and clothing and also income by trade and from rent. There is no evidence that they included extensive woodlands either at that time, or later (Evans, 1973, p. 11; Rackham, 1986, p. 83). The monks, for example, used peat both for fuel and for firing bricks (Chapman, 1907, pp. 30, 67; Evans, 1973, p. 13). Of the major works carried out in the cathedral in the thirteenth and fourteenth centuries, the building of the choir and of the nave roof was at the expense of bishops Northwold (1234–1251) and Hotham (1321–c.1338) and the building of the Octagon and Lady Chapel (1321–1350) was at the expense of the prior. Only the accounts for Northwold's work and for the Octagon survive and, unfortunately, only the latter detail both the receipts and expenditure. However, it may be pertinent to note that the survey of the bishop's estates known as the Old Coucher was completed in 1251, the year in which his building campaigns on the cathedral and elsewhere were concluded. These certainly drew heavily upon his financial resources and no doubt his woodlands also, as for example the new roof of the nave built c. 1240 using over 200 mature oaks (Esling et al., 1989; Simpson, 1996a, unpublished).

There is no record that Exeter received any gifts of timber from royal forests but most of the builders' requirements could be provided it seems from the manorial estates of the bishop or of the dean and chapter in south Devon. The Exeter fabric accounts for midsummer 1332, for example, record preparations for the construction of the nave roof (Hewett, 1985, p. 46; Mills, 1988). First the payments for the felling and transport of large oak timber on heavy horse-drawn wagons from the bishop's woods at Chudleigh (SX 8780) are detailed, then the same for boards to cover the rafters and finally the purchase of lead and bundles of moss on which to lay it (Erskine, 1983, pp. 80, 249). It seems that the boards came directly from Chudleigh and so were presumably sawn *in situ* as in Worcester cathedral's woods and elsewhere in the seventeenth and eighteenth centuries (Hendry et al., 1984; Fearn and Simpson, 1997). At other times the master carpenter purchased building timber from Langford and other parts of Devon but it is not always clear if it was from markets or estates of the laity. The dean and chapter's Stoke Wood, just to the north of Exeter (SX 930962) during the whole period covered by the accounts (1279–1353) provided only coppice and underwood, particularly for fuelling limekilns and for making the hurdles which were used to deck the builders' scaffolding. At Lincoln cathedral part of such a hurdle made of sallow has been reused in the early thirteenth century as part of the centring of the vaulting of a small chamber in the north-east transept (Litton and Simpson, 1996, p. 203). Scaffolding poles at Exeter and elsewhere usually came from local sources. They were sometimes of poplar but alder was prefered, probably because it gave lighter and more slender poles (Erskine, 1983, pp. 291–292, 321; Salzman, 1952, p. 323).

However, it was poles of imported conifer that particularly had these qualities and which were therefore most sought after even into the present century, before the introduction of tubular steel scaffolding. In the Middle Ages, Norway spruce and Scots pine were imported as boards, as poles for scaffolding and for making ladders, and 'spars' for roof rafters (Chapman, 1907, pp. 18, 47, 54–55). Conifer timber was being imported from Norway by the late twelfth century (PPRS, 1914, pp. xxi, 116 and 199) and later, from the early thirteenth centuries, together with oak, from the southern shores of the Baltic. Much of the oak came as boards which were, at least at first, riven not sawn from the log. The vault of the lantern of Ely cathedral is largely constructed of them, imported through King's Lynn (Chapman, 1907, p. 72). Baltic oak can be identified by tree-ring analysis (Eckstein et al., 1986). The boards have been riven from very long-lived trees having many very narrow rings – like the quality of English oak that was available to the Lincoln carpenters in the twelfth century (Fig. 4.1). The timber came from the valleys of the rivers which rise in the mountains of eastern and central Europe and end their journeys at the Baltic. These forests were first settled by Neolithic farmers practising slash and burn agriculture in the fifth millenium BC but there were probably still areas of primary forest and certainly forest where human intervention had been minimal or non-existent over long periods of time (Milisauskas, 1986, p. 3; Brown, 1997, pp. 211–214). The trees grew here in conditions very competitive for light and nutrients and developed tall trunks without significant branches except at forest canopy (Crumlin-Pedersen, 1986, pp. 141–142; Savill and Spilsbury, 1991). The lack of knots or shakes in the timber made it easy to split into long, wide planks and the rays gave the surfaces attractive patterning. Documentary and dendrochronological sources indicate that timber was being imported from northern and north-eastern Europe in considerable quantity to eastern ports of Scotland and England (Simpson, 1996c, p. 90). Oak timber was also being imported from Ireland from early in the thirteenth century, mostly to western and south-western ports of Britain. Irish timber was used in the building of the east end of Salisbury cathedral c. 1224 (Simpson, 1996b, p. 17) and Irish boards were bought for Canterbury cathedral in 1254 and to make an altar table at Exeter cathedral in 1316 (Blore and Harvey, 1945, pp. 29, 35; Salzman, 1952, p. 245; Erskine, 1981, pp. 89, 138).

Around the early sixteenth century the seaboard nations of western Europe began to develop their naval power and one of the consequences of this was that the Baltic, because of its raw materials – not only timber but iron, pitch, hemp, saltpetre for gunpowder and many other things – became an arena for political intrigue. Henry VIII was not only intent upon safeguarding his overseas supplies of timber from the likes of the Dutch, who had few forests of their own, but also in maximizing the revenue and the timber that English forests could supply (Green, 1870, p. 183). Following his dissolution of the monasteries in the late 1530s, a surveyorship of their woods was created and in 1542 John Mynne was appointed 'Master of the Woods'. His alternative title 'Master of the King's Wood Sales' betrays the true purpose of his appointment. With the home supply of naval timber in mind, Parliament in 1543 passed the first important timber preservation act. It stipulated that wherever woods were cut down a minimum of 12 young trees must be left on every acre and also the minimum size that mature trees ('standards') should be before they could be felled (Albion, 1926, pp. 121, 154; Statute 35 Hen. VIII, c. 17). The legislation codified good practice for woodland management and effectively prevented the conversion of wooded land to other uses. Its influence can be seen in the leases that the post-Reformation deans and chapter of Lincoln drew up with their tenants. On the rare occasions that a tenant was allowed to fell timber the relevant sections of the act were spelled out in the lease (Cole, 1917, pp. 51–52). The provisions of the act were strengthened by Parliament in 1570 (Statute 13 Eliz. c. 25) and 10 years later Lord Burleigh, Queen Elizabeth's Secretary of State, attempted to control the felling of timber by demanding that bishops and secular owners of large estates send him details of their fellings during their incumbencies (Lemon, 1856, pp. 645–654; Heal, 1980, p. 287).

This measure was to some degree a response to the excessive exploitation of the woodland resources of his estates by the bishop of London. Bishop Cox of Ely was another who allegedly derived a substantial income from this source. Although most cathedral manors were let out to tenants their woodlands were usually kept in hand by their ecclesiastical landlords because they were assets which could be readily turned into cash. This assumed a greater significance following an Act of Parliament in 1559 (Statute 1 Eliz. c.19) by which the Crown during the vacancy of a See could exchange impropriated rectories and other spiritual revenues which it had acquired on the dissolution of the monasteries for the temporal estates of the bishops. Ely, one of the richest Sees in the country, was one of the first to suffer. When Richard Cox was appointed towards the end of that year he found that he had been deprived of most of his Norfolk and Suffolk estates in exchange for spiritualities which brought in a reduced income. Greater efficiency in management, the renegotiation of leases, the sale of farm produce and of timber were the principal means by which the bishops endeavoured to maintain their real incomes in the face of not only an unsympathetic queen and parliament but also price inflation (Heal, 1980, pp. 32, 194, 267). The sale of timber, it would seem, was a method particularly favoured by Bishop Cox. Unfortunately there is no return from Ely to Burleigh's request of 1579 because the bishop retired the same year and the estates came under the administration of the Crown, until his successor was appointed nearly 20 years later. Meanwhile the Crown received the revenues of his estates (Heal, 1973, pp. 206–210).

At this time most cathedral chapters still retained one or more woods from which timber was taken as required for new work and for repairs which were becoming increasingly necessary in their cathedrals and buildings of the close. Extensive rebuilding of the medieval roofs of Lincoln cathedral following the fall of the central spire in 1548 was probably carried out using trees from the 'Minster Wood' at Harby, just across the county boundary in Nottinghamshire, about 6 miles west of Lincoln (SK 8870). The manor of

Harby had been a property of the Common Fund since at least the fourteenth century when its wood was known as boscus Hallewoode.[3] The eight massive tie-beams across the top of the tower demonstrate that timber of the same dimensions and quality was still available to the cathedral carpenter in the mid sixteenth century as to his predecessor in the thirteenth century.

During the English civil war Parliament saw the ecclesiastical estates as potential sources of revenue to be used in much the same way as by Henry VIII a century before. The estates of the bishops were put up for sale by Act of Parliament in November 1646. In 1649 the sale of Crown estates was legalized by Parliament and the estates of the deans and chapters were pledged for an immediate loan of £300,000 from the financial community in the City to finance a military expedition to Ireland; the lands were surveyed and valued in preparation for their sale (Firth and Rait, 1911, p. 81; Madge, 1938, pp. 65, 107; Gentles, 1973, p. 615). Harby Wood was only 18 acres and its survey is fairly brief (Venables, 1884, p. 249 – see Table 4.1). Within 10 years the timber was all felled and the land sold (Williamson, 1956). However, Lincoln cathedral and the buildings of the close remained largely intact, unlike Worcester cathedral where lead and timber from the roofs were sold off although fortunately the dean and chapter still had plenty of timber on their estates for their restoration.

Figure 4.3 shows two of the manors of the Bishops of Worcester which were sources of timber for the cathedral. Wolverley in the royal forest of Kinver is first mentioned in a charter of the mid ninth century in terms which suggest that compartmented wood-pasture was already in operation there. Henry III granted the bishop timber from here and from the Forest of Dean to build the roof of the choir in 1232 (Rackham, 1980, p. 134; Fearn and Simpson, 1997).[4] Most of the manor of Eymore, on the east bank of the Severn just west of Kidderminster, was taken up with woods (SO 780795). *Parco de Eymore* is first recorded in a charter of 1278 and was one of the principal sources of timber for the cathedral to the eighteenth century (Simpson *et al.*, 1994, unpublished; Fearn and Simpson, 1997). The 1649 survey of it begins with a similar inventory and valuation as the Lincoln document (Table 4.1). It is, however, very much longer and sets out in great detail how the woods were managed in the early seventeenth century and probably for quite some time before that (Cave and Wilson, 1924, pp. 55–59). Moreover, a map of the wood made for the dean and chapter by John Broome in 1756 illustrates many of the physical features described in the earlier survey (Fig. 4.4) and a contemporary terrier adds further details.[5]

The wood was in the care of a woodward who was also the tenant of the only residence within its bounds which are defined in the document by reference to topographical features. He was responsible at his own expense for the upkeep and repair of the woodbank, which on the north side was followed (at that time) by the county boundary between Staffordshire and Worcestershire, and for any internal hedges, ditches and quicksetts. He also acted as gamekeeper, for the dean and chapter valued the wood for its 'deer hunting, hawking and other games' as well as for its timber and underwood. There was no common pasture within the manor and only sheep were allowed

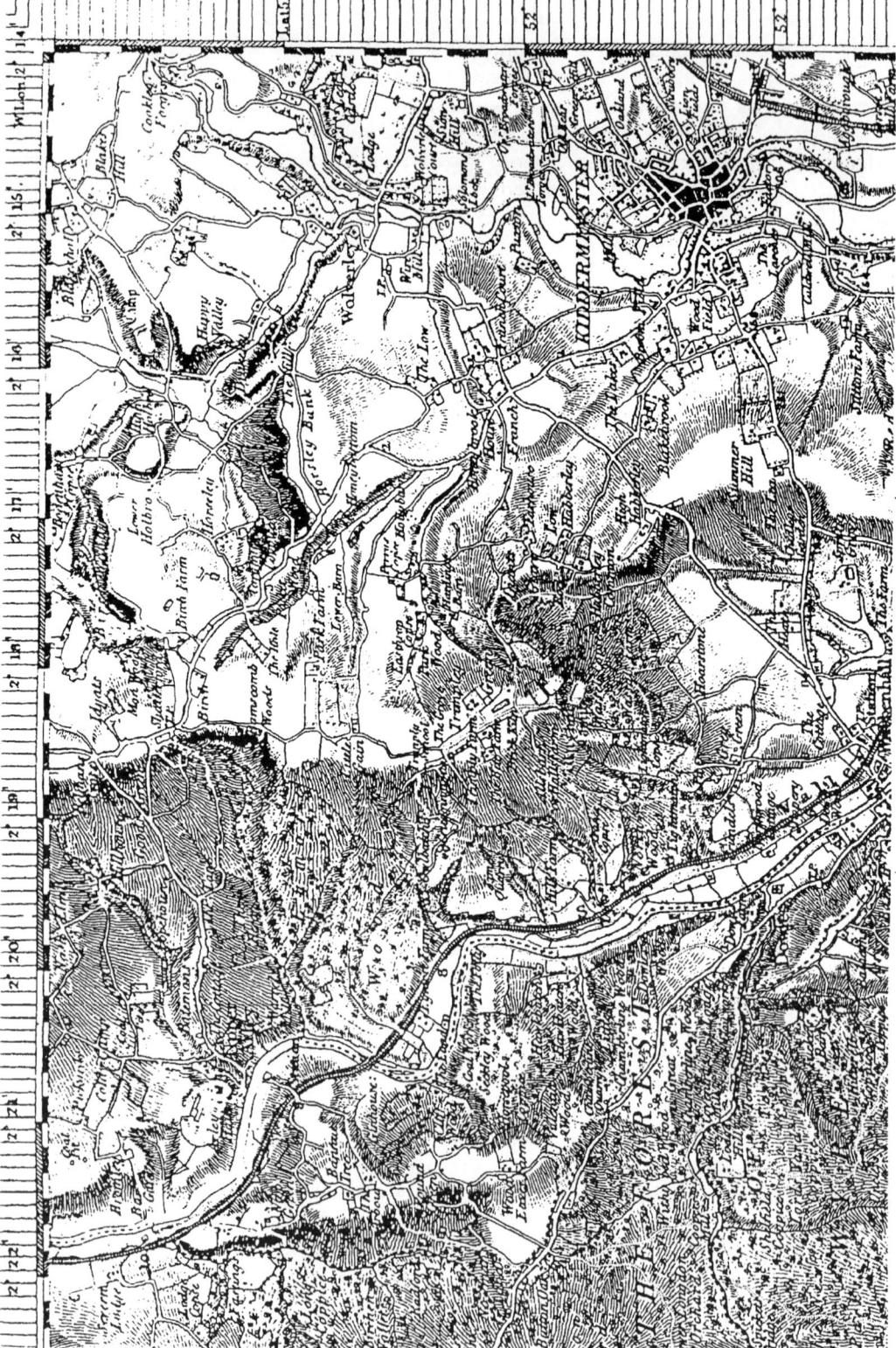

Fig. 4.3. (Opposite) A section of the Ordnance Survey map (Sheet 50) of 1832 showing Wolverley manor (top right), Eymore Wood (left of centre) and the coppiced woods of the Wyre Forest (bottom left). The map has many indications of industrial activities which were established in the area at that time.

Table 4.1. Parliamentary Survey, 1649–50, Harby Wood, Nottinghamshire.
'A survey of the wood commonly called ye Minster wood, lying and being in the Hamlett of Harby in the said County, formerly used for the repair of the Cathedrall Church of Lincolne being in present possession.'

	Numbers of trees	Rates p. tree	Total of rates £	s	d
Imprimis two faire Tymber oak Trees of assize called the ffower sisters, and the great faire Tree	002		15	00	00
Tymber oake Trees of assize	026	£3	78	00	00
Tymber oake Trees of assize	039	50s	97	10	00
Tymber oake Trees of assize	046	40s	92	00	00
Tymber oake Trees of assize	059	30s	88	10	00
Tymber oake Trees of assize	066	20s	66	00	00
Tymber oake Trees of assize	101	15s	75	15	00
Tymber oake Trees of assize	110	12s	66	00	00
Tymber oake Trees of assize	120	7s	42	00	00
Tymber oake Trees of assize	160	5s 6d	44	00	00
Sapling oake Trees	180	4s	36	00	00
Sapling oake Trees	200	2s 6d	15	00	00
	1109		715	15	00

There are small underwoods of Hazell and Sallow, by est. Eight acres, more or Lesse, valued in Grosse at ... ... £16. *Anuall Vallue*

The soyle of the said wood (being wett and morist Ground cont. in the whole by est. 18 acres, more or Lesse, being estimated as grubbed upp and worth p. ann. £3 12s

to graze in winter and swine with snouts ringed at acorn time. The wood was compartmented into 21 sales or coppices and one was to be felled each year. Whatever standards were required had to be felled at the same time and in the same coppice. The woodward had to heed both folk lore and state law in this process. He had 'to observe chiefly the wind and the moon' – no timber was to be felled when the wind was in the east, north-east or north, nor within 5 days before and after the change of the moon – and he had to leave 'so many of the likeliest young trees and so many samplers (saplings) as the law requires' (Statute 35 Hen. VIII, c. 17). He had to square up the timber and make up the 'shides, kidds and faggots' (various kinds of fuel – see Rackham, 1980, p. 142) before 10 June. Everything then had to be taken to one of the two quays on the Severn bank and loaded on to barges and taken to the Prior's slip at the College of Worcester. The surveyors end their report with a summary of their conclusions which includes the following evocative description:

> The said wood or park is very well preserved from spoil of cattle, and from stealing as by view thereof appears. The timber trees are very handsome, tall and

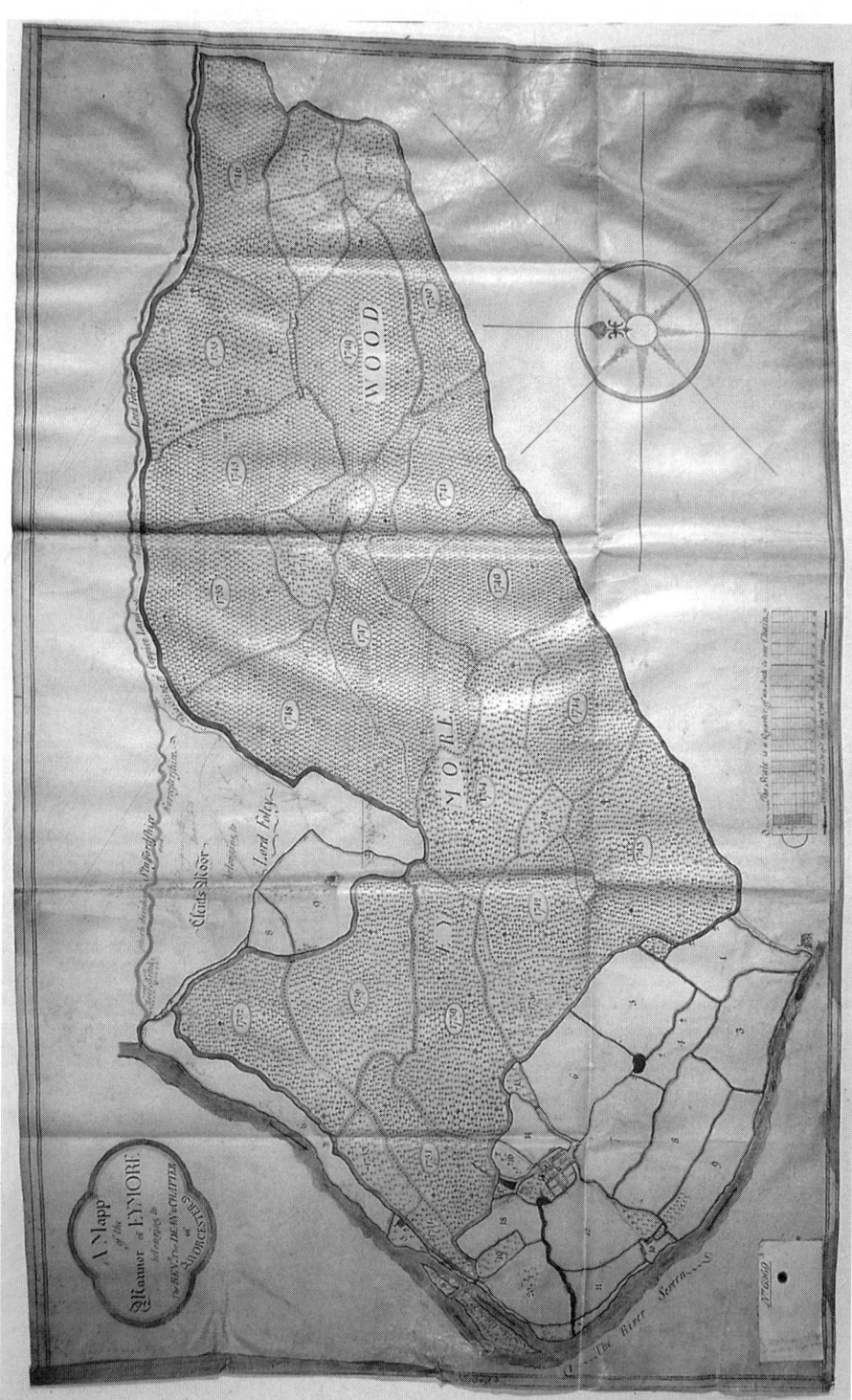

Fig. 4.4. Map of Eymore Wood drawn by John Broome in July 1756. The boundary between Worcestershire and Staffordshire follows the north edge of the wood and the bend of the river Severn is the west boundary of the manor.

smooth, of lively growth, and few or none decaying having but few boughs thereon and those near the tops for the greatest part. The underwoods are the worse for the timber trees there growing.

(Cave and Wilson, 1924, p. 59)

The description clearly refers to that part of the wood with mature trees, probably spaced at around 40 to the acre and ready for felling (Rackham, 1975, p. 47). The management system is clearly one of coppice-with-standards but the average distribution of timber oaks from the figures given is only five per acre (see Table 4.2), which was below the legal limit (Statute 35 Hen. VIII, c. 17).

The dean and chapter seem to have had varying fortunes in avoiding the depredations of the Parliamentary commissioners in their woodlands. Eymore Wood was bought by the tenant/woodward who restored it to them after the Interregnum. In 1661 the Clerk of the Works at Worcester was authorized by the chapter to make a payment of £10 to the widow of another of its woodwards for managing 'to preserve the timber at ye Berrow [in the parish of Cradley, Herefs – SO 7347] from cutting down in the bad time of sequestration and spoil' (Page and Willis-Bund, 1913, p. 171).[6] However, in the manor of Hardwick, in the eighth century estate of *Wican*, just to the west of Worcester, the surveyors found the manor house demolished 'and all the Timber Trees cut down by the enemies of the Parliament in time of the wars, and nothing to be seen but grass and weeds' (Cave and Wilson, 1924, p. 60; Hooke, 1982, p. 97; Hooke, 1985, p. 109). A similar fate seems also to have befallen much of the 280 acre, late eighth century, Cornewood in the manor of Newnham (SO 6270) in the parish of Knighton where the surveyors found 1690 timber oaks in 1649 (Cave and Wilson, 1924, pp. xvi, 108; Hooke, 1985, pp. 15, 68, 70). They proposed that these be sold off at 10s a tree and the area of managed woodland reduced

Table 4.2. Parliamentary Survey, 1649–50, Eymore Wood, Worcestershire.
'One wood or Park called Eymore containing in the whole according to Land measure, viz. 16½ ft. to the perch, 301ac. 2rd., but the same being deduced into Wood measure, viz., 18ft. to the perch is 253ac. 1rd. 14p., divided into 21 several coppices, falls or sales, which being particularly viewed by us we Value:'

	Numbers of trees	Rates p. tree	Total of rates £	s	d
Timber oake trees	250	40s	500	–	–
Timber oake trees	300	26s 8d	400	–	–
Timber oake trees	800	15s	600	–	–
Timber oake trees	208	9s	92	12	0
	1558		1593	12	0
We Value the underwoods with the Saplings and Black poles therein growing to be worth in present			700		
The Soil or ground, the Wood being taken off, the same being healthy barren ground upon a rock, 301 ac. at 12d per ac. is worth per annum			15	1	0
Total			2308	13	0

to 70 acres. Although the manor was restored to the dean and chapter in 1660 the wood does not feature again in fabric accounts or other chapter records as a source of timber for rebuilding the cathedral (Fearn and Simpson, 1997).

The destruction and exploitation of forests during the civil wars, the losses of shipping in the Dutch Wars (1652–1674) and the increasing demands of the building trade (aggravated particularly by the rebuilding of London after the fire of 1666), are the principal reasons given for a severe shortage of great timber and plank of English oak in the last quarter of the seventeenth century (Albion, 1926, pp. 217–218; Yeomans, 1992, p. 12). Consequently, it appears there was a significant increase in the amount of foreign timber imported, particularly pine (*Pinus sylvestris*) from Norway and the Baltic and this now began to come in through Irish Sea ports to western and north-western counties where it was used by entrepreneurs for their commercial enterprises, by builders for new houses of the gentry and even in small quantities at Worcester cathedral, where oak was still plentifully available (Tyson, 1996, pp. 161–169).[7] However, London and the east coast ports probably still received the majority, as in medieval times. Lincoln cathedral has very detailed fabric accounts from the year of the Restoration (Litton and Simpson, 1996, p. 198). The first record of the use of imported deal boards in any quantity is in the contract drawn up in 1674 by the dean with the local builder William Evison for a new cathedral library to the designs of Sir Christopher Wren (Wren Society, 1940, pp. 76–77). From the early 1680s deal replaced the traditional oak battens for covering the cathedral roofs. Oak for the repair of the principal roof timbers, however, probably still came largely from local woodlands, from Saxilby and Ingleby, adjacent parishes to Harby, for example.[8] The first imported pine baulks at Lincoln were used by the Cambridge architect James Essex to rebuild the roof of the Chapter House in 1762 (Yeomans, 1992, pp. 52–54). The roofs of many medieval English cathedrals have considerable late eighteenth and nineteenth century repairs in Baltic pine which could easily provide tie-beams, for example, of the dimensions of the original medieval ones.

The dean and chapter of Worcester obtained timber from Powick, Ombersley (SO 8463), Ankerdene (SO 7456), Berrow (SO 7347) and Eymore for the restoration of the cathedral in the 1660s. The last four of these sources were woodlands in their own estates. The roof of the Chapter House was the first to be rebuilt using timber from some or all of them (Fearn and Simpson, 1997). Their later history has not been traced in any detail but the first (1832) edition of the One Inch Ordnance Survey map (Fig. 4.3) shows quite clearly that Eymore became part of the landscape of the Industrial Revolution in the eighteenth century as witnessed by coal mines, mills, the Staffordshire and Worcestershire Canal and the numerous coppice woods, particularly just across the Severn in the Wyre Forest. Underwood was required to make charcoal to smelt iron, as pit-props and to fuel brick and pottery kilns, and timber was needed for industrial buildings and transport projects such as bridges, canal locks and wooden railways for the mines and quarries (Trinder, 1973, pp. 50, 116–118). Whether Eymore Wood continued to be managed in the traditional way and to have the principal outlets for its wood products downstream in Worcester or whether advantage was taken of the new industrial markets to the north is as yet unresolved. However, it remained the property of the dean and chapter until 1861 by which time the common availability of cheap coal was undercutting the wood-fuel market (Page and Willis-Bund, 1913, p. 171).

References

(Ordnance Survey map references are given for woods named in the text. Existing woods have six-figure references and four-figure references are given for woods which no longer exist or which cannot be precisely located.)

Albion, R.G. (1926) *Forests and Sea Power: The Timber Problem of the Royal Navy, 1652–1862.* Harvard University Press, Cambridge, Massachussetts.

Blore, W.P. and Harvey, J.H. (1945) Recent discoveries in the archives of Canterbury Cathedral. *Archaeologia. Cantiana* 58, 28–39.

Brown, A.G. (1997) *Alluvial Geoarchaeology: Floodplain Archaeology and Environmental Change.* Cambridge University Press, Cambridge.

Cave, T. and Wilson, R.A. (eds) (1924) *The Parliamentary Survey of the Lands and Possessions of the Dean and Chapter of Worcester Made in or About the Year 1649 in Pursuance of an Ordinance of Parliament for the Abolishing of Deans and Chapters.* Worcester Historical Society, London.

Chapman, F.R. (1907) *The Sacrist Rolls of Ely.* 2 vols. Cambridge University Press, Cambridge.

Cole, R.E.G. (ed.) (1917) Chapter Acts, Lincoln Cathedral, 1547–59. *Lincoln Record Society* 15.

Crumlin-Pedersen, O. (1986) Aspects of wood technology in medieval shipbuilding. In: Crumlin-Pedersen, O. and Vinner, M. (eds) *Sailing into the Past.* Proceedings of the International Seminar on Replicas of Ancient and Medieval Vessels, Roskilde, 1984. Roskilde.

Eckstein, D., Bauch, J., Klein, P. and Wazny, T. (1986) New evidence for the dendrochronological dating of Netherlandish paintings. *Nature* 320, 465–466.

Edwards, J.F. and Hindle, B.P. (1991) The transportation system of medieval England and Wales. *Journal of Historical Geography* 17(2), 12.

Erskine, A. (1981) The Accounts of the Fabric of Exeter Cathedral, part 1, 1279–1326. *Devon and Cornwall Record Society* n.s. 24.

Erskine, A. (1983) The Accounts of the Fabric of Exeter Cathedral, part 2, 1328–1353. *Devon and Cornwall Record Society* n.s. 26.

Esling, J., Howard, R.E., Laxton, R.R., Litton, C.D. and Simpson, W.G. (1989) List 29: Nottingham University Tree-Ring Dating Laboratory Results. *Vernacular Architecture* 20, 39.

Evans, S. (1973) *The Medieval Estate of the Cathedral Priory of Ely.* Ely.

Fearn, K. and Simpson, W.G. (1997) Worcester Cathedral as a source of forest and woodland history. In: Barker, P. and Guy, C. (eds) *Archaeology at Worcester Cathedral: Report of the Seventh Annual Symposium.* Worcester, pp. 6–16.

Firth, C.H. and Rait, R.S. (eds) (1911) *Acts and Ordinances of the Interregnum, 1642–1660.* v.2.

Foot, N.D.J., Litton, C.D. and Simpson, W.G. (1986) The high roofs of the east end of Lincoln Cathedral. In: Heslop, T.A. and Sekules, V.A. (eds) *Medieval Art and Architecture at Lincoln Cathedral* (BAA Conf. Trans., 8), pp. 47–74.

Gentles, I. (1973) The Sales of Crown Lands during the English Revolution. *Economic History Review,* 2nd series 26, 614–635.

Green, M.A.E. (ed.) (1870) *Calendar of State Papers Domestic, 1601–1603,* London.

Heal, F. (1973) The Tudors and church lands; economic problems of the bishops of Ely during the sixteenth century. *Economic History Review,* 2nd series 26, 198–217.

Heal, F. (1980) *Of Prelates and Princes: A Study of the Economic and Social Position of the Tudor Episcopate.* Cambridge University Press, Cambridge.

Hendry, G.A.F, Bannister, N. and Toms, J. (1984) The earthworks of an ancient woodland. *Bristol and Avon Archaeology* 3, 47–53.

Hewett, C.A. (1985) *English Cathedral and Monastic Carpentry.* Phillimore, Chichester.

Holdsworth, C.J. (ed.) (1974) Rufford Charters, 2. *Thoroton Society Record Series* 30.

Hooke, D. (1982) The Anglo-Saxon landscape. In: Slater, T.R. and Jarvis, P.J. (eds) *Field and Forest: An Historical Geography of Warwickshire and Worcestershire.* Norwich, 79–104.

Hooke, D. (1985) *The Anglo-Saxon Landscape, The Kingdom of the Hwicce.* Manchester University Press, Manchester.

Kitching, C.J. (1986) Re-roofing old St Paul's Cathedral, 1561–66. *The London Journal* 12 (2), 123–133.

Lemon, R. (ed.) (1856) *Calendar of State Papers Domestic, 1547–1580.* London.

Litton, C.D. and Simpson, W.G. (1996) Dendrochronology in cathedrals. In: Tatton-Brown, T. and Munby, J. (eds) *The Archaeology of Cathedrals.* Oxford.

Madge, S.J. (1938) *The Domesday of Crown Lands: a Study of the Legislation, Surveys and Sales of Royal Estates under the Commonwealth.* Routledge, London.

Milisauskas, S. (1986) *Early Neolithic Settlement and Society at Olszanica.* University of Michigan Press.

Mills, C.M. (1988) Dendrochronology in Exeter and its applications. Unpublished thesis for the degree of PhD, University of Sheffield.

Page, W. and Willis-Bund, J.W. (eds) (1913) Worcestershire, 3. *Victoria County History.*

PPRS (1914) *Publications of the Pipe Roll Society.* London.

Rackham, O. (1975) *Hayley Wood; Its History and Ecology.* Cambridgeshire and Isle of Ely Naturalists' Trust, Cambridge.

Rackham, O. (1980) *Ancient Woodland: Its History, Vegetation and Uses in England.* Edward Arnold, London.

Rackham, O. (1986) *The History of the Countryside.* Dent, London.

Salzman, L.F. (1952) *Building in England Before 1540.* Oxford.

Savill, P.S. and Spilsbury, M.J. (1991) Growing oaks at closer spacing. *Forestry* 64, 373–384.

Simpson, W.G. (1996a) *Ely Cathedral: The Nave Roof Archive Report.* Unpublished report, Historic Buildings Research Unit and Tree-Ring Dating Laboratory, University of Nottingham for English Heritage.

Simpson, W.G. (1996b) Documentary and dendrochronological evidence for the building of Salisbury Cathedral. In: Keen, L. and Cocke, T. (eds) *Medieval Art and Architecture at Salisbury Cathedral* (BAA Conf. Trans., 17), pp. 10–20.

Simpson, W.G. (1996c) Master-builders: fresh research on cathedrals and other medieval buildings by the Historic Buildings Research Unit. In: Wilson, R.J.A. (ed.) *From River Trent to Raqqa: Nottingham University Archaeological Fieldwork, 1991–95.* University of Nottingham, Nottingham, pp. 87–92.

Simpson, W.G., Howard, R.E. and Guilding, E. (1994) *The Survey, Dating and Analysis of the Roof of the Nave, Worcester Cathedral.* Unpublished report, Historic Buildings Research Unit and Tree-Ring Dating Laboratory, University of Nottingham for English Heritage.

Trinder, B.S. (1973) *The Industrial Revolution in Shropshire.* Phillimore, Chichester.

Tyson, B. (1996) Some Cumbrian builders, 1670–1780. *Trans. C. & W. A. & A. Soc.* 96, 161–186.

Venables, Rev. E. (1884) The Vicars' Court, Lincoln, with the Architectural History of the College and an Account of the existing buildings. *Associated Architectural Societies Reports* 17, 235–250.

Williamson, D.M. (1956) *Lincoln Muniments.* Friends of the Cathedral, Lincoln.

Wren Society (1940) *Publications of the Wren Society* 17, 76–77.

Yeomans, D. (1992) *The Architect and the Carpenter.* RIBA, London.

Notes

1. Grants of timber to Ely are recorded in the Close Rolls as follows: *Rotuli Litterarum Clausarum* ii (1224–1227), 106; *Calendar of Close Rolls* i (1227–1231), 46; *Calendar of Close Rolls* ii (1231–1234), 468; *Calendar of Close Rolls* iii (1234–1237), 12, 45, 129; *Calendar of Close Rolls* v (1242–1247), 508; *Calendar of Close Rolls* vi (1247–1251), 34.
2. Some of this timber may have come from royal forests – see, for example, Holdsworth, 1974, pp. 341–343 (mss 677–680A) which concern permission granted by Edward I to the abbot of Rufford to cut down and sell 40 acres of trees in Sherwood Forest in 1304. There were three buyers of the trees, one of whom was a monk, and the land was sold to two different buyers.
3. Lincolnshire Archives Office: D & C. Dij 71/3/31-43 – Common Fund.
4. *Calendar of Close Rolls* ii (1231–1234), 64, 92.
5. Worcester County Record Office: 5403/20.b.009:1BA John Broome's 'Mapp of the Mannor of Eymore' made for the Dean and Chapter of Worcester, 1756. BA 2602.b/009.1.75 'A Terrier to a Mapp of the Mannor of Eymore', 1756. The writer is grateful to Dr C.R. Salisbury for his photograph of the map and to the WCRO and the Church of England Record Centre for permission to publish it here.
6. Worcester Cathedral Library: A73 Receiver-General Accounts, f. 53.
7. Worcester Cathedal Library: A73 Receiver-General Accounts, f. 66.
8. Lincolnshire Archives Offices: D & C. Bj/1/9 – Fabric Accounts.

CHAPTER 5
The rise, decline and extinction of spring wood management in south-west Yorkshire

Melvyn Jones
Sheffield Hallam University, City Campus, Sheffield S1 1WB, UK

Writing in his *Trees and Woodland in the British Landscape,* Oliver Rackham (1976) pointed out that the reasons for the decline of coppicing in Britain had still to be critically studied. He repeated the point in the second edition of the book (Rackham, 1990). It is equally true that, despite some detailed regional and local studies (e.g., Linnard, 1982; Jones, 1986a; Rackham, 1986; Squires and Jeeves, 1994), the rise of coppicing in different regions, the evolution of regional variations in coppice management and the varying importance, over time and space, of markets for the timber and underwood, and the subsequent decline of coppicing, have been inadequately researched. This is a surprising circumstance in view of the fundamental importance of coppice management in local economies over a long period of time. At the national level Collins' (1992) recent paper on British woodlands during and after the charcoal iron era has gone some way to meeting Rackham's criticism, but at the regional and sub-regional levels there is still much to be learnt about the nature and timing of changes in woodland management practices.

This chapter presents evidence concerning changes in woodland management in south-west Yorkshire over the last 1000 years. For the purposes of this chapter south-west Yorkshire is defined as the present metropolitan districts of Sheffield and Rotherham (Fig. 5.1). The area contains great topographic variety. In the west the landscape has developed on the Millstone Grit Series, but most of the area lies on the Coal Measures. The folding, faulting and tilting of these rocks and their subsequent erosion have given rise to a 'belted' landscape with a succession of broken but bold sandstone cuestas in the west and a succession of broad shale vales in the east. On the eastern margins of the area, the Coal Measure rocks dip below a Magnesian Limestone escarpment. The geomorphology and the underlying geology have exerted a significant influence on the development of the settlement pattern, on economic development in general, and on the clearance and exploitation of the woodland resource in particular (Jones, 1993, unpublished).

From late Saxon times, four distinct phases of woodland management can be distinguished, during which coppice management gradually became

© CAB INTERNATIONAL 1998. *European Woods and Forests: Studies in Cultural History* (ed. C. Watkins)

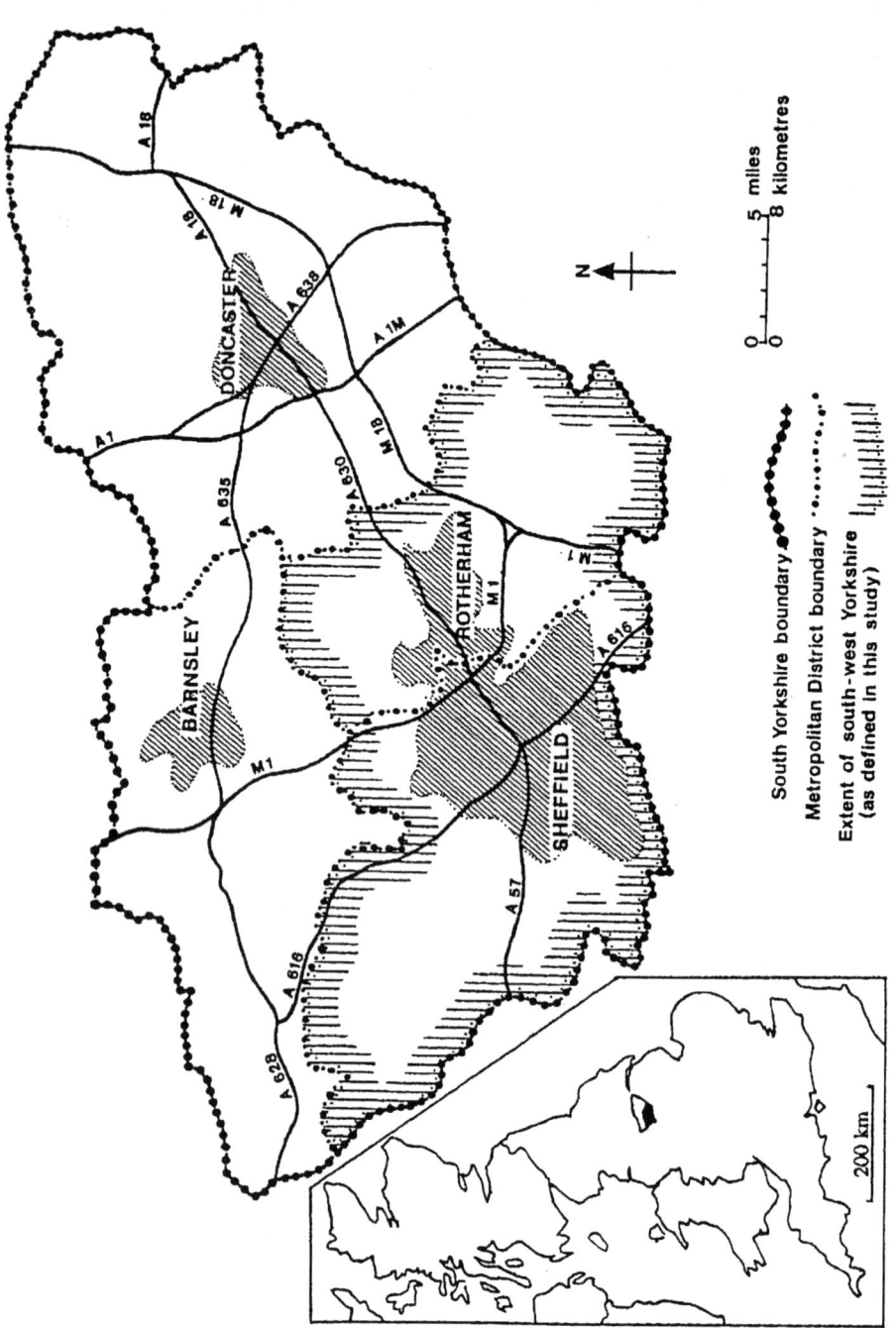

Fig. 5.1. South-west Yorkshire: location.

dominant in the woods of south-west Yorkshire, reached a high degree of sophistication, exploited a succession of markets, and then, gradually, went into almost terminal decline (Fig. 5.2).

Phase 1: the Wood Pasture Tradition

The Domesday survey of 1086 provides the first opportunity for an analysis of woodland cover and woodland management in south-west Yorkshire. Rackham (1980) has calculated that the Domesday survey covered 27 million acres of land of which 4.1 million were wooded, that is 15% of the surveyed area. His figure for the West Riding of Yorkshire is 16%. My own calculation for south-west Yorkshire is just under 20% (Jones, 1989). What all this means is that in the eleventh century the country generally, and south-west Yorkshire specifically, were relatively sparsely wooded even by twentieth century standards.

Before looking at how woodland in south-west Yorkshire was managed according to the Domesday survey it is helpful to be aware of the difficulties of interpretation of the Domesday information. Almost all woodland in south-west Yorkshire was recorded in terms of its length and breadth (in leagues and furlongs); a few areas of woodland were given in terms of acres. The problem is to convert the Domesday measurements into modern areal units. A Domesday acre was probably smaller than the modern acre and a number of studies have used the device of multiplying Domesday acres by 1.3 to make up the difference. The length by breadth measurements are a little more problematic. It is generally assumed that a Domesday league was 12 furlongs or a mile and a half. This assumption would give, of course, the dimensions of woodlands

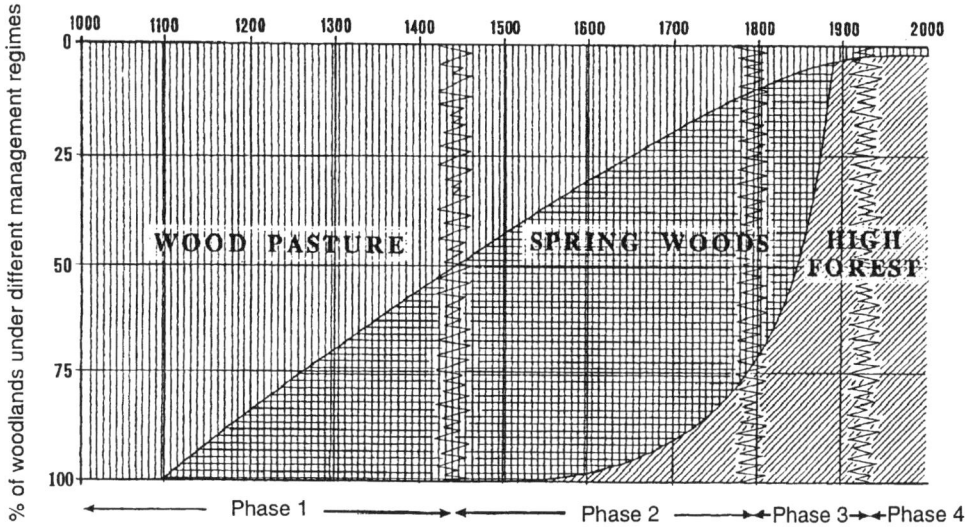

Fig. 5.2. Phases of woodland management in south-west Yorkshire.

that were perfectly square or rectangular, but this was never the case in reality. Moreover, in many cases the linear dimensions must have been arrived at by lumping together two or more woods. To overcome these problems Rackham uses a 'form factor' (Rackham, 1980). This is a statistic used to compensate for the oversimplification of the Domesday entries. After comparing Domesday entries with medieval woods of known size which are known not to have changed in the intervening period, Rackham suggests a form factor of 0.7, i.e. if the length is multiplied by the breadth then it must be assumed that the woodland area would be about 70% of the figure arrived at. Using the formulae for Domesday areal and linear entries outlined above, what is now Sheffield Metropolitan District had about 20,400 acres of woodland in the eleventh century which represented 22.5% of the land area. The corresponding figures for what is now Rotherham Metropolitan District were 11,600 acres or 16.6% of its area (Jones, 1995).

We can now turn to Fig. 5.3 which shows the distribution and types of woodland in south Yorkshire as recorded in Domesday. With the exception of the giant manor of Hallam, places with woods are shown by circles varying in size according to the extent of the woodland. There are noticeable variations in the distribution of woodland. In the west in what is now Sheffield, Barnsley and the western half of Rotherham Metropolitan District, woodland was relatively extensive, with a substantial number of communities having more than 1000 acres of woodland.

In contrast, in the eastern part of the county, in what is now the eastern part of Rotherham Metropolitan District and the whole of Doncaster Metropolitan District, the picture was very different. In those areas woodland was more scattered, amounts in individual communities were much smaller than in the west, and in 19 places no woodland was recorded at all. This suggests early clearance and continuous occupation and cultivation by a relatively dense population for thousands of years. It can be seen that this part of the county includes the belt of Magnesian Limestone country already referred to, which has long been regarded as the most fertile and attractive area for early settlement in south Yorkshire.

The types of woodland in south Yorkshire at Domesday also suggest a shortage of woodland in parts of the east of the county and a relative abundance in the west. Turning to Fig. 5.3 again, the key shows that Domesday woodland in the county was described in four different ways: *silva, silva modica, silva minuta* and *silva pastilis*. *Silva* is simply woodland; the meaning of *silva modica* is not clear; *silva minuta* is coppice; and *silva pastilis* is wood pasture. Of the 111 places in which woodland was recorded in south Yorkshire, seven had coppice woods and 102 had wood pastures. All seven areas of coppice wood were in the eastern half of the county, all outside south-west Yorkshire as defined in this paper, with five in the Magnesian Limestone belt, underlining the view that it was an area of dense population. On the other hand, although wood pasture was found in all parts of the county it was very extensive and the only type of woodland found in the west, including the whole of south-west Yorkshire.

To sum up, using the methods of calculation outlined above, at Domesday

Spring wood management in south-west Yorkshire

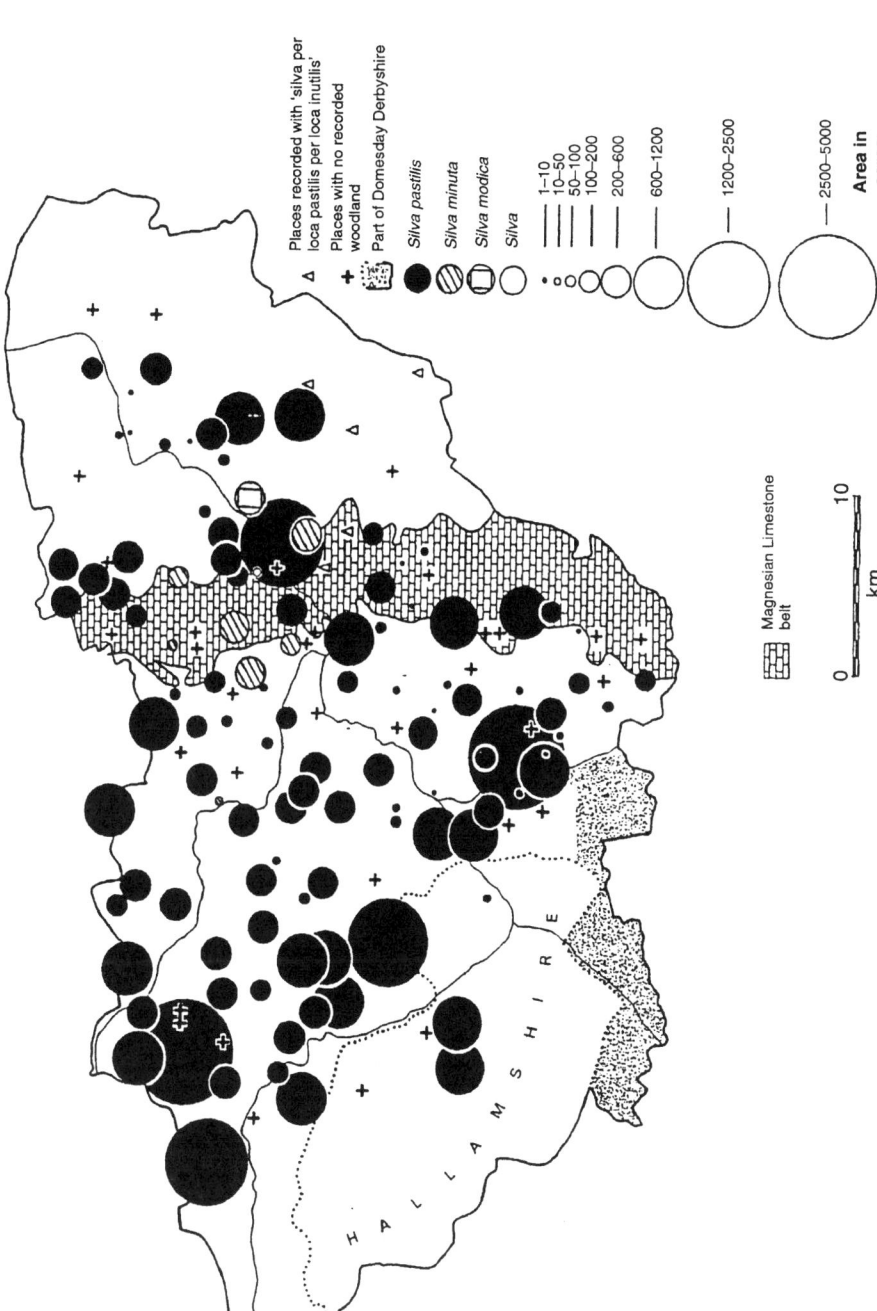

Fig. 5.3. Woodland in south Yorkshire according to the Domesday survey of 1086. The dotted line shows the probable ancient boundaries of Hallamshire. Within these boundaries is recorded a giant manor containing Sheffield and Attercliffe and 16 un-named hamlets. The manor contained *silva pastilis* 4 leagues in length and 4 in breadth, i.e. 16,128 acres using a form factor of 0.7.

about one-fifth of the area of south-west Yorkshire (19.9%) was covered by woodland. The population was small enough not to exert extreme pressure on the woods, and animals were allowed to graze over large areas of woodland. This does not mean that coppicing was unknown in the region. It was probably widespread and of ancient origin, but it must have been carried on within the wood pasture system on a less systematized basis compared with the situation that was to develop in the late medieval period and beyond. The wide distribution in south-west Yorkshire, particularly in the western half of the region, of the place-name element -*storth*, the Old Norse term for coppice, suggests that coppicing was practised at the time of the Domesday survey.

For at least three centuries following the Domesday survey, the wood pasture tradition in the area continued to be dominant, on the extensive wooded commons, on the wooded chases at Rivelin and Wharncliffe, and in some 20 deer-parks (Jones, 1996). This extensive, multi-purpose system of management, however, was gradually replaced by coppice management in order to conserve wood supplies which were becoming depleted as the population grew and woodland was cleared for agriculture. Having said that, however, evidence available at the present time suggests that coppice management did not replace wood pasture in south-west Yorkshire as the dominant form of woodland management until a comparatively late date.

A number of documents illustrate the dominance of wood pasture. For example in 1161 the monks of the Cistercian abbey of Kirkstead in Lincolnshire were granted land in Kimberworth to mine iron ore and operate two bloomeries and two forges. Their fuel was to be dead wood from the wooded commons not from coppice woods. In the same year the monks of Ecclesfield Priory were given the right to pasture their flocks every year from January to Easter in a large wood in the valley of the River Don, the surviving parts of which constituted a compartmented coppice wood four centuries later. The twelfth century document suggests a wood pasture regime. Nearly 200 years later in 1332, in the *inquisitione post mortem* of Thomas de Furnival, the lord of the Manor of Sheffield, 11 localities were listed under the heading of pastures in woods, moors and commons, including Greno Wood, Beeley Wood and Bowden Housteads Wood, all of which by Elizabethan times were coppice woods in which animals were unwelcome.

Phase 2: Spring Woods and Fuel Supplies

By the late Middle Ages wood pasture was in rapid retreat and coppice with standards had emerged as the dominant form of woodland management in south-west Yorkshire. These coppice with standards woods were usually referred to as spring woods, *spring* being the Anglo-Saxon name for coppice.

In the fifteenth century, leases of spring woods invariably mention charcoal making, and there can be little doubt that the increasing dominance of coppicing was closely related to the expansion of iron making and the related nail-making and edge tool trades. For example, one lease dated 1462, refers to seven woods in Norton parish, then part of north Derbyshire (Hall,

1914). The document records that John Cotes and John Parker had been granted permission by William Chaworth, knight and lord of Norton, 'to fell ... cole [i.e. to make into charcoal] and carye the said Woddes' preserving for the owner 'sufficiaunt Wayvers after the custom of the contre'. The mention of wavers, the young timber trees, shows that the woods in question were coppices with standards. Wavers are also mentioned in another document, written in 1496, which refers to Hutcliff Wood in the Sheaf valley. The wood at that time was the property of Beauchief Abbey and the document records that 'the abbot of Beucheff' had granted permission 'to cooll [i.e. to make into charcoal] ii certen wodds that is to say hudclyff and the brood medowe Abutt', the woods to be left 'weyverd workmonlyke'. The lease also refers to a bloom hearth (a primitive furnace) and a dam (the local name for a pond at a water-powered industrial site). By the end of the sixteenth century spring wood management was general. In an undated document written for the 7th Earl of Shrewsbury, the major landowner in south-west Yorkshire who succeeded to the title in 1590 and died in 1616, 49 spring woods were listed, and, significantly, they were all listed as belonging to the Earl's forges.

Between the last quarter of the sixteenth century and the middle of the eighteenth century in the woods along the extreme western and southern boundaries of south-west Yorkshire, was another woodland industry also making a fuel for smelting ore and sustaining the management of local woods as spring woods. This time the ore was lead and the fuel was called whitecoal (Kiernan, 1989). A number of local landowners, including the Earls of Shrewsbury, the Strelleys of Beauchief and Ecclesall and the Brights of Ecclesall and Whirlow, are known to have been very active in the lead trade during that period. They obtained ore from the Derbyshire Peak District, smelted it in water-powered smelters called ore-hearths near their Coal Measure oakwoods and then transported the lead to the Humber for sale in the London market. Whitecoal was small lengths of wood, dried in a kiln until all the moisture was driven out. According to Linnard (1982), charcoal and whitecoal were mixed together in lead smelting because 'charcoal made too violent a fire, and wood alone was too gentle'. The former presence of whitecoal making in woods in south-west Yorkshire is betrayed by the presence of characteristic depressions, about 4–5 m in diameter with a noticeable spout at one end. The spouts always slope downhill. These are the remains of the whitecoal kilns. In 1657 Ecclesall Woods, extending to more than 130 ha, was let to a lead smelter who was permitted to make both charcoal and whitecoal. He also leased two water-powered ore hearths. The sites of about 140 whitecoal kilns still survive in Ecclesall Woods (P. Ardron, pers. comm.).

By the seventeenth century the iron industry in south-west Yorkshire had achieved a high degree of sophistication and was increasingly characterized by a large measure of vertical integration and horizontal combination (Hopkinson, 1963). By the 1650s, the most powerful ironmaster in the region was Lionel Copley (Goodchild, 1996) who entered into a succession of agreements with local landowners to fell and coal their spring woods. The surviving deeds illustrate contemporary coppice practice. For example, in a deed dated 30 December 1657, the 2nd Earl of Strafford of Wentworth Woodhouse entered into a 10-year agreement with Copley, to fell the underwood and trees in 13 woods on the Earl's estate in south-west Yorkshire. Under the contract Copley was to cut 1000 cords of wood (a pile of wood 4 ft × 4 ft × 8 ft) each year for charcoal making. The agreement dealt with the felling and use (predominantly for conversion into charcoal) of 'young timber trees, Lordings, Blackbarks, powles, coppices and Springwoode', together with 'the Bark thereof'. Copley was given permission to 'pill [peel the bark] fell and cutt downe and to cutt into Cordwood and Cord all and every the said trees'. The lessee was instructed to make sure that 'all the said Springwoode [is] well and sufficiently weavered', i.e. young timber trees (wavers) had to be left standing to mature into standards. He was also to ensure that the coppice was 'workmanlike cutt downe ... and the stowens [stools] thereof neare to the roote so as best preserve the future growth and next springing thereof'. Copley was further instructed to 'burne the Ramell thereof upon such part of the before mentioned parcells of ground where the same ramell now groweth as may be least prejudiciall to the weavers and Springwood which shall be left to grow'. The 'ramell', or *ramelia* as it was known in Latin in medieval documents, was the smallest brushwood and other waste material from cutting with axe, saw and billhook. It was important to protect the new growth in a coppice wood from grazing animals, particularly sheep, cattle and deer, and Copley was told that he must not take timber out of hedges and dykes around woods or from partitions in woods.

Various aspects of spring wood management in south-west Yorkshire are also well illustrated in two schemes devised by Thomas Wentworth, 1st Marquis of Rockingham, who inherited his estates in 1723. In 1727 he devised what he called 'A Scheme for making a yearly considerable Profit of Spring Woods in Yorkshire', and in 1749 what he described as 'A Scheme for a Regular Fall of Wood for 21 years ...'. In the 1749 scheme a 21-year cycle was used so that the woods coppiced in 1749 would be cut again in 1771. This meant that the Marquis's 876 acres of woodland in south-west Yorkshire would produce a regular crop of 40 acres of underwood a year for 21 years at which point the crop in the first wood to be cut, would be ready to be cut again. The Marquis was very specific about the timber trees: there were to be five reserves (mature timber trees called black barks) and 70 wavers (sapling timber trees) per acre.

The Marquis was in the fortunate position of having both ironstone and extensive woodlands on his estate and he linked the mining of the former with the charcoaling of the latter. In 1745 the Marquis wrote that:

> and whereas it is the Iron Men that keep up the Price of the Wood, Especiall care must be taken that the Iron Stone be never let for a longer time than the Woods are agreed for

A year later the Marquis was able to write, with obvious satisfaction:

> a new term of 11 years to Michaelmas 1758 was agreed upon for the Iron Stone & for all the Cord Wood for the same term at nine Shillings Per Cord

But this was the beginning of the end. By 1780 the lead ore hearths had been replaced by coal-fired cupola furnaces and by the end of the eighteenth century iron furnaces were converted or were rebuilt to use coke (Hey, 1977) and new ironworks were, from the first, coke fired. The market for whitecoal disappeared and that for charcoal was much reduced.

Phase 3: Decline and Extinction of Spring Wood Management and the Rise of High Forest

Nationally, coppicing did not decline in the immediate aftermath of the loss of the market for charcoal for iron smelting, and one author has recently suggested that the first half of the nineteenth century may be regarded as 'the golden age of traditional English woodmanship' (Collins, 1992). In the case of south-west Yorkshire this was not the case. Its golden age was in the seventeenth and eighteenth centuries when production of charcoal and whitecoal were at their height. During that period the 2nd Earl of Strafford and his successors on the Wentworth estate even exported the south-west Yorkshire coppicing tradition to their Irish estates in County Wicklow where the woods served a wide range of markets across Leinster and also Whitehaven shipbuilders (Jones, 1986a).

In south-west Yorkshire some markets for charcoal remained in the nineteenth century and others expanded. Most importantly, charcoal was used in making blister steel in so-called cementation furnaces where successive layers of bar iron interbedded with layers of charcoal were heated up to high temperatures for up to 8 days. Over 250 such furnaces with their characteristic conical chimneys were built in the Sheffield area. Another industry based on charcoal present in the region was gunpowder manufacture; in the nineteenth century there were works at Worsbrough Dale and at Wharncliffe. Charcoal was also used as blacking by moulders in the many iron foundries in the area. Charcoaling also led to the development of industries involved in the extraction of wood acid and wood spirit and two local firms became substantial purchasers of cordwood for this purpose in the nineteenth century.

Other important local markets for the produce of spring woods which continued to be important in the nineteenth century were for oak bark for tanning, and timber and poles for use as pitwood; for both of these markets the oakwoods of south-west Yorkshire were admirably suited. As elsewhere (Collins, 1992), the smaller coppice crafts, such as fencing, scaffolding, building materials, ladders, brushes, clogs and baskets, were sustained by the demands of the rapidly growing urban settlements and industrial villages.

However, the surviving coppice-using industries could not sustain spring wood management on the existing scale. Markets for coppice poles began to decline for a variety of reasons. Locally, the early adoption of coal as a domestic fuel instead of wood affected those coppice woods where firewood markets were important. The growth of factory production and rapid and reliable distribution by rail, meant that factory-made products often in metal where previously wood had been used, hit many local craftsmen badly. As early as 1847 the agent on the Earl of Scarbrough's estate on the eastern margins of south-west Yorkshire advocated the clear-felling, grubbing up and conversion to agriculture of the largest wood on the estate because it was far from railways (Beastall, 1974). Even where the factory products were still wooden, regional specialization and large-scale organization meant that the previously ubiquitous self-employed craftsmen found it difficult to compete, and gradually disappeared. The demand for bark from tanners also declined as they took advantage of cheaper imported bark and chemical substitutes. The presence of valuable coal and ironstone seams under many local woods, and the rapid physical growth of Sheffield led to the eventual destruction of centrally located woods and the further contraction of the coppicing tradition. In the second half of the nineteenth century the increasing amount of trespass from the growing urban populations seeking recreation in the surrounding countryside induced landowners to sell their woods to the local authorities or to present them as gifts for use as parks and recreation grounds.

By the 1890s coppicing was nearing its end. Income from local coppices had declined sharply during the second half of the nineteenth century. As a result more and more woods were gradually converted into high forests. In essence they were becoming plantations, and forestry was replacing woodmanship. This was achieved by natural regeneration, singling the multi-stemmed coppice stools in order to allow the best stem to grow into a standard tree, by clearing away altogether the oldest stools and sickly trees, and in their place planting young trees to be grown for timber on a long cycle. Many of the newly planted trees were not native to south-west Yorkshire, trees such as beech (*Fagus sylvatica*), sweet chestnut (*Castanea sativa*), and sycamore (*Acer pseudoplatanus*). Amongst the broadleaved trees, conifers, mainly *Pinus sylvestris* and *Larix decidua*, were often planted as a 'nurse crop'. For example, in October 1898 the Duke of Norfolk's forester (on the same estate, before the nineteenth century, the word forester was not used; the official in charge of the estate woodlands was the woodward) began systematically to plant timber trees in large sections of local spring woods. In Hesley Wood he planned to plant 100 acres (40 ha) with ash, elm, sycamore, birch, lime, sweet chestnut and beech 8 ft apart and 'filled up' at 4 ft intervals with larch. Another 40 ha were to be planted in the same way in Smithy Wood, 50 ha in Greno Wood, 25 ha in Beeley Wood, 16 ha in Bowden Housteads Wood, 10 ha in Hall Wood and 8 ha in Woolley Wood. Conifers were to be planted in all the woods mentioned above, but when he came to Shirecliffe Wood he noted that 'being situated nr Sheffield & therefore affected by smoke etc Larch & conifers would not grow well'. On 16 November he placed his first order for young trees: 20,000 larch, 10,000 sycamore, 5000 beech, 2000 birch and 2000 sweet chestnut.

On Earl Fitzwilliam's Wentworth estate the transition from spring woods to high forest had started in the first half of the nineteenth century. This is well illustrated by changes in the extensive Ecclesall Woods. The first known map of Ecclesall Woods compiled sometime in the seventeenth century named 17 woods covering 381 acres (155 ha). These woods constituted compartments or falls in a coppice regime. Between 1715 and 1774, 17 named woods were combined in various ways to provide 22 falls. Rotation lengths varied, the average length of coppice cycle during the 1715–1774 period being 24 years. After 1774 individual woods or compartments were not itemized in the account books but the entries indicate unequivocally the continued practice of coppicing, and it is clear that the woods were coppices with standards. From 1775 until 1847 sales from Ecclesall Woods were only entered under the title of a 'fall of wood' or 'falls of wood' or 'fall of coppis [sic]' or simply 'fall' or 'falls'.

Yet today there is virtually no sign of a coppice structure in the woods, and one of the most marked characteristics of the present-day woods is their mixed composition. Growing side by side are trees and shrubs native to the site and the surrounding region, such as sessile oak (*Quercus petraea*), birch (*Betula pendula* and *Betula pubescens*), rowan (*Sorbus aucuparia*), ash (*Fraxinus excelsior*), alder (*Alnus glutinosa*), hawthorn (*Crataegus monogyna*), hazel (*Corylus avellana*) and holly (*Ilex aquifolium*), together with introduced species such as beech (*Fagus sylvatica*), sweet chestnut (*Castanea sativa*), sycamore (*Acer pseudoplatanus*), hornbeam (*Carpinus betulus*), Scots pine (*Pinus sylvestris*) and larch (*Larix decidua*). This mixed composition has been characteristic of the woods for most of this century – the British Association's *Handbook and Guide to Sheffield* (Porter, 1910) described Ecclesall Woods as 'A mixed wood of birch, chestnut, larch, oak, etc. with undergrowth of bracken' – and for an unspecified period before 1900. Obviously planted trees and their descendants have played a significant role in forming the character of the woods as we know them today.

The first known reference to planting in Ecclesall Woods is as early as 1752. In that year ashes and elms were planted in a large glade or laund within the southern part of the woods which had previously been used as a hay meadow by one of the estate's tenant farmers. No further planting was recorded in Ecclesall Woods until 1824–1826. There was then a gap of 3 years before a prolonged period of planting activity began in 1830 which lasted until 1845. After that date planting in Ecclesall Woods was recorded on only four further occasions, in 1885, 1886, 1889 and 1899. However, non-site-specific payments for planting in the woods on the Wentworth estate were recorded annually from 1864 until 1874, in 1878, and annually from 1885–1899.

The planting activity in Ecclesall Woods from 1824 appears to be entirely consistent with a gradual transition from coppice with standards to high forest management (see Fig. 5.4a). It is likely that three processes were taking place or being encouraged simultaneously that would eventually change the composition and structure of the woods and produce a different balance of products to match changing market requirements:

1. Planting up of gaps ('filling up the falls', 'to fill up the vacancies') in the various coppice compartments at the end of a coppice cycle. Evidence from the account books and the woods themselves by the beginning of this century show that by this process the woods were becoming mixed woods containing both broadleaves and conifers.
2. Natural regeneration.
3. Storing coppice, i.e. the retention of coppice poles past the normal coppice

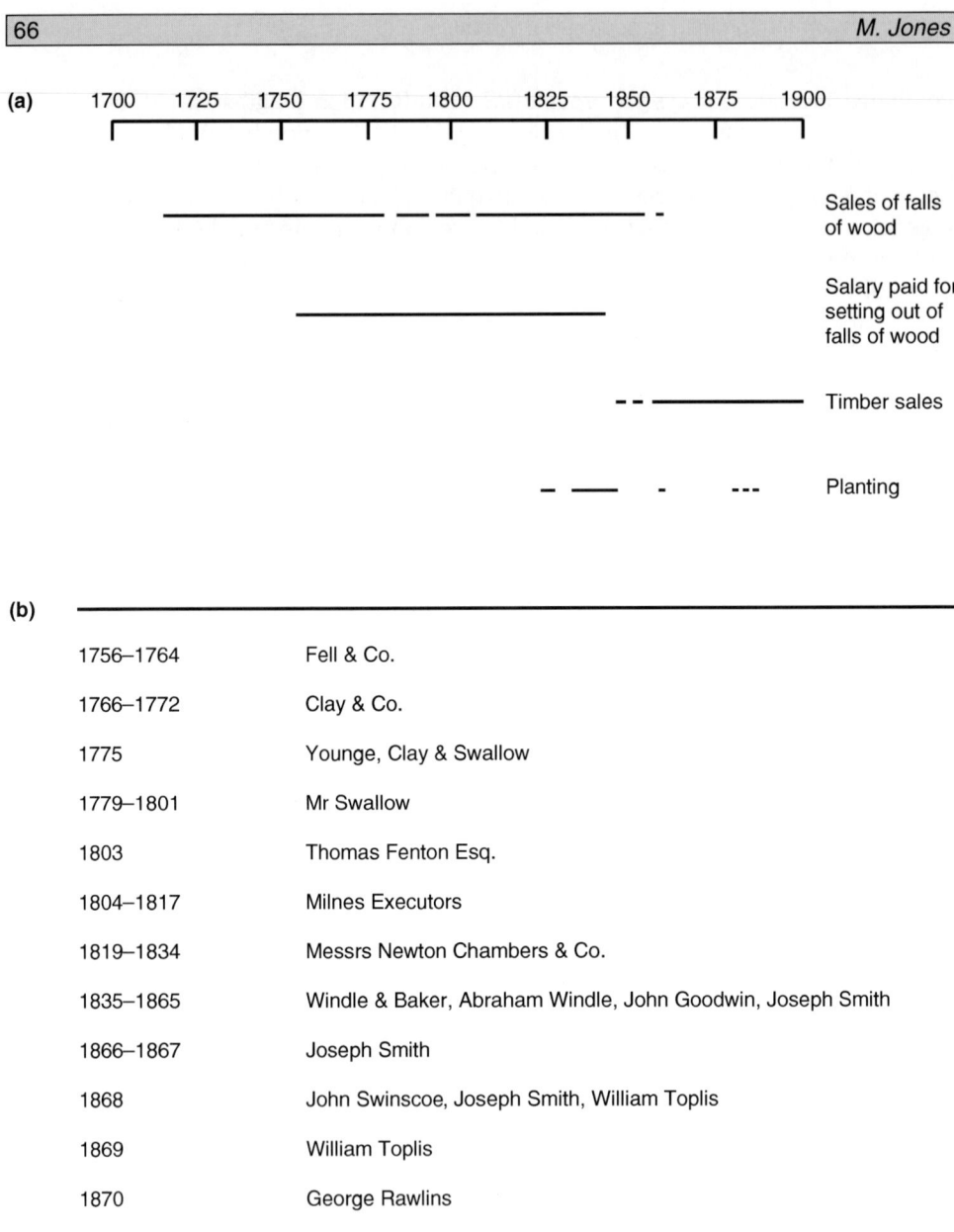

Fig. 5.4. (a) Indicators of changing management practices in Ecclesall Woods, 1715–1901. Source: Sheffield Archives, Wentworth Woodhouse Muniments A700–744, A257–485, MP42 and MP44.
(b) Purchasers of wood and timber from Ecclesall Woods, 1756–1901. Source: Sheffield Archives, Wentworth Woodhouse Muniments A257–485 and A700–784.

rotation length. On each stool only one well-formed vigorous stem would have been left to grow on to produce timber-sized material.

Falls of wood or coppice were recorded in the account books from 1756 to 1847. After that date further falls of wood were recorded from 1848 to 1852 and from 1856 to 1859. After 1859 sales of falls of wood were never recorded again in the 41 years up to 1901. In 1848 a timber sale was recorded for the first time. A further timber sale took place in 1851, and then continuously from 1853 until the records end in 1900–1901, with the exceptions of 1897 and 1898.

Clearly, the period from the mid 1820s until the late 1850s was a transitional period during which intensive planting took place at selected locations within the woods, the last falls of coppice occurred and the change from a coppice to a high forest structure was effected. It must be remembered that this transition was not solely dependent upon planted trees; the early timber sales must have been dominated by stored coppice and single-stemmed trees that had originally been set out as standards during the previous coppice with standards regime. Early sales of planted trees must have been dominated by thinnings. This is partly confirmed in the account books: in 1836 £10 15s 10d was paid 'for poles sold out of the Plantations in Ecclesall'; in 1843 £2 8s 9d was paid for 'Fir poles sold from Ecclesall Wood'; and in 1855 a small consignment of larch poles was sold from the woods. The main purchasers of wood and timber from Ecclesall Woods reflect the changing nature of the woods from coppice with standards to high forest (Fig. 5.4b). Most of the early purchasers – John Fell (1756–1764), Henry Clay (1766–1772), Richard Swallow (1779–1801) and Newton Chambers (1819–1834) – were local ironmasters with ironstone mining, and in the last two cases, coal mining interests. Fell's blast furnace at Chapeltown was fuelled by charcoal until the 1770s (Hey, 1977) and therefore supplies of cordwood from coppices for charcoal making was a major requirement. John Swallow succeeded Fell at Chapeltown and, although the blast furnace was coke fuelled by 1779, Swallow had ironstone pits and collieries nearby, and coppice poles and small timber for pit props was still a major requirement. Timber, from standards, would also be consumed in mining operations. Newton Chambers, who opened their Thorncliffe Works, also at Chapeltown, in 1793, also needed wood and timber in considerable amounts for their extensive coal and ironstone mining operations, and for sleepers for the extensive railways in and about the works. By the 1860s, by which time Ecclesall Woods were being managed as high forest, local industrial buyers had disappeared and had been replaced by timber merchants. John Swinscoe, for example, who was one of three purchasers of timber from the woods in 1868, was acting on behalf of the firm of James and Joseph Swinscoe who were described in White's Sheffield Directory of that year as hardwood merchants. William Toplis, who was the sole purchaser at 27 of the 30 annual timber sales between 1869 and 1900–01 (there were no sales in 1896 and 1897), was a Chesterfield timber merchant.

Phase 4: Coniferization, Neglect, Amenity and Revival of Woodmanship

On the Wentworth estate the former spring woods, converted to high forest, have continued to be managed commercially as broadleaved plantations. Elsewhere, coniferous plantings have supplanted broadleaves and large woods such as Wharncliffe and Greno have been almost completely coniferized. A large number of former coppice woods have also come into the ownership of the local authorities either as gifts or by purchase. Roe Wood was gifted to the City Council by the Duke of Norfolk in 1897 and Wincobank Wood was given by the same owner in 1904. Bowden Housteads, purchased from the Duke of Norfolk in 1914, and Ecclesall Woods, bought from Earl Fitzwilliam in 1927, are just two examples of former coppice woods converted to canopy woods and then sold to Sheffield City Council. Rotherham Borough Council also made important purchases from the trustees of the Earl of Effingham's estate in the 1920s and more recently from the Fitzwilliam estate.

However, vandalism and neglect have taken their toll in the more accessible private and publicly owned woods and there are development pressures on some woods. By the 1980s, many woods were over-mature, lacked a shrub layer and contained much poorer ground floras than had been the case in the past (see Fig. 5.5). A significant number of breeding birds and butterflies were no longer found in woods where they had been common in the past and local residents were increasingly afraid of walking in the woods because they were dark and gloomy, and engendered a fear of personal attack. There was a real danger by the 1980s that south-west Yorkshire's woodland heritage would be squandered.

Since then there have been some encouraging developments. Awareness of their cultural importance has been raised to a much higher level than hitherto and interest in their economic as well as recreational potential has been re-awakened. In Sheffield and Rotherham detailed studies of the ancient woods were commissioned (Jones, 1986b, unpublished; Jones 1989–1992, unpublished). These have highlighted the historical and ecological significance of the surviving woods and in many cases their neglected state. In both Sheffield and Rotherham woodland officers have been in post for a number of years and there is now a wider appreciation of the value of the south-west

Fig. 5.5. (Opposite). Diagrammatic representation of a wood in south-west Yorkshire in 1650, 1890 and 1980. The diagram shows a typical ancient coppice (spring) wood on a steeply sloping parish edge site. In 1650 the wood is a coppice with standards with only native trees and shrubs. Except where bounded by a watercourse, the wood is protected by a bank or wall or both. By 1890 significant changes in the composition and structure of the wood have taken place. The wood is being converted from a spring wood into a high forest plantation by singling and planting. Species non-native to the site now dominate the wood. By 1980 the wood has changed character again. It is now not managed as a commercial wood; it is a recreational amenity. It is even-aged and over-mature and has only a sparse shrub layer and the ground flora is impoverished compared to the situation in 1650 when the felling of timber and the cutting of coppice were regular activities.

Yorkshire's woodlands and the need to actively manage them. Detailed management plans have been drawn up for a number of woods and these are in the early stages of implementation. Additionally, four coppicing experiments associated with the re-introduction of charcoal making and small coppice crafts are being undertaken in four Sheffield woods (Jones and Talbot, 1995).

A major influence on local attitudes to woodland management in the last few years has been the South Yorkshire Forest Project. This project, established in 1991, is a partnership between Barnsley, Rotherham and Sheffield Councils, Sheffield Development Corporation, the Countryside Commission and the Forestry Commission. Its aim is to 'develop multipurpose forests which will create better environments for people to use, cherish and enjoy'. Although not just concerned with ancient woodlands, among its objectives are commitments to protect areas of historical, archaeological and ecological interest (i.e. former spring woods), to increase opportunities for access and recreation, and to encourage the development of timber-based industries, employment opportunities and woodland products. Following a year of public consultation, the South Yorkshire Forest Plan was published in August 1994. This establishes a policy framework and a strategic approach to woodland management throughout the South Yorkshire Forest area – for private as well as publicly owned woods – and will guide developments well into the next century.

Acknowledgements

I would like to thank Olive, Countess Fitzwilliam's Wentworth Settlement Trustees and the Director of Sheffield City Libraries for permission to quote from the Wentworth Woodhouse Muniments in Sheffield Archives.

References

Beastall, T.W. (1974) *A North Country Estate*. Phillimore, Chichester.
Collins, E.J.T. (1992) Woodlands and woodland industries in Great Britain during and after the charcoal iron era. In: Metaille, J.P. (ed.) *Protoindustries et Histoire des Forêts, (Les Cahiers de l'ISARD* 3). University of Toulouse, Toulouse, pp. 109–120.
Goodchild, J. (1996) Lionell Copley: a seventeenth century capitalist. In: Jones, M. (ed.) *Aspects of Rotherham: Discovering Local History*. Wharncliffe Publishing, Barnsley, pp. 16–22.
Hall, T.W. (1914) *Descriptive Catalogue Forming the Jackson Collection*. J.W. Northend, Sheffield.
Hey, D. (1977) The ironworks at Chapeltown. *Transactions of the Hunter Archaeological Society* 10, 252–259.
Hopkinson, G.G. (1963) The charcoal iron industry in the Sheffield region 1550–1775. *Transactions of the Hunter Archaeological Society* 8, 122–151.
Jones, M. (1986a) Coppice wood management in the eighteenth century: an example from County Wicklow. *Irish Forestry* 43(1), 15–31.
Jones, M. (1986b) *Sheffield's Ancient Woods: Some Notes on Their History and Past*

Management with Special Reference to Woods Owned by the City Council. Unpublished report for Sheffield City Council.

Jones, M. (1989) *Sheffield's Woodland Heritage*. Sheffield City Libraries, Sheffield.

Jones, M. (1989–1992) *Inventory Survey of Ancient Woods in Rotherham Metropolitan Borough.* Progress reports 1 (1989), 2 (1990) and 3 (1992). Unpublished reports for Rotherham Metropolitan Borough Council Planning Department.

Jones, M. (1993) *South Yorkshire Forest: Historic Landscapes Study.* Unpublished report for South Yorkshire Forest Project.

Jones, M. (1995) *Rotherham's Woodland Heritage*. Rotherwood Press, Rotherham.

Jones, M. (1996) Deer in South Yorkshire: an historical perspective. In: Jones, M., Rotherham, I.D. and McCarthy, A.J. (eds) *Deer or the New Woodlands?* (*Journal of Practical Ecology and Conservation*, Special Publication No. 1).

Jones, M. and Talbot, E. (1995) Coppicing in urban woodlands: a progress report on a multi-purpose feasibility study in the City of Sheffield. *Journal of Practical Ecology and Conservation* 1, 46–52.

Kiernan, D. (1989) *The Derbyshire Lead Industry in the Sixteenth Century*. Derbyshire Record Society, Chesterfield.

Linnard, W. (1982) *Welsh Woods and Forests: History and Utilization*. National Museum of Wales, Cardiff.

Porter, W.S. (ed.) (1910) *Handbook and Guide to Sheffield*. J.W. Northend, Sheffield (for the British Association for the Advancement of Science).

Rackham, O. (1976) *Trees and Woodland in the British Landscape*. Dent, London.

Rackham, O. (1980) *Ancient Woodland: Its History, Vegetation and Uses in England*. Edward Arnold, London.

Rackham, O. (1986) *The Woodlands of South-East Essex*. Rochford District Council, Rochford.

Rackham, O. (1990) *Trees and Woodland in the British Landscape,* 2nd edn. Dent, London.

South Yorkshire Forest (1994) *Forest Plan.* SYF Project, Sheffield.

Squires, A. and Jeeves, M. (1994) *Leicestershire and Rutland Woodlands Past and Present.* Kairos Press, Newtown Linford.

CHAPTER 6
The continuous conflict between sustainable management regulations and over-utilization of woodland caused by local demands in Austria from the thirteenth century onwards

Elisabeth Johann
Austrian Forest Society, Vienna, Austria (Institute for Forest Policy and Forest History, University of Freiburg, 17 Bertoldstrasse, D-79085 Freiburg, Germany)

Introduction

Over many centuries natural conditions and human activities have influenced and changed ecosystems. With the increase in population and rise of technology, people's claim on the environment has increased too. Many ecological and economic crises have arisen due to unexpected changes in natural conditions which have impaired the life of the local population. During the pre-industrial period these crises were mainly of regional or local importance. Nevertheless their traces are still visible. This chapter is based on a project examining the relationship between humans and nature in the historical development of the Hohe Tauern National Park region of Carinthia. I will explore the connections between the people living in this area and the wooded landscape and hope to gain knowledge which will help to focus public consciousness on current ecological problems.

Dimensions of Sustainability — Sustainable Management

Many authors have referred to historical changes in the definition of *sustainability*. In this connection Sagl (1993) has emphasized that, despite these changes, the core meaning of the word has changed little. At its heart, sustainability contains a bundle of notions including the passing of time, of steadiness and continuity, of development, of potential value, and current and future use. Within forestry, sustainability concerns the yield of wood and other products in relation to time. The principle of sustainability applies to the total complexity of relations between humans and forest based on the available 'usefulness' for human existence. However, the volume and importance of the functions that make up this 'usefulness' depend on different factors which are conditioned by site and time.

Utilization of the Landscape Potential

Following Ottitsch (Glück and Ottitsch, 1995) the possibilities offered by nature for various land uses may be considered a limited budget available in a particular space. Within a certain time period the budget concept refers to a specific spatial unit (Leser, 1991). The landscape potential, which is given by natural factors and cannot be increased, appears on the income side of this budget. On the expenditure side we find the various land uses and their intensity. The total expenditure should not exceed the available income.

The allocation of the various budget items on the expenditure side, that is land exploitation, is determined by the different fields of interest of the parties involved in land use. The land user has always tried to maximize personal profit. This has led to the land use intensity exceeding the landscape potential. The exodus of capital has been greater than was available in a particular unit of time and space. Since it is not possible to import capital from other spatial units this in turn meant that the level of benefits to be had were reduced. For example, stand density of Alpine forests was reduced by intensive logging and branch litter utilization which led to subsequent agricultural use by fire cultivation or forest grazing, with the effect of increased soil erosion, floods and avalanches.

Utilization Conflicts

At the heart of this study is the question of how former society related to its surroundings. In this case, society means a multitude of relations between individuals based on evolving circumstances. The shaping and structure of the society is assisted by particular ideas and measures of order provided in general by the government. Together with other influencing factors, these rules and regulations act as limits within which a certain economic framework can develop. The way a society tries to establish a relationship with its environment depends on the social and economic position of the population. Within a certain type of society there are different demands with regard to the type and volume of wood exploitation, and there are also differences between interest groups within that society (Fig. 6.1).

The following conflicts are emphasized: first, unstable conditions in forest and domestic policies (epidemics, crop failures, wars and revolutions); second, the national interest of the sovereign or state (the promotion of industry at the expense of forest management); third, increase in population (surplus demand); and fourth, the different interests of the user of the forests with regard to

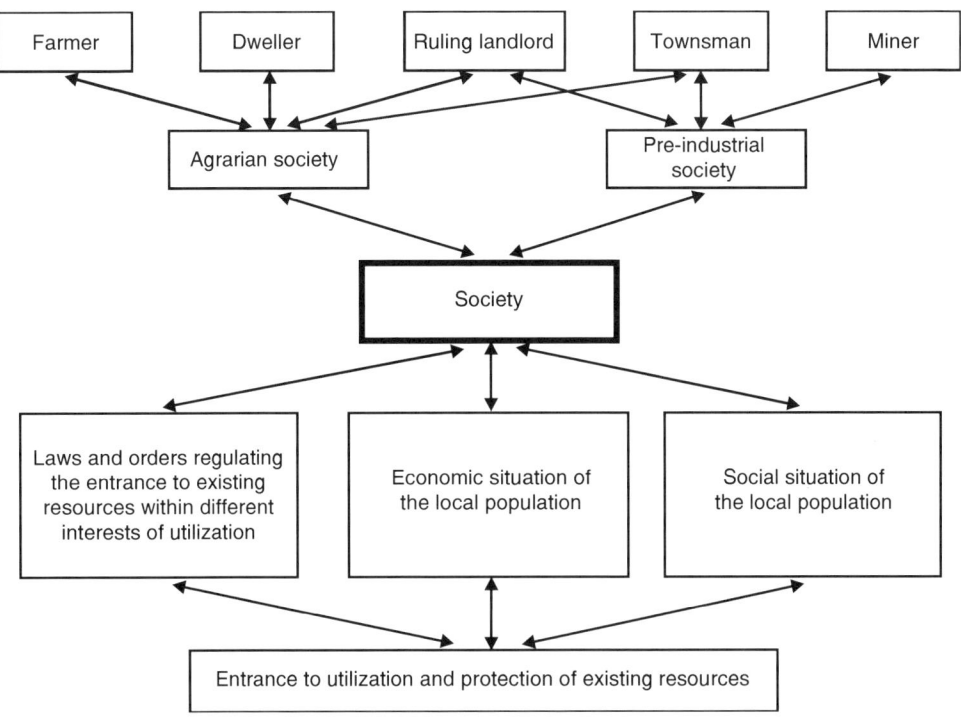

Fig. 6.1. Interactions of society with environmental conditions in a given region.

expected wood production. The conflict between those wishing to produce timber ('proper forestry') and those wishing to exploit the forest for agricultural purposes ('minor utilization') is particularly important.

The influence of humans on the forest varies from place to place in intensity, time and duration according to the course of events. In the Alpine region, however, there has, from the beginning of permanent human settlement up until the start of the nineteenth century, been a continued dependence by agriculture on the utilization of the forest. Access to the forest resources was usually reserved for the established population. As this population depended on these resources for its survival, the continued guaranteed availability of these natural resources in the locality was of great concern (Figs 6.2–6.3 and Table 6.1). During the pre-industrial period, however, additional demands placed on the local resources by non-local interests brought about the potential for additional conflicts of interest. In order to help control these conflicts of interest, regulations were drawn up to control access to the resources (see Box 6.1).

Forest Management Strategies to Resolve Utilization Conflicts

A decrease in the availability of wood, which in some areas reached the level of wood poverty, was the principal reason for worries about the conservation of

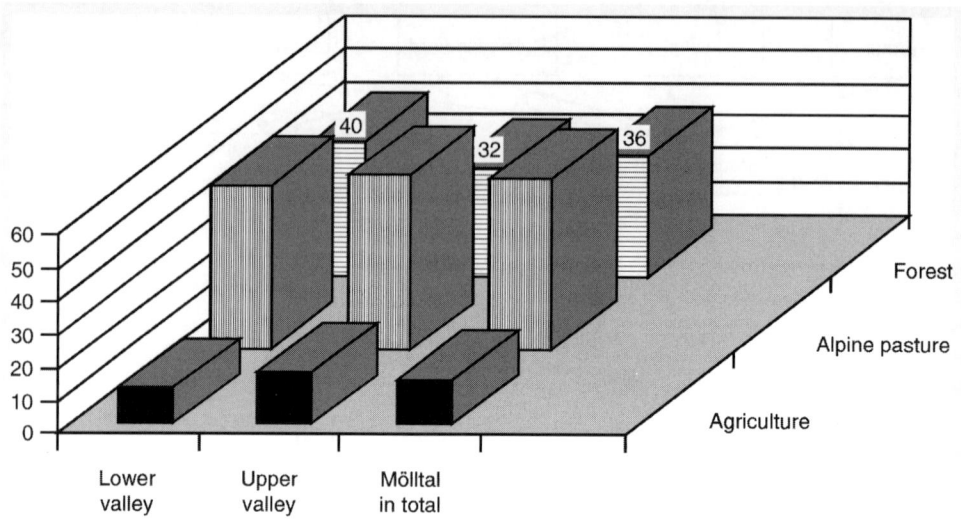

Fig. 6.2. Distribution of the cultivated area with regard to different species (as a percentage of the total cultivated area).

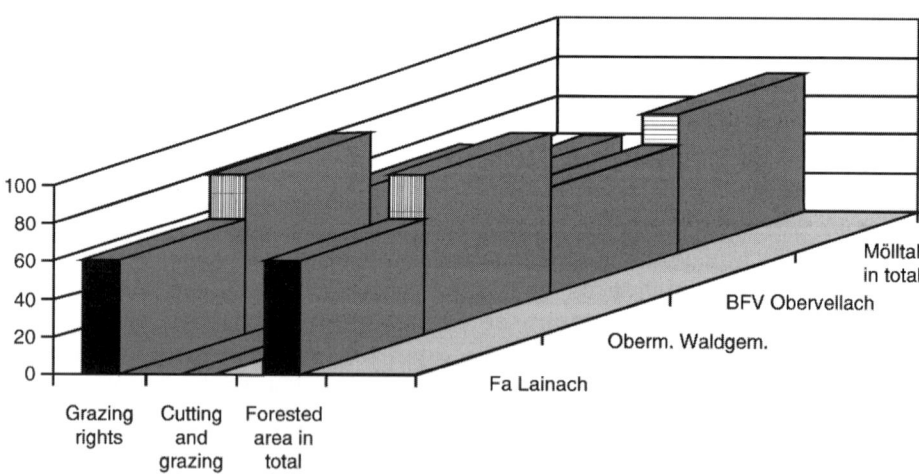

Fig. 6.3. Forested areas burdened with rights in the Mölltal Valley (Carinthia) in 1926 in relation to ownership (as a percentage of total forested area). 1. Private forests (FA Lainach): 60.6% of the forested area; 2. Cooperative forests (Obermölltaler Waldgemeinschaft): 83.3% of the forested area; 3. State forest (BFV Obervellach): 66.0% of the forested area; 4. 'Rights' area in total: 71.0% of the forested area.

the forest and its products. This situation encouraged discussion about the best ways of balancing out the different forest uses. There was thus a general movement away from devastating forest exploitation to regular forest management with the ordered use and care of sustainable young wood. The origin of the term 'sustainable forestry' at the end of the seventeenth century

Table 6.1. Development of livestock in the region of 'Mölltal'.

Year	Horses	Cattle	Sheep	Goats	Pigs	Total	Fowl	Hives
1832	525	7,433	11,202	2,000	2,500	23,760		
1857	655	14,313	16,738	7,634	3,288	42,628		
1880	758	14,565	17,171	4,716	1,212	38,442		1,620
1900	812	13,752	11,978	4,386	1,326	32,254	7,751	3,021
1923	901	9,032	15,154		2,108	27,195	9,005	
1930	925	10,480	10,075	1,553	3,412	26,445	9,919	1,660
1934	806	10,124	10,600	2,159	4,230	27,919	13,562	2,755

is partly explained by the fear of a general wood poverty in central Europe at that time. This had been brought about through a combination of forest devastation and excess cutting, but was also due to an increased standard of living and, later on, the beginning of industrialization. Thus 'sustainable

Box 6.1. Standards to regulate access to the resources and the fields of conflicts.

Legal basis

Local regulations ('Weistümer') since the thirteenth century
The sovereign's right (mountain right, hunting right, Montan forest reserve) in particular since the sixteenth century
Regional forest regulations, particularly from the sixteenth to eighteenth centuries
National forest laws (eighteenth–nineteenth centuries)
Forest Law 1852
Several game laws (particularly since the seventeenth century)
Legislation regarding harmless diversion of mountain streams 1884

Interest of the sovereign

Economic interest
Increase of government tax revenue through promotion of mining, particularly precious metals, iron and salt

Promotion of industrial development by sustainable securing of energy supply

Interest of regional culture
Segregation of forest reserves and protective forests together with restriction of exploitation (prohibition of clear cuttings) for population protection

Segregation of national parks and natural reserves, water protection areas

Private interest
Exclusive shooting rights

Interest of landowners

Free disposal of forest property with minimum restriction on kind, time, place and volume of exploitation and, with other legitimate users, access to other forms of:
- forest utilization
- grazing utilization
- forest farming
- utilization of litter
- utilization of resin
- utilization of various forest products

Demand for wood required for housekeeping (so-called 'Hausholzbedarf') free of charge since colonization (forest rights) and grazing free of charge (grazing rights)

Continued overleaf

Box 6.1. Continued.

Interest of old-established farmers	Interest of non-landowners
Preservation of possibly undivided property	Covering the need of:
Sustainable securing of the privileges entitled to the farm, to cover the need of:	• fire and construction wood • litter • grazing ground • alpine pasture
• wood (firewood, wood for construction and handicrafts) • litter (leaf and branch litter) • grazing ground • alpine pasture	
Interest of the manorial system	**Interests of forest users**
Restriction of rights for wood utilization for entitled users with regard to quality and quantity	*Forest rights* Demand for disposal of construction or firewood free of charge or against remuneration, as so-called household wood requirements
Transformation of rights to obtain wood free of charge into charged rights to obtain wood	
Restriction for grazing rights with regard to number and kind of livestock, time and place	*Grazing rights* Demand for the permission to practise grazing rights free of charge or charged, at least only for small cattle
Transformation of grazing rights free of charge into charged grazing rights	Demand for free of charge utilization of secondary products like mushrooms and berries
Conflicts with regard to the so-called secondary utilization like mushrooms, resin, leaf litter and chopped branches	Demand for free access to the forest
Conflict with regard to utilization of forest areas for recreation purposes	

forestry' was used as a key to reconstruct the devastated woodlands of central Europe and to increase the supply of wood. It was also regarded as the instigation of 'proper forestry'. Forestry can be defined to include every measure of forest technology and management necessary for the conservation, tending and exploitation of a productive forest. Thus forestry includes both technical–ecological and socio-economic components (Mantel, 1990).

The requirements for forest utilization without conflicts are:

1. Stability with regard to socio-economic conditions (living conditions).
2. Stability with regard to political conditions.
3. Continuity with regard to the type and volume of forest utilization.
4. Continuity with regard to forest beneficiaries.
5. Technical progress.

The main movement towards forest protection began at the end of the eighteenth century. One of the main reasons for this was public concern over the damage caused by floods and soil erosion following deforestation. This resulted in national and international support for the legal protection of forests and support for long-term forest ideals (see Box 6.2). As a result of the various management strategies introduced, after the passing of 150 years the formerly devastated area became fully stocked with high forest and most of the tree species produced utilizable timber. Even, however, at the outset,

Box 6.2. Main contents of the most important laws and orders.

Regulations on local level
Particular forest regulations (e.g. closure against goat-grazing by the local administration of Triest in 1150)
General common regulations since the fourteenth century (e.g. Pfunds 1303)
- characteristics of the landed real property
- regulations according to forest controls
- utilization systems and utilization restrictions ('Hausholzbedarf') in regard to the annual demand for wood required for housekeeping
- prescription for harvesting of wood suitable to time and space
- prescription for shifting cultivation methods
- regulations for forest grazing rights
- protection of special tree species, young plantations and young wood
- regulations for timber transport
- regulations for the sale of wood

Regulations on a regional level: mining regulations – forest regulations
(Earliest mining regulation from 1213: region of Trentino (Italy); restrictions with regard to shifting cultivation since 1207; silvicultural instructions for foresters since 1308 (Viennese Forest); forest regulations since 1492: in the Tyrol in upper and lower Inntal valley)
- protection of forest land
- control of exploitation (restrictions with regard to the number of utilizers, felling methods, cutting area, amount of cut, utilization period, cutting rights)
- forest supervision
- silvicultural methods
- management systems to guarantee a spatial yield regulation
- the sovereign's right
- preservation of the sustainability of forest stands
- forest exploitation and reafforestation

Regulations on a national level
- The Forest Law from 3 December 1852: maintenance of the extension of the existing forested area and improvement of forest stands
- Order from 5 July 1853 in regard to the regulation and detachment of cutting rights, grazing rights and other forest rights
- Law from 7 July 1883 in regard to the division of community-owned forest land or the regulation of cutting rights
- Legislation regarding harmless diversion of mountain streams, 30 June 1884
- Forest Law from 12 July 1962
- Forest Law from 1 January 1976

certain negative ecological effects of the forest management programme were recognized. The strategies were seen as being an artificial intervention in the 'natural balance' of the landscape and the environment (Fig. 6.4) Therefore, based on current scientific knowledge and a more comprehensive definition of sustainability, and not least, due to the pressure of public opinion, since the second half of the nineteenth century in some places forestry has strived for ecological targets too.

Progress in agricultural technology and the replacement of charcoal by lignite in industrial processes reduced the local pressure on forest stands. At the same time, people living in industrialized regions became physically and psychologically separated from the land that sustained them. Consequently, forests were no longer regarded as a main resource for forest products but as places for recreation and public health. People saw forests as undisturbed nature and wanted to preserve them for that reason. There was increasing support for the idea of national parks which would preserve the forests and help conserve nature. At the beginning of the twentieth century a local landowner gave 4000 ha of land in the region of Großglockner to the Austrian and German Alpine Society. This was the first step in the establishment of the today's Hohe Tauern national park in this formerly over-utilized valley.

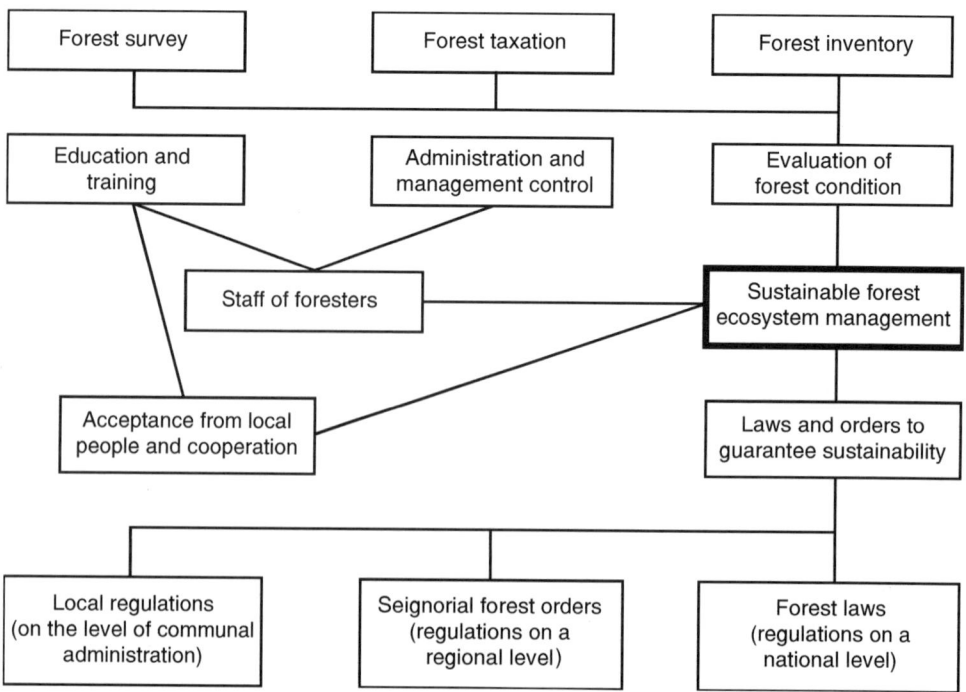

Fig. 6.4. Prerequisites for a sustainable management.

Conclusion

Global concern and discussion on deforestation has led to renewed interest in the history of woodlands. This is particularly true for the factors which underlie the process of deforestation and reasons for the subsequent expansion of forests. In this chapter the major historical factors underlying the development of the woodlands and forestry in the Hohe Tauern national park have been examined. Although unsustainable land management systems were able to destroy the natural resources of cultural landscapes and resulted in soil erosion, floods and avalanches, they did not result in any irreversible impact on the productive potential of the region. The study also shows that a general knowledge of the economic development of a region is an essential prerequisite for an understanding of its woodland history.

References

Glück, P. and Ottitsch, A. (1995) *Entwicklung eines Rauminformationssystems Wienerwald. Evaluierung der Wienerwalddeklaration.* Institut für Forstliche Betriebswirtschaft und Forstwirtschaftspolitik, Universität für Bodenkultur, Wien, pp. 34–38.

Leser, H. (1991) *Landschaftsökologie: Ansatz, Modelle, Methodik, Anwendung.* Hartmut Leser; Ulmer, Stuttgart, p. 37.

Mantel, K. (1990) *Wald und Forst in der Geschichte.* M. and A. Schaper, Hannover, pp. 89–110.

Sagl, W. (1993) Dimensionen der Nachhaltigkeit. In: *20. Tagung der Fachgruppe Wald – und Holzwissenschaft 27 und 28 Okt.* Universität für Bodenkultur, Wien, pp. 1–20.

CHAPTER 7
An ecological revolution? The 'Schlagwaldwirtschaft' in western Germany in the eighteenth and nineteenth centuries

Christoph Ernst
Universitaet Trier, B III Geschichte, D-54286 Trier, Germany

The directive of the Supreme Court, the German Reichskammergericht, issued in 1764 to the electoral government in Trier was unambiguous: 'The felling of trees by sections which was recently introduced in the Idar Forest is to be discontinued immediately' (Landeshauptarchiv Koblenz [LHAK] 56/1642). In 4 years' time, the lawyer of the suing eight communities had argued, the forest would be 'completely hewn'. The Supreme Court shared this assessment, and for this reason prohibited the successive clearing of sections that had been established a few years before.

This then is the focus: the 'Schlagwaldwirtschaft' mentioned by the judges, which is a system based on growing and felling trees section by section. In this sense the core principle was to be found in coppice woods with and without standards as well as in high forests. I am concentrating on the Eifel and the Hunsrueck regions. Both lower mountain ranges extend north and south of the river Moselle. The Moselle flows into the Rhine near Koblenz, while Cologne is found further north. In the eighteenth century the region in question belonged to the Electorate of Trier and secular sovereigns, in the nineteenth century it was part of the Prussian Rhine province. Approximately half of the territory was covered by woods, of which the state owned three-quarters, mainly going back to property of the kings, which was transferred to regional princes in medieval times (for regional studies see Schmithüsen, 1934; Bauer, 1962; and Schwind, 1983; for an introduction to German forest history research see Radkau and Schäfer, 1987, and, in English, Radkau, 1996).

This approach to forest management has not been investigated in detail in Germany until now, although it is as significant as the improved three-field agrarian system, and in addition to that, the Schlagwaldwirtschaft has determined forestry practice in Germany until recently. Moreover it is an excellent example of how the condition and expansion of the forests are frequently influenced by human activity and especially in what ways conceptions of nature are related to concrete transformation. To link the historical discourses on nature with processes in nature is generally acknowledged as a missing link in environmental history (Simon, 1993; Abelshauser, 1994).

The social construction of the forest (Bird, 1987; Evernden, 1992; Simmons, 1993; Olwig, 1996) can be described precisely whenever changes are about to happen. Hence this contribution concentrates on the qualitative and quantitative transformations in dealing with the forest in the eighteenth and nineteenth centuries in the following four steps. First of all I describe the idea of 'Schlagwaldwirtschaft'; second I deal with the motives that led to its implementation. Third, I will explain how these deliberations were realized. The fourth point deals with ecological and social consequences that arose with the application of the method. Do these fields of investigation show features of an ecological revolution? This is the central question which is to be discussed finally. By 'ecological revolution' I follow the distinguished American environmental historian Carolyn Merchant who defined it as 'major transformations in human relations with non-human nature' (Merchant, 1987, p. 265).

The Schlagwaldwirtschaft

The Schlagwaldwirtschaft is known to have made use of wood from the forest. It regulated how and in which order expanses of forests were to be cleared. To be able to achieve the clearings in the long run the reafforestation of cleared woodlands had to be effected. Thus, a forest had to be evenly divided into sections. If you wanted to use 30-year-old firewood from a woodland area, then you had to measure off a district for each of the following 30 years. In this way, one-thirtieth of the forest could be used annually without affecting the greater substance of the forest. This method equally protected the property and the annual rates of return.

In the eighteenth and nineteenth centuries economists and forest researchers elaborated the central features of this new forest management model. First, it was to secure the 'sustainability' of the timber supply. In 1713 the leading official of a Saxon coal-mine had exhorted the 'sustainable use of the forests' (Carlowitz, 1713, p. 105). From this time on 'sustainability' was something like the pivotal point of forestry. In it, the economist Pfeiffer in 1781 saw a way to realize 'the perpetual maximum utilization of the forests' (Pfeiffer, 1781, p. 15). Second, the method was based on scientific findings. In field experiments the physiological features of trees and their conditions of location for optimal growth were examined. Generally speaking, the significance of the forest for the climate, water balance, and formation of the country was pointed out in the early nineteenth century. Third, it was part of general tendencies towards centralization in so far as the central authorities controlled and managed the system; they thereby replaced regional and local practices. Fourth, it was crucial that more timber could be produced in this way, timber being well known as the essential resource of pre-industrial society. Hence foresters primarily saw a forest as a 'timber-production forest' – and little more (Hasel, 1985; Mantel, 1990).

From time to time in the Middle Ages, the pre-eminence of wood exploitation was documented, particularly in the sixteenth century, when people and manufacturing were dependent upon a well-balanced wood supply, that is in towns and in the vicinity of salt-works, mines and iron works. Still, the 'wood-production forest' was clearly to be distinguished from the common method of forest management. It is true that forests had produced firewood and building timber before. Above all, however, they served as pasture, and as a reservoir of land and fertilizer for the agrarian ecosystem: it was mainly an 'agricultural forest of sustenance'. This diversity of utilization by the rural population endangered the forest stand in the eyes of the foresters because – according to their new standards – it was unorganized, unscientific and unsustainable. The agrarian model was particularly devalued because the agricultural ways of utilization required expanses of land, nutrients and fruit which were hence not available for wood production.

Motives for Implementation

This new 'paradigm of forest management' (Selter, 1995, p. 291) was elaborated since, with the increase of population, the demand for firewood, building and construction timber rose – all these being resources to which there was no alternative in Germany until late in the nineteenth century. Apart from that, the industrial wood supply had to be secured, for instance, for the iron works in the Hunsrueck which – as the largest commercial consumer – expanded their production from 1750 onwards. Drastic price increases as a result of a rising and at the same time liberalized demand induced a wood boom (Rubner, 1967, pp. 60, 82), which followed a pattern similar to those agrarian crises and booms examined by Wilhelm Abel (1974). In times of a growing demand for wood, the 'Schlagwaldwirtschaft' made it possible, as it is frequently said in relevant sources, 'to turn the wood into silver'. For instance, nine-tenths of the electorate's forest returns came from sales of wood from the forest stands. These profits increased sharply after the turn of the century and contributed approximately 10% to the gross income of the treasury. Consequently, the forest and its intensive management were of an overriding economic importance to the state. Winfried Schenk has shown this for other regions in Germany, too, using financial documents, which astonishingly have only been occasionally analysed in Germany so far (Schenk, 1996).

Thus, it becomes clear why regulations for forest protection, or rather regulations for wood protection, were taken up from the eighteenth century onwards; in fact they were not new, but of renewed relevance. People wanted to meet the allegedly arisen, but at least the feared, shortage of firewood and building timber, in order to preserve the forest – not entirely unselfishly – for the 'descendants', as they were termed in the sources. The emergence of barren land was to be restricted and finally turned around.

In some regions of the Eifel, deforestation was in fact widely advanced. Numerous forest descriptions indicate this. However, these sources ought to be studied critically, as they are written from an authoritarian perspective. The forest map (Fig. 7.1) shows the 'Kondelwald' in the south of the Eifel in 1760 (see Ernst, 1995, for more detailed maps). It is particularly valuable because a distortion in perspective of the forest condition can be excluded. It was commissioned by the two rulers of the Sponheim territory to record the value of their district as accurately as possible for a planned division. As can easily be seen from the map, at that time only half of the original forest area was still wooded. It is possible to recognize the degradation in the south and east from their light colour, in contrast with the dark tree symbols in the north; the level of afforestation is revealed statistically again on the bottom of the map. Overgrazing near settlements, the favourable location near the Moselle, and the effects of war had aggravated the condition of the forest in this extreme regional example of a state-owned forest, whereas this destiny was no doubt shared by numerous communal forests. But nevertheless the other state-owned forests of the Grafschaft Sponheim were deforested to only about 15% (see Bauer, 1981 for other regional forest maps). Therefore one has to differentiate and be attentive in interpreting the vociferous complaints about the wood 'shortage' in sources and contemporary literature. They must not hide the fact that this scenario of disaster was to legitimize the authoritarian intervention and the method of forest management in question (Radkau, 1983, 1986; Allmann, 1989; Schäfer, 1992).

What Did These Authoritarian Measures Consist of in Concrete Terms?

Several central requirements had to be fulfilled for the 'wood-production forest' to supersede successfully the 'agricultural forest of sustenance'. First of all, a legal framework was required. In the eighteenth and nineteenth centuries detailed forest laws were enacted to extend and codify the medieval regulations and those from the sixteenth century more precisely. In order to introduce and implement this method of forest management on a long-term basis and to be able to punish offenders, people had to rely on a scientifically trained and functioning forestry administration. It proved to be difficult to control the forest administration in the face of corruption. In addition to that, damage caused by snow, fire and fauna endangered the forest stands. The latest research in Germany has revealed a further point; namely that before forestry could act according to its own criteria, it had to emancipate itself from those two fields that had strongly influenced the forest, if not dominated it completely: hunting and agricultural exploitation. As long as rulers and their households only saw the forest as a large game preserve, wood production was not only out of fashion, but rather difficult because princely numbers of deers imperilled the natural regrowth (Eckardt, 1976). The agricultural exploitation of forests on the other hand had to be restrained and finally got rid of for the reasons mentioned earlier. This disconnection turned out to be lengthy and extremely problematic. In the nineteenth century, for instance, the feeding of cattle in barns during the summer, which is known to have been an essential matter of concern to agrarian innovators, was only possible in many places

The Schlagwaldwirtschaft in western Germany

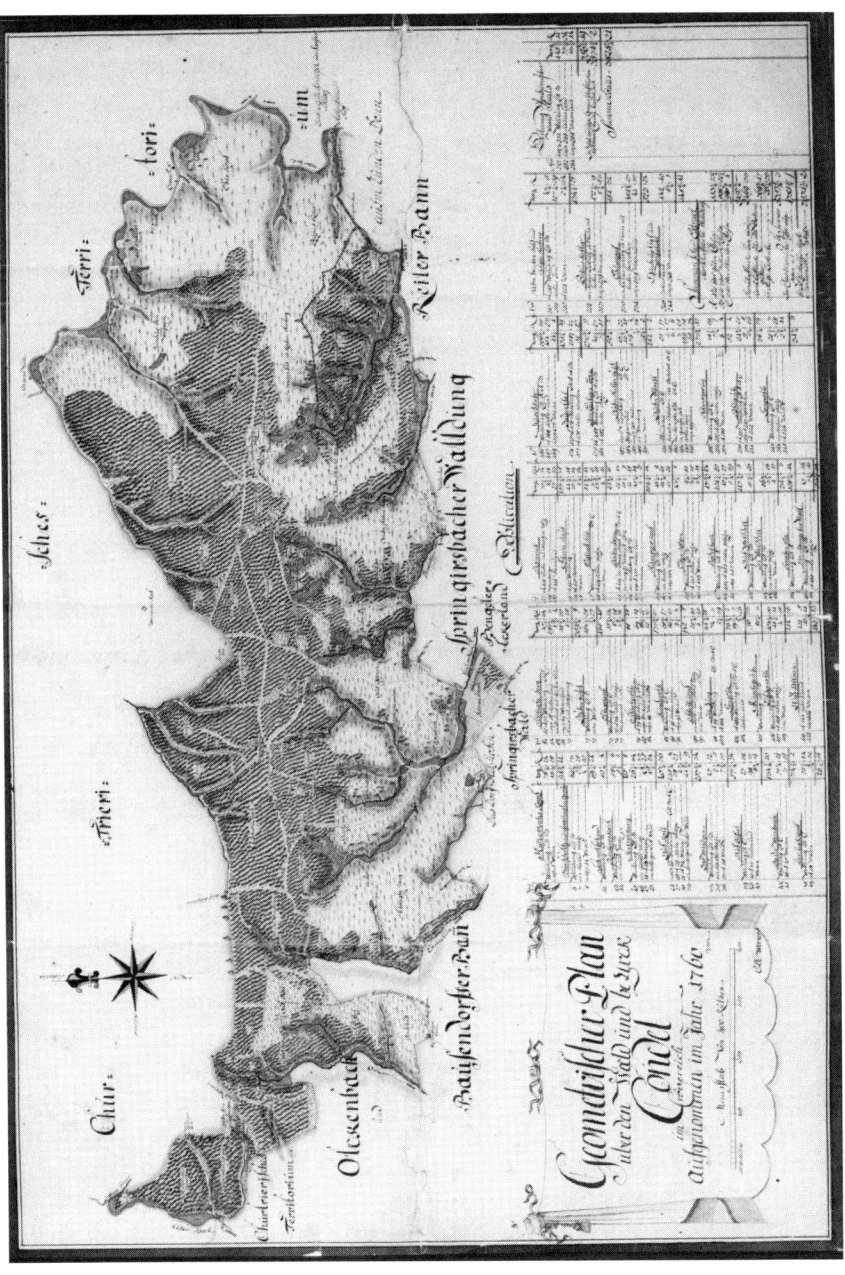

Fig. 7.1. The Kondelwald/Eifel, 1760 (Stadtarchiv Trier 54 K 5104).

where leaves and grass could be fetched from the forest to substitute for (the missing) hay. The consequences for the forests were disruptions of growth and – again – lack of natural regrowth: improvements in agriculture and forestry could only be successful when attempted together (Selter, 1995, p. 371).

Despite all these difficulties, one contemporary was mistaken when he stated with regard to the 'Schlagwaldwirtschaft' of forests: 'They do not adopt characteristics of human systems' (LHAK 1 C 1416: fol. 56v–57r). Quite the opposite is the case. In 1787 all state-owned forests in the Electorate of Trier were included in the table (reproduced as Fig. 7.2) that probably goes back to the draft outline of a forest scientist named Burgsdorf. It showed forest stands for 35 years to come with data on the amounts that were orientated exactly towards the annual felling; wood supplies were therefore reserved for the future. The sheet including the Hermeskeil forest is shown. The time of felling was laid down for each of the 31 compartments, with the following felling planned for 11 years later. After the French occupation of the Rhineland, lasting from 1794 to 1814, the Prussian forest administration prolonged the plan of forest stands up to our time and included communal forests as well as private forests in the operations.

What Ecological and Social Consequences Were Involved in the Schlagwaldwirtschaft?

The division of the 'wood-production forest' into sections and the replanting of trees by age-groups forced the disconnection of the wooded country ecosystems: the spatial and chronological cooperation of utilization and tree species grew into a spatial and chronological discontinuity. It has been emphasized that the stability of the ecosystem 'forest' was thereby reduced, a fact that did not escape contemporary attention: for instance, at the time the ruler's lawyer ascribed the storm damage of the 1760s in the Hunsrueck forests to tree-felling by sections; and also damage caused by erosion and floods were at times connected with the forest condition and its management in the early nineteenth century. Obviously, the concept of 'sustainability' was understood too onesidedly from an economic point of view; hence a 'biological sustainability' of the forest that was orientated towards ecological criteria like naturalness, health of the forest and diversity of species, had fallen into oblivion (Bode and von Hohnhorst, 1994) – a diagnosis that is particularly applicable to pine monocultures nowadays. Such monocultures emerged from 1840 onwards when the degraded soils in the Eifel could only be reafforested with pine, which in comparison to the natural vegetation of beech, hornbeam and oak is rather undemanding on soil nutrients. This led to a massive change in the existing kinds of trees. However, this was not inconvenient for forestry later, for the pine is known to produce more wood in the same period of time than other trees (Wenzel, 1962).

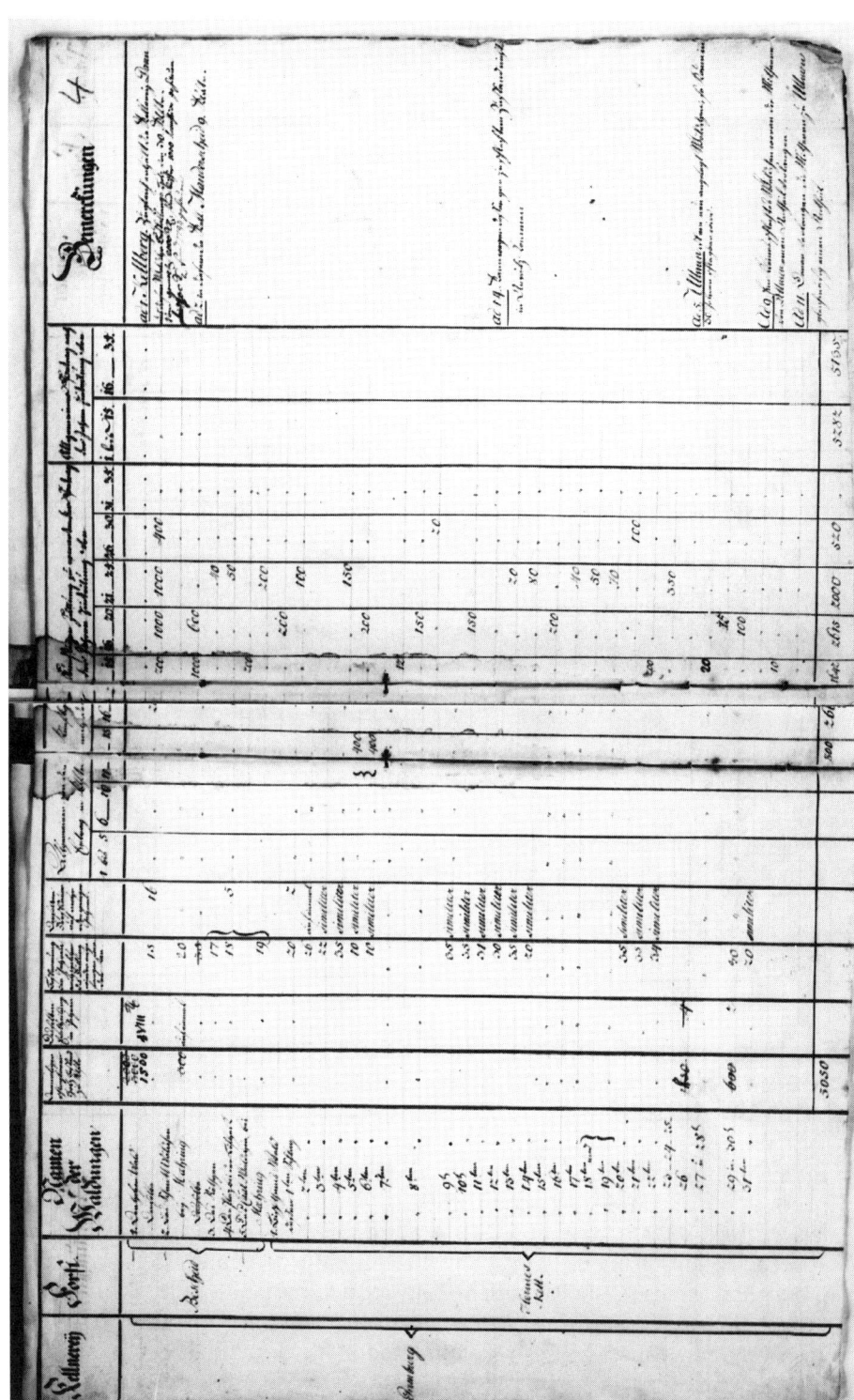

Fig. 7.2. State-owned forests in the Electorate of Trier (obererzstiftische Wälder) and their scheduled management 1787/1791 (LHAK 1 C 18820, fol. 4).

Conflicts about forest property and exploitation can be verified from the late Middle Ages onwards (Epperlein, 1993). Mass infringements of the forest laws are generally acknowledged as indicators of the social situation of farmers and peasants. Struggles for the rights of disposal of forest resources are documented particularly during the Peasant Revolt in the sixteenth century and the Revolution in 1848. From time to time only the military could secure the stands and measures of afforestation.

Until now, it has been overlooked that alongside these 'classical lines of conflict' so-to-speak, the 'Schlagwaldwirtschaft' in the Hunsrueck was the object of numerous lengthy trials that were initiated against the treasury by the communities. These trials produced some of the rare kind of sources which allows one seriously to investigate the farmer's perspective. Many committees examining the forests on behalf of the courts confirmed that the Schlagwaldwirtschaft reduced wood pasture and – which was generally accepted – rights of disposal. With the latter a compromise was intended: in 1790, an official from Reinsfeld summarized that the sections in the Osburger Hochwald had been felled so professionally that 'they give the most promising prospect for the future' (LHAK 56/3810: fol. 129r). He had to concede, however, that less waste and dead wood could be obtained, to which, until recently, the local communities were entitled: 'But of course there is waste wood nowadays when the forests are exploited so economically and sustainably, and no trees are left standing, much less than in the past when the forests had no value.' In various judgements from this time on, the Schlagwaldwirtschaft was only permitted if it did not endanger the communities' entitlement to gather firewood from the forests. In the last 30 years of the eighteenth century these entitlements were quantified exactly and withdrawn from the annual wood 'harvest' of state-owned forests. Only if something remained, was the treasury allowed to dispose of it. The courts' appeal led to a socially more wholesome management. Not so much 'environmental protection', but effective legal protection must have been at the centre of the reasoning of the Electoral courts and the Supreme Court in this matter.

What Features of an Ecological Revolution can be Revealed in the Forests of Hunsrueck and Eifel in the Eighteenth and Nineteenth Centuries?

In this period of time, economists and forest researchers developed the concept of Schlagwaldwirtschaft, that is, a system based on growing and felling trees section by section. Its features were: sustainability, a scientific approach, central management and the primary orientation towards wood production. Despite similar small-scale local forerunners, the Schlagwaldwirtschaft stood out from earlier methods of forest management. The decisively new aspect of this revolutionized method was that it was put into practice extensively. At first, only the state-owned forests were included, but from 1800 onwards the communal forests also gradually followed suit. In some ways, in terms of property rights, all forests became equal in the face of Schlagwaldwirtschaft. By separating woodlands from agricultural lands, the ecological consequences reached a new dimension: in comparison to past times, the forest biotope was constructed entirely according to different social interests. Since the majority of the population was confronted with the modification of their rights to use the forests, this process also had far-reaching social consequences.

This ecological revolution, understood as a 'major transformation in human relations with non-human nature', between 1750 and 1850 shaped the beginnings of an early 'environmental policy' in dealing with forests. It had two components: on the one hand it aimed to increase woodland management, in order to provide the population, trade and proto-industry with timber. Material interests accelerated the endeavours for efficiency. On the other hand, systematic wood protection was integrated to sustain and expand the ecological and economic potential in the long term. Now and then, however, non-material interests were also taken into consideration.

This process, exemplified by the proceedings in Hunsrueck and Eifel, obviously refers to a common challenging situation: during this period of time the Schlagwaldwirtschaft and related environmental policies were introduced in almost all German territories with some chronological variation. But they did not revolve so much around the poles of 'dependency on nature' and 'exploitation of nature' as we would imagine today. Rather, their concern was more subtly diversified in the social construction of nature. The result of constructing nature not only mentally, but also in concrete terms, is the shape and shaping of a cultural landscape, the way we know it today.

References

Abel, W. (1974) *Massenarmut und Hungerkrisen im Vorindustriellen Europa. Versuch einer Synopsis.* Hamburg, Berlin.
Abelshauser, W. (ed.) (1994) *Umweltgeschichte. Umweltverträgliches Wirtschaften in Historischer Perspektive.* Acht Beiträge, Göttingen.
Allmann, J. (1989) *Der Wald in der Frühen Neuzeit: eine Mentalitäts – und Sozialgeschichtliche Untersuchung am Beispiel des Pfälzer Raumes 1500–1800.* Berlin.
Bauer, E. (1962) *Der Soonwald im Hunsrück. Forstgeschichte eines Deutschen Waldgebietes.* Freiburg.
Bauer, E. (1981) *Unsere Wälder im Historischen Kartenbild. Beitrag zur Geschichte des Forstkartenwesens in Rheinland-Pfalz.* Grünstadt.
Bird, E.A.R. (1987) The social construction of nature: theoretical approaches to the history of environmental problems. *Environmental Review* 11, 255–264.
Bode, W. and von Hohnhorst, M. (1994) *Waldwende. Vom Försterwald zum Naturwald.* München.
Carlowitz, H.C. von (1713) *Silvicultura oeconomica, oder hausswirthschaftliche Nachricht und naturmässige Anweisung zur wilden Baum-Zucht.* Leipzig.
Eckardt, H.W. (1976) *Herrschaftliche Jagd, bäuerliche Not und bürgerliche Kritik. Zur*

Geschichte der fürstlichen und adligen Jagdprivilegien vornehmlich im südwestdeutschen Raum. Göttingen.

Epperlein, S. (1993) *Waldnutzung, Waldstreitigkeiten und Waldschutz in Deutschland im hohen Mittelalter: 2 Hälfte 11. Jahrhundert bis ausgehendes 14.* Jahrhundert, Stuttgart.

Ernst, C. (1995) Ein neuer Umgang mit Natur? Der Kondelwald im 18. Jahrhundert. In: Klaus Freckmann (ed.), *Sobernheimer Gespräche 3. Das Land an der Mosel – Kultur und Struktur.* Köln, pp. 21–32.

Evernden, N. (1992) *The Social Creation of Nature.* Baltimore and London.

Hasel, K. (1985) *Forstgeschichte. Ein Grundriß für Studium und Praxis.* Hamburg and Berlin.

Mantel, K. (1990) *Wald und Forst in der Geschichte. Ein Lehr – und Handbuch.* Alfeld, Hannover.

Merchant, C. (1987) The theoretical structure of ecological revolutions. *Environmental Review* 11, 265–274.

Olwig, K.R. (1996) Environmental history and the construction of nature and landscape: the case of the 'landscaping' of the Jutland heath. *Environment and History* 2, 15–38.

Pfeiffer, J.F. von (1781) *Grundriss der Forstwissenschaft, zum Gebrauch dirigierender Forst- und Kameralbedienten, auch Privatguthsbesitzern.* Mannheim.

Radkau, J. (1983) Holzverknappung und Krisenbewußtsein im 18. Jahrhundert. *Geschichte und Gesellschaft* 9, 513–543.

Radkau, J. (1986) Zur angeblichen Energiekrise des 18. Jahrhunderts. Revisionistische Betachtungen über die 'Holznot'. *Vierteljahrschrift für Sozial- und Wirtschaftsgeschichte* 73, 1–37.

Radkau, J. (1996) Wood and forestry in German history: in quest of an environmental approach. *Environment and History* 2, 63–76.

Radkau, J. and Schäfer, I. (1987) *Holz. Ein Naturstoff in der Technikgeschichte.* Reinbek.

Rubner, H. (1967) *Forstgeschichte im Zeitalter der Industriellen Revolution.* Berlin.

Schäfer, I. (1992) *'Ein Gespenst geht um': Politik mit der Holznot in Lippe 1750–1850. Eine Regionalstudie zur Wald- und Technikgeschichte.* Detmold.

Schenk, W. (1996) *Waldnutzung, Waldzustand und regionale Entwicklung in vorindustrieller Zeit im Mittleren Deutschland. Historisch-Geographische Beiträge zur Erforschung von Kulturlandschaften in Mainfranken und Nordhessen.* Stuttgart.

Schmithüsen, J. (1934) *Der Niederwald des Linksrheinischen Schiefergebirges. Ein Beitrag zur Geographie der Rheinischen Kulturlandschaft.* Bonn.

Schwind, W. (1983) *Der Wald in der Vulkaneifel in Geschichte und Gegenwart. Übernutzung der Vulkaneifel bis 1840.* Göttingen.

Selter, B. (1995) *Waldnutzung und ländliche Gesellschaft. Landwirtschaftlicher 'Nährwald' und neue Holzökonomie im Sauerland des 18. und 19. Jahrhunderts.* Paderborn.

Simmons, I.G. (1993) *Interpreting Nature. Cultural Constructions of the Environment.* London.

Simon, C. (1993) *Environmental History Newsletter*, Special issue No. 1 (1993): Umweltgeschichte heute: Neue Themen und Ansätze der Geschichtswissenschaft – Beiträge zur Umwelt-Wissenschaft.

Wenzel, I. (1962) *Ödlandentstehung und Wiederaufforstung in der Zentraleifel.* Bonn.

CHAPTER 8
'A solemn and gloomy umbrage': changing interpretations of the ancient oaks of Sherwood Forest

Charles Watkins
Department of Geography, University of Nottingham, Nottingham NG7 2RD, UK

> It is also necessary to take down some of the Randle Pike oaks in the park; for they are loosing money every year; for when an oak, or other Timber tree, once looseth the verdure of his head, it's like an old man going down to his Grave, and leaves nothing behind him but his Bones
> (Observations by John Amery on the Park at Grove, near Retford, north Nottinghamshire c. 1762[1])

Introduction

The surviving remnants of Sherwood Forest are found at Birklands and Bilhaugh near Edwinstowe, Nottinghamshire. Although only a few hundred hectares in extent, they have in recent years been recognized as of great nature conservation value and have been categorized as a Site of Special Scientific Interest. The surviving ancient oaks are valued by nature conservationists for providing particularly valuable dead wood habitat and they are found in important fragments of lowland heathland. In addition to their ecological interest, the trees are an enormous tourist attraction. The Sherwood Forest Visitor Centre, run by Nottinghamshire County Council, receives around one million visitors a year; the Forestry Commission has established a Sherwood Forest Initiative to help restore aspects of the Sherwood Forest landscape.

The cultural meanings of trees have been widely discussed (Thomas, 1983; Daniels, 1988; Schama, 1995). In this chapter I explore changing interpretations of a small number of ancient oaks by archaeologists, historians, writers and tourists in the eighteenth and nineteenth century. Several layers of meaning are uncovered; the trees were ascribed different values by different people at different times. As the trees get older, and die, they accumulate additional meanings yet lose others. Their position in Sherwood Forest has afforded them a particular range of cultural values. The linkage with the owners of the Dukery estates is explored, as is their inheritance of the legend of Robin Hood.

© CAB INTERNATIONAL 1998. *European Woods and Forests: Studies in Cultural History* (ed. C. Watkins)

Remnant Royal Rights

Sherwood Forest was first mentioned in 1154 and is thought to have covered the whole of Nottinghamshire north of the River Trent until 1227 (Cox, 1905, p. 204; Crook, 1979). After that date, the legal boundaries of the forest were restricted to an area roughly 20 miles long and 8 miles wide, largely on the dry sandy soils of the Bunter sandstone (Boulton, 1959). The forest was characterized by a shifting mosaic of unenclosed oak and birch woodland and heathland subject to grazing, and enclosed arable land and woodland. Some of the arable land was in the form of temporary enclosures known as 'brecks' (Chambers, 1932; Fowkes, 1977). Most of the land was in private ownership, although large areas were subject to common grazing and other common rights until these rights were extinguished by enclosure (Cowell, 1998). By the end of the seventeenth century the area of woodland in Sherwood had reached its lowest point (Watkins, 1981)

In the late eighteenth century the surviving Royal Forests were considered by many to be an anachronism redolent of 'old corruption'. A Royal Commission was established to examine the extent of surviving Crown rights (James, 1981). This published 17 reports between 1787 and 1793. The fourteenth report, published in 1793, examined the 'state and condition' of the Crown's 'woods, forests and land revenues' in Sherwood Forest. The report was compiled by Charles Middleton, John Call and John Fordyce of the Land Revenue Office.[2]

The commissioners drew on a wide variety of evidence in compiling their report on Sherwood. This included perambulations of the forest preserved in Records kept in the Tower of London and in the Court of Exchequer[3] and various surveys of the forest, including one made in 1609. In addition to documents kept in London, and 'wishing for farther and more certain Information, we thought it necessary personally to view the Forest, and make Enquiries on the Spot.'[4] They interviewed John Gladwin, who was steward of the courts of the Forest and George Clarke, woodward of Birkland and Bilhaugh in November and December 1791. They learnt from Mr Gladwin that

> there was lying in the *Swan Inn*, in *Mansfield*, a voluminous Collection of antient Papers, supposed to relate to this Forest, and locked up in a Chest, which had formerly been kept in the Castle of *Nottingham*; but whether these Papers were of any Consequence, or what they contained, was not known. We found also that there were some antient Forest Books and Papers in the Possession of different Gentlemen residing in the neighbourhood of the Forest.[5]

Extracts were taken from these various documents, and reproduced, together with the reports of the interviews with Gladwin and Clarke, in 32 appendices to the report.

The report describes how the greater part of the Forest had been granted to private individuals over many years and that 'the Rights of the Crown had been lost Sight of', and were, prior to their enquiries, 'very imperfectly known'. The nine forest keepers 'knew very little of their Walks' and 'never acted' other than in 'receiving their trifling salaries'. No deer survived in the forest, other than those in Thorney Woods and those were claimed by Lord Chesterfield. Moreover, 'not the Right of Common only, but the Soil itself, of every part of the Forest, was claimed by neighbouring Proprietors.'[6] The report concluded that the property of the Crown in the forest:

> appears to consist of Three Kinds; of the Soil in *Birkland* and *Bilhagh*, where neighbouring Inhabitants have Rights of Common; of Timber in a State of Decay, and exposed to constant Injury; and of Forestrial Rights over an extensive District, which tend to obstruct Improvement, and lessen the Value of private Property, without bringing, at present, any Profit to the Crown.[7]

What of the surviving woodland and the ancient oaks? The surveyors were keen to establish which woodland remained Crown property and the value of any trees. Their earlier reports on the Forest of Dean and the New Forest had shown 'how loose the Management in Forests has been' and they were not surprised to find that at Sherwood 'situated at a Distance from any Dock Yard ... many Abuses had been suffered to prevail ... in the Management of Timber'.[8] They found that the 'Timber Trees, admitted to belong to the Crown, were going fast to Decay'[9] and made special enquiries about the way in which the trees were managed.

The woodward of the two main surviving areas of woodland at Birkland and Bilhagh was George Clarke. When interviewed Clarke stated that these woods were about 2000 acres in extent and confirmed that the trees in these woods were the sole property of the Crown and that 'no Timber or Wood is cut there for the Use of any private Person.'[10] He had been appointed woodward in 1772 and his job was to see that no 'Waste be committed on the Wood and Timber belonging to the Crown' and provide information about the quality and quantity of timber that might be harvested when required to do so by the Surveyor General of His Majesty's Woods. The woodward had to account for the disposal the 'Lops, Tops, Bark and Offal Wood'. His salary was £20 per year together with the income, worth about £10 a year, from all windfallen trees.[11]

The commissioners used documentary evidence together with a visit to the woods to establish their condition. They found that

> The Trees in *Birkland* and *Bilhagh* are of great Size, and now of great Age. It appears by a Return to a Commission for Enquiring into the State of the Timber in this Forest, and the Abuses committed in it, dated 3d *January* 1598, 'That the Trees in Birkland were then of 300 Years Growth, and more;' and those in Bilhagh were 'of 200 Years and more' – The far greater Part of those Trees are now in a State of Decay, and it is not easy to find such as have not some Defect in the Heart, where Trees first begin to fail. This Difficulty gives rise to the greatest Abuse which we have found to prevail in this Forest.[12]

Each year five trees were taken by the four Verderers and the Steward of the Forest Courts as 'Fee Trees' in payment for the posts they held. George Clarke describes how these officers

> or those to whom their Trees are sold, choose whatever Tree they like best; and it is a common Practice to bore the Trees first, to see if they are sound; and if a Tree after being bored is not liked, other Trees are tried in the same Manner, until the Party finds one he approves of.[13]

The surveyors do not hold back on the criticism of this method. They point out that the very Officers whose duty it was to prevent abuses in the forest frequently sold their fee trees to individuals who were able to choose which tree they wanted. The purchaser 'to guard against the Danger of buying [a tree] that is unsound, bores the best-looking Trees to the Heart with an Auger, rejecting every one in which there is any Mark of Decay'. They note that 'as this happens every year' many of the better trees had been damaged in this way. Indeed they found that 'each Time that we viewed this Forest, we found some which had been recently bored'.[14]

In addition to the damage through boring the trees, the commissioners found that a system of enumerating the trees had been in use which itself damaged the trees. George Clarke reported that in 1775 he had been directed to mark and number the trees by 'cutting off a Piece of the Bark about Five Inches Square, and stamping the Crown, the Number of the Tree, and the Name of the Forest, on each Tree, with an Iron Instrument, on the solid Wood, many of which Numbers and Marks are now partly grown over by the Bark' but the order was changed before the marking was finished 'from an Apprehension the Trees might be injured from that Mode of marking; and he was directed to mark the remaining Trees by shaving off a Part of the Bark, so as to make a smooth Surface, on which the Marks and Numbers were stamped.'[15] The commissioners reported that 'the greater Part of the Trees' had suffered from this 'Mode of Marking' which had been introduced to help preserve them.[16]

The commissioners compared the results of various previous surveys of the timber in Birklands and Bilhaugh and found that 'the Number and Value of the Trees appear, for two centuries, to have been in a State of continual Decrease'.[17] A survey of the 'Woods and Tymber' of 1608 by Sir John Bentley, John Whitehall and William Deane found 49,900 trees of which 23,100 (46%) were 'Tymber Okes' and 26,800 (54%) were 'Okes not Tymber' which were also classed as 'decayed'. There was no underwood[18]. Seventy-two years later, a survey was carried out by Peter Brunsden 'Calker of Deptford' and John Bowyer 'Purveyor' of the navy. They found that the total number of trees had fallen to 33,996, of which 1400 (4%) were suitable for 'His Majesty's Shipping' and 32,596 (96%) were for the 'Country's use'. Many of the trees not fit for the navy were 'frow' (misshapen) and 'shaken' (having cracks in the growing timber) and had many 'dead Knotts'. At this date 8621 trees (25% of those not fit for the navy) were described as 'Young thriving Trees' but many of these were also cracked.

This 1680 survey contains a rare reference to the cutting of oak branches: it ascribes the poor condition of 8060 oak trees at Bilhagh to 'oft lopping'. This suggests that, in this section at least, pollarding was frequent, or had been so in the past. Various forest inhabitants claimed a right to cut timber in the seventeenth century although this seems not to have been the case in the eighteenth, as evidenced by a number of presentiments to the quarter sessions against labourers who had 'stolen' firewood or damaged trees (Cowell, 1998). Although the residents of Edwinstowe retained a right to collect firewood in the late eighteenth century it appears that pollarding had been largely stamped out by that time.

In 1788/89 the woodward George Clarke made a survey of 'the Oak Timber and Wood in *Birkland* and *Bilagh*' in 12 books. By this date there were 10,117 oak trees left of which 1368 contained 'Timber fit for the navy', 2213 were 'Trees fit for Coopers Use', 4572 'were Trees for Carpenter Use' and 1964 were 'Trees fit only for Cordwood and Firewood'. In addition there was in Birkland 'a Quantity of Birch Wood, containing about Four or Five hundred Cords' but no birch in Bilhagh[19] (Fig. 8.1).

Further evidence of bad management of the oak trees and their poor quality was derived by the commissioners from an investigation of the timber and wood sales accounts for 1700–1777, after which no timber had been felled. This showed that the expenses from felling the trees exceeded the income by over £9000 and that forest officers had made many fraudulent claims. The accounts also show the uses to which the timber was put and its condition. Sales of 2 March 1702 of 'dotard and decayed trees' and of 16 October 1711 of 'sear, dead, broken and decayed trees' provide confirmatory evidence to the seventeenth century surveys of the continued existence of decayed trees. A storm of 1 February 1714 resulted in a sale in 1716 of 'Trees blown down, broken and shattered, or otherwise damaged'. The only sales of timber for the navy in this period were between 1762 and 1777 when five sales totalling 682 loads of oak timber were made.[20] One load usually consisted of 50 cubic feet of timber, and there were about two loads in a good sized oak (James, 1981, p. 148), so these sales roughly equalled 340 oak trees.

In addition to the decline in the number of trees, there was a lack of regeneration of young oak trees. This was probably due to grazing. George Clarke told the commissioners that the parishioners of Edwinstowe:

> claim a Right to the Acorns, when they fall, and take in Swine to feed on them, at certain Rates *per* Head, according to the Plenty or Scarcity of Food. They also depasture their Sheep in those Woods, but not Horses, or Horned Cattle, for which the pasturage is not sufficient; and do not take in any Sheep belonging to others to feed there. No other Persons enjoy any Right of Common in those Woods.[21]

The surveyor employed by the commissioners concluded that 'the Land, in its present State, to be worth very little for Pasturage',[22] but that if the timber were felled, and the land enclosed, the land at Birklands would be worth 8 shillings an acre and that at Bilhagh 12 shillings. Although worth little, in the surveyor's view, the area was clearly grazed at this time. The survey reported by Hayman Rooke for 1790 shows that Joseph Peatfield had a sheep pen in Warsop Quarter of Birkland, Elizabeth Peatfield had a sheep pen in Clipstone Quarter and John Denton had a sheep pen in Budby Quarter (Fig. 8.2).

Further evidence for grazing is provided in the commissioners' report where they comment on the lack of young regenerating oak trees. Moreover, they contemplated the necessity of enclosing the forest, as they had in the New Forest and in the Forest of Dean, if the Sherwood woods were to continue to produce timber. They found that:

> There are, at present, no young Trees coming up in Birkland and Bilhagh, to supply the Place of those very antient Trees that are now upon it; and if a Right

ABSTRACT of a Survey of the Oak Timber and Wood in *Birkland* and *Bilhagh*, in the Forest of *Sherwood*, made in the Years 1788 and 1789, by George Clarke, by Order of *John Robinson*, Esquire, Surveyor General of His Majesty's Woods, contained in Twelve Books; viz.

BOOKS.	Number of Trees in each Book.	Timber fit for the Navy.		Trees fit for Coopers Use.	Trees for Carpenters Use.	Trees fit only for Cordwood and Firewood.	Value of Trees in each Book, with the Tops and Bark.		
		Trees.	Round Measure.						
			Loads.	Feet.				£.	s.	d.
Nº 1.	Nº 1 to 711	276	422	33	121	341	73	1,453	18	6
2.	712 to 1,431	255	342	39	114	271	180	1,249	16	5
3.	1,432 to 2,378	151	320	33	207	347	242	1,528	17	9
4.	2,379 to 3,322	88	168	4	165	449	242	1,113	7	9
5.	3,323 to 4,260	140	228	7	127	430	241	1,276	2	3
6.	4,261 to 5,204	156	241	9	138	441	209	1,272	15	—
7.	5,205 to 6,148	116	165	21	117	395	316	1,080	16	3
8.	6,149 to 6,461	43	47	17	40	131	101	396	—	3
9.	6,462 to 7,406	138	222	26	280	454	73	2,160	—	—
10.	7,407 to 8,351	103	141	39	241	502	100	1,699	18	9
11.	8,352 to 9,295	66	128	17	350	445	83	2,144	3	2
12.	9,296 to 10,117	39	70	3	313	366	104	1,771	19	3
		1,368	2,498	48	2,213	4,572	1,964	17,147	15	4

N. B.—Besides the above there is a Quantity of Birch Wood, containing about Four or Five hundred Cords, growing in *Birkland*, but none in *Bilhagh*.

Geo. Clarke,——Novʳ 17, 1791.

Fig. 8.1. Extract from House of Commons (1793).

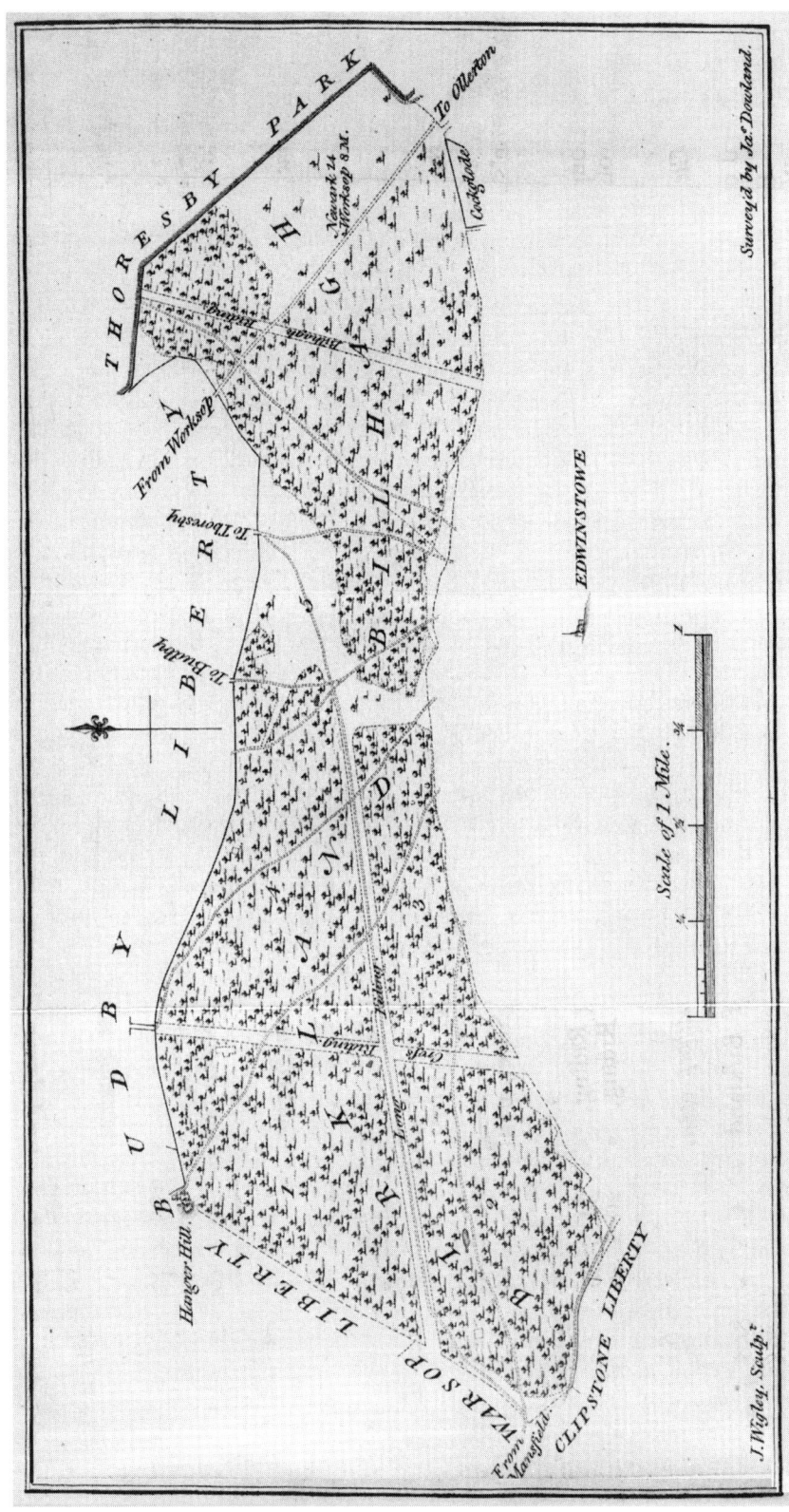

Fig. 8.2. Plan of Birkland. From Rooke (1799).

to inclose, for the Growth of Timber, should be obtained under such Conditions as we have proposed in our Reports on those Forests which it seems expedient to retain, it would be necessary, immediately, to build Houses for Wood-Bailiffs, there being none upon it at present, and to pay the Wages of those Officers for at least Fourscore Years, before any Supply to the Navy could be expected from it. The accumulated Expence on so small a Quantity of Land, in such a Length of Time, would, we apprehend, probably exceed the Value of the Produce.[23]

This lack of a satisfactory potential financial return, together with the fact that the 'Situation of the Forest is remote from any of His Majesty's Dock Yards' and the remaining area owned by the Crown was too small to 'retain it as a Nursery of Timber for the Navy'[24] were the reasons the commissioners recommended that the Birkland and Bilhagh should be sold. They suggested that the Duke of Portland should be given first refusal on this sale as his tenants had rights of common in Birkland and Bilhagh.

What can we conclude about the 'official' view of the condition of the ancient Sherwood oaks at the end of the eighteenth century? The commissioners had the job of finding out what Crown rights remained in Sherwood Forest and, if any remained, what should be done with them. The evidence they collected to do this task, and the way in which they interpreted this evidence allows us to examine the way in which they thought of and used the oak trees of Birkland and Bilhagh. The commissioners themselves saw the trees as a potentially valuable asset to the Crown which should be managed to the benefit of the nation. However, their precise calculations showed that the small remaining area of Crown woodland at Birklands and Bilhagh together with the need to wait 80 years before valuable timber would be available, meant that it was in their terms uneconomic for the Crown to enclose and manage the woodland. The distance from naval dockyards was an additional problem. The language used by the commissioners is one which emphasizes the need to overcome past abuses and provide a reasoned and orderly use of the woodland. The timber trees and land are seen as a resource which should be efficiently managed.

The abuses which the commissioners uncovered were widespread. Forest officers claimed excessive expenses and officers were appointed who 'for many Years, have been chosen for no other reason than to entitle them to their Fees'.[25] The officers were destroying the very trees they were paid to protect because the trees themselves were used as a medium of exchange. And to gain the maximum value of a fee tree, it was bored to the centre to check it was not hollow. The trees were used and described in similar terms to forest deer: they are used by the Crown to make payments of gifts, the 'best-looking Trees [are bored] to the Heart with an Auger',[26] waste wood is described as 'offal'.[27]

The remoteness from naval dockyards, together with the enormous power of local landowners who were the principal forest officers, had resulted in a massive decline in the power of the Crown in this remnant of Sherwood Forest. It was only after a thorough examination of surviving documents that the commissioners were able to demonstrate that the Duke of Portland's claim to own the soil of Birkland and Bilhagh was unfounded in law. And yet it was to the Duke of Portland that the commissioners recommended the land should be sold. The report emphasizes that it is under Crown ownership and control that the abuses it uncovers have been allowed to unfold and develop. It sees private ownership as the only satisfactory way of encouraging the future management of Birkland and Bilhagh. The commissioners' surveyor has informed them that the existing pasturage in Birkland and Bilhagh was worth 'very little' and the implication is that it would be best for the land to be inclosed and 'cleared of Timber'.[28] The evidence collected by the commissioners does not indicate pollarding was practised in the eighteenth century, although the survey of 1680 with its reference to trees suffering from 'oft lopping' indicates this may have been true earlier. The poor timber quality of many of the trees is emphasized throughout the report. Many trees are described as hollowed, withered, decayed, mishapen or seared and therefore of little economic value. The only hint that there may be some other sort of value given to these old trees is in George Clarke's evidence to the commissioners when he tells how in 1775 he was asked by the Surveyor General to fell 300 trees, but when 117 loads (about 55 trees) had been cut 'an Order came to him to cut no more, which he has heard and believes was occasioned by the Interference of the Duke of *Newcastle*, who wished the old Timber to be left standing'.[29]

Ducal Landowners

The concentration of aristocratic landownership in the northern part of Sherwood Forest resulted in the area becoming known as the Dukeries in the late eighteenth century. The four main estates were Clumber (Dukes of Newcastle), Welbeck (Newcastle and Portland), Worksop Manor (Norfolk) and Thoresby (Kingston). The families owning these estates were interconnected and titles varied due to vagaries of succession and issue. The domination of such a large area by four aristocratic estates, however, resulted in a consistent cultural signature of parks, mansions, plantations and modern agriculture (see Seymour, Chapter 9, this volume). This ducal landscape was established over many years by the purchase and appropriation of tracts of land, some former Crown land. The new landscape of improvement was wrested from the poor sandy soils of Sherwood Forest and from smaller landowners. By the 1790s the rump of Sherwood Forest was an odd island of Crown land in a sea of private property.

Many of the ancient trees were incorporated into parks. The most famous example was the Greendale Oak at Welbeck. This enormous tree was already famous in the mid seventeenth century and its dimensions are described in John Evelyn's *Sylva* (Evelyn, 1670). It achieved its greatest fame, however, through having a large hole cut through its trunk. This came about as follows. The 2nd Duke of Newcastle died at Welbeck in 1711. His daughter married Lord Harley, the connoisseur, who succeeded as 2nd Earl of Oxford in 1724.[30] The opening 'is said to have been made in consequence of an after-dinner bet by the owner, who declared that he had a tree in his park with a sufficiently large trunk to allow an aperture to be cut through which a coach and six could be driven' (Rodgers, 1908, pp. 292–293). To celebrate the event a set of five drawings was published showing views from every angle and in plan (Figs 8.3 and 8.4). Moreover the Countess of Oxford 'had a cabinet made of the oak that was cut from the heart of this tree which still survives. On the cabinet are inlaid representations of the Greendale Oak, and of a carriage and six horses being driven through the opening' (Rodgers, 1908, pp. 292–293). In the spring of 1725 Mr Thomas, the Earl of Oxford's Chaplain, wrote in his diary of 27 April:

> on Tuesday about eleven in the morning, my Lord, Mr Morley, Mr Hobart and myself rode out that way where the famous tree is, called the 'Grindall' Oak, which has lately had a passage cut through it, large enough for any coach to drive through, and accordingly seldom any pass that way without going through it; there were several of the most heavy and cumbersome branches cut off at the same time to ease the tree, but there are still reckoned to be about nine tons of timber remaining upon it. We rode through the body of it.
> (Quoted by Rodgers, 1908, pp. 292–293)

Here, we see that the Greendale Oak was celebrated for its great size, and its curiosity value was increased by the opening which allowed the owner, his friends, and anyone travelling along the road, to pass through the tree and experience the outside view and interior spaces. The tree was captured by the artist and popularized in prints. A version appeared in Hunter's edition of John Evelyn's *Sylva*. The dimensions of the tree and the amount of timber it contains were recorded by the Duke's chaplain. Parts of the timber were converted to furniture and displayed as a trophy in the mansion house.

As well as appropriating the interior of trees, the landowners cut long, wide, straight rides through the remaining oak woodland at Birkland and Bilhagh. One was cut through Birkland and Bilhagh in 1703, and in 1706 another was made from Thoresby House through Bilhagh Wood.[31] In 1709, the same year he obtained permission to enclose Clumber Park,[32] the Duke of Newcastle (Lord Warden of the Forest) cut a new ride through Birklands: 'a very broad Riding was cut through the Whole of *Birkland Wood*, from one End to the other; and the Timber, which was valued at £1,500 was given to his Grace; but the Expences attending the Fall, amounting to £.118. 17s. 2d. were charged to the Crown'.[33] The rides are shown on a map of 1799 (Fig. 8.2). These rides allowed neighbouring ducal owners to take full advantage of the Crown Forest for hunting: riders could traverse the woodland quickly and view the deer clearly. In the case of the Duke of Newcastle, his local power was so great that he was able to charge the Crown for the cost of making the ride and take the profit from the trees felled. The rides demonstrated to all and sundry the status and power of the landowners. As with the increasingly popular long avenues of trees planted in parks and along access routes to mansions, the straight-sided rides indicated that the woodland had become part of a larger planned, controlled and subordinated landscape.

Towards the end of the eighteenth century, the old Sherwood oaks attracted the attention of Major Hayman Rooke. Rooke, a retired army officer, lived at Woodhouse Place just outside Mansfield. He has been described as 'the

Fig. 8.3. The Green Dale Oke near Welbeck, 1727. From Nottingham University Manuscripts Department.

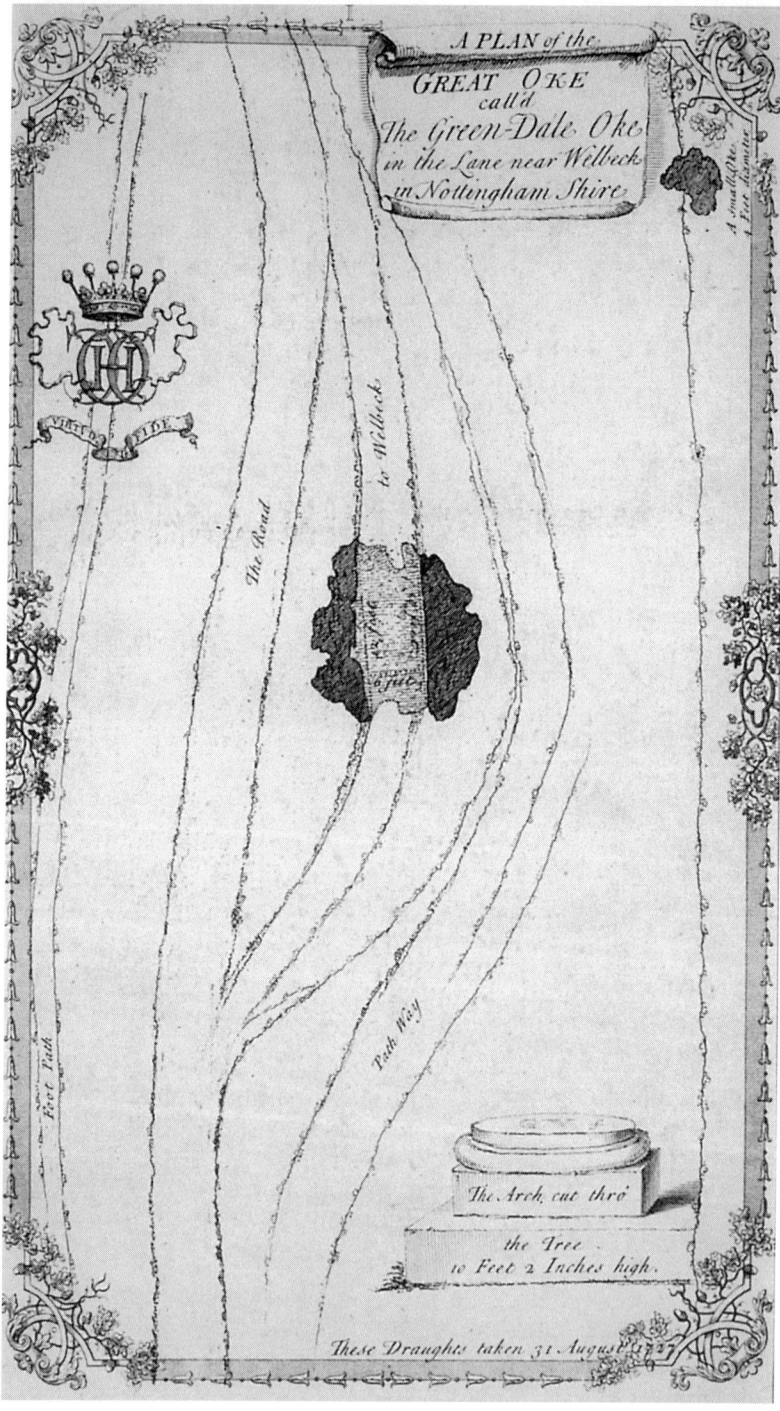

Fig. 8.4. A Plan of the Great Oke call'd the Green Dale Oke in the lane near Welbeck in Nottingham Shire, 1727. From Nottingham University manuscripts department.

real pioneer of archaeology in Nottinghamshire' and was a member of the Society of Antiquaries (Butler, 1954, pp. 3–4; Sherrat, 1965). A man of wide interests he excavated a Roman Villa at Mansfield Woodhouse, and published a meteorological register from 1785 to 1805. Towards the end of the eighteenth century he turned his attention to the ancient oaks of Sherwood. In 1790, over 60 years after the opening was cut in the Greendale Oak, Rooke published his *Descriptions and Sketches of some Remarkable Oaks, in the Park at Welbeck..., a Seat of His Grace the Duke of Portland, To Which are Added, Observations on the Age and Durability of that Tree, With Remarks on the Annual Growth of the Acorn* (Fig. 8.5). The volume was dedicated to the Duke of Portland and Rooke notes that 'These sketches [were] taken under favour of his repeated hospitality at Welbeck' (Rooke, 1790). Nine years later he published more extensive observations on the forest in his *A Sketch of the Ancient and Present State of Sherwood Forest in the County of Nottingham* which incorporated material from the Commissioners Report on Sherwood Forest of 1791 and the results of some of his archaeological work (Rooke, 1799).

Rooke's publications allow us to view the Sherwood oaks through the eye of a scholarly late eighteenth century antiquarian. The trees are celebrated for a variety of reasons. There is wonder at their size, form and dimensions; there is keen interest in their age; there is fascination with their Royal and other historical associations. These aspects are combined with a desire to dissect the trees with a scientific purpose: to gain knowledge of their origin and age. Rooke is imbued with an Enlightenment understanding, and so he is keen to marshall statistics about the decline in the number of trees and the scale of destruction. For these he draws heavily on the Report of the Commissioners of 1793. But he combines this statistical approach with an attempt to link the oaks both with the classical world and with the ancient Britons and more particularly the Druids:

> The venerable and majestic Oak seems to claim superiority over all other trees. It was styled by the ancients *Jovis Arbor*; and the Celtic statue of Jupiter was a tall oak. Our ancestors, the ancient Britons, held the oak sacred; and their priests the Druids, who took their name from the British *Derw*, an oak, and esteemed the mistletoe of that tree above that of all others, consecrated groves of oaks as one species of temple worthy of their religious ceremonies.
>
> (Rooke, 1790, p. 5)

He then connects this historical understanding of the ancient trees with the experience of contemporaries visiting the oaks 'Were we, even now, to enter a grove of stately oaks, seven or eight hundred years old, whose spreading branches form a solemn and gloomy umbrage, I think we could not behold them without some degree of veneration' (Rooke, 1790, pp. 5–6).

Rooke draws on classical and modern authorities to ascertain the possible ages of oaks. He notes that 'It has generally been thought, that the age of an oak seldom exceeds three hundred years' but considers this to be 'certainly an erroneous calculation'. He alludes for evidence to his assertion to 'some old writings in the Duke of Portland's possession' which describe the 'planting' of

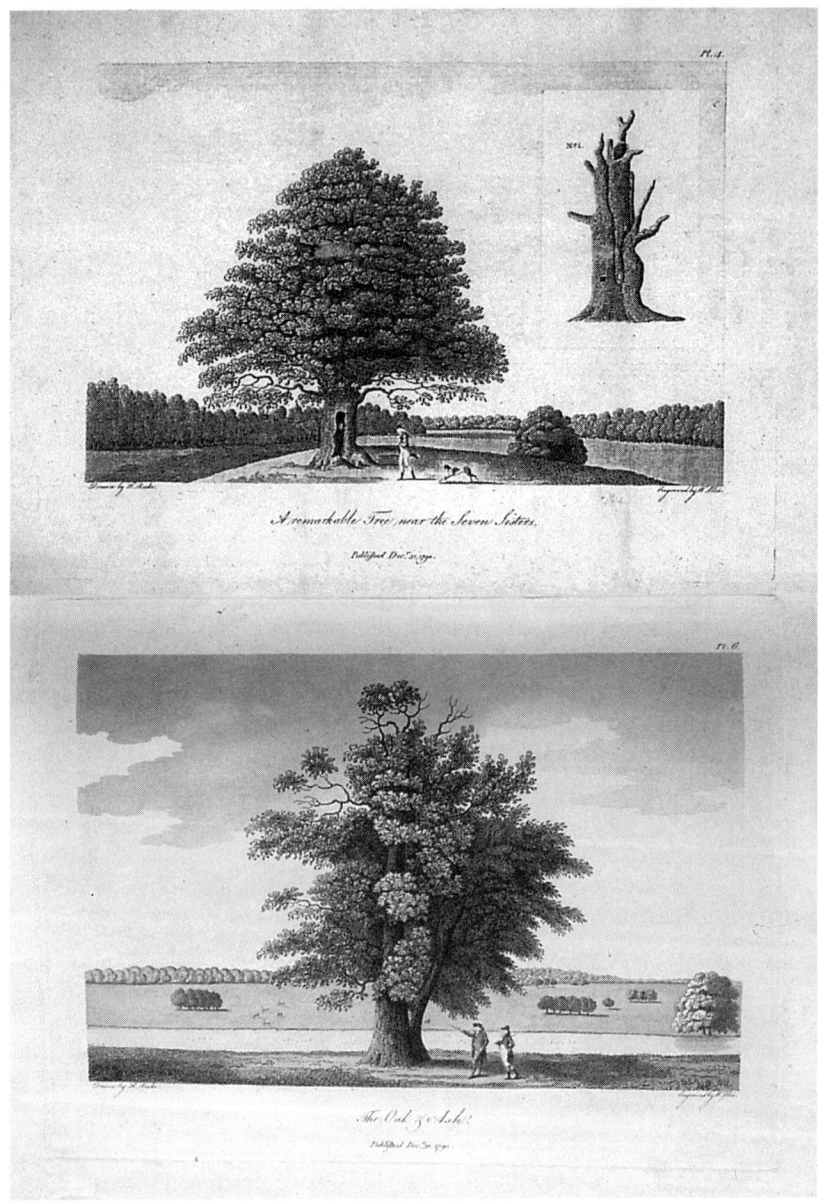

Fig. 8.5. *A Remarkable Tree near the Seven Sisters* and *The Oak and the Ash*. From Rooke (1790).

oak trees 500 years previously. Unfortunately he does not identify these old writings. However, he does draw on his archaeological and antiquarian knowledge to attempt to date the trees. He considers the principle of the annual growth rings of trees and quotes directly from Evelyn (1670, pp. 159–160) that:

> It is said, that the trunk or bough of a tree being cut transversely, plain and smooth, sheweth several circles or rings, more or less orbicular, according to the external figure, in some parallel proportion one without the other, from the centre of the wood to the inside of the bark, dividing the whole into so many circular spaces ... It is commonly, and very probably, asserted, that a tree gains a new ring every year.

But Rooke is able to go beyond paraphrasing John Evelyn. Using his own local knowledge and observation he recounts (pp. 16–17) how:

> There are now and then opportunities of knowing the ages of oaks almost to a certainty. In cutting down some trees in Birchland ... letters have been found cut or stamped in the body of the tree, marking the king's reign, several of which I have in my possession. One piece of wood marked J.R. (James Rex) was given me by the woodman, who cut the tree down in the year 1786. He said, that the letters appeared to be a little above a foot within the tree, and about one foot from the centre; so that this oak must have been near six feet in circumference when the letters were cut. A tree of that size is judged to be about one hundred and twenty years growth. If we suppose the letters to be cut about the middle of James the First's reign, it is 172 years to the year 1786, which, added to 120, makes the tree 292 years old when it is was cut down. The woodman likewise says, that the tree was perfectly sound, and had not arrived to its highest perfection. It was about 12 feet in circumference.

Rooke uses his archaeological imagination to apply a novel means of dating the ancient trees, by making use of Crown ownership marks buried deep within the body of the boles of the trees. Rooke's celebration of the ancient oaks is wide ranging. He is a detective who dissects the trees in order to understand their origin. But he is also a publicist who espouses a particular political cause. Rooke supported the aristocratic Whiggish landed interest. He organized a special celebration in 1788 of the centenary of the Glorious Revolution. He identified even the old and decrepit Sherwood oaks with the greatness of the British navy and elided their Druidic ancestry with the new plantations of the aristocracy:

> It is with pleasure we see that efforts are making to adorn this ancient Forest in a manner truly patriotic and worthy of imitation; the many respectable Persons, whose Mansions and Parks border on the Forest, have made, and continue to make, large Plantations in honour of the splendid Victories gained by our gallant Admirals
>
> (Rooke, 1799, pp. 18–19. See also Daniels, 1988)

What Rooke does not do, however, is to provide a particular aesthetic with which to appreciate the ancient trees. His views are those of the establishment, and are heavily influenced by John Evelyn whose *Sylva*, first published in the seventeenth century, was reissued in many editions in the eighteenth century and became a staple of libraries of the landed gentry. It was in the last decades of the eighteenth century, however, that another way of seeing and appreciating the Sherwood oaks came into prominence.

The Picturesque and Robin Hood

The picturesque became an increasingly powerful and influential way of understanding the landscape from the late eighteenth century onwards (Daniels and Watkins, 1994). The aesthetic theory of the picturesque is complex and takes many forms. Copley and Garside (1994, p. 1) note that it is 'a notoriously difficult category to define'. What is clear, however, is that popularizers of the picturesque such as William Gilpin (1791) and Uvedale Price (1794) presented a new vocabulary with which polite society could enjoy the landscape. Ancient trees were especially valued. While Evelyn and Rooke celebrated old trees for their dimensions, age, and classical and patriotic associations, Gilpin and Price, drawing on Italian landscape painting, were more concerned with their appearance. In his *Remarks on Forest Scenery*, Gilpin (1791, p. 7) 'laments' the 'capricious nature' of picturesque ideas which 'in many instances' 'run counter to *utility*'. He argues "What is more beautiful, for instance, on a rugged foreground, than an old tree with a *hollow trunk*? or with a *dead arm*, a *drooping bough,* or a *dying branch*? (Gilpin, 1791, p. 8). Later he extolls the 'blasted tree' as having a:

> fine effect both in natural, and in artificial landscape. In some scenes it is almost essential. When the dreary heath is spread before the eye, and ideas of wildness and desolation are required, what more suitable accompaniment can be imagined, than the blasted oak, ragged, scathed, and leafless; shooting it's [sic] peeled, white branches athwart the gathering blackness of some rising storm?
>
> (Gilpin, 1791, p. 14)

Price placed the picturesque as a category lying between Burke's sublime and beautiful. Again, the ancient tree is a signature of the category. In his *Essay on the Picturesque* he compares the 'tameness of the poor pinioned trees (whatever their age) of a gentleman's plantation drawn up strait and even together' with old pollarded trees which 'stretch out their limbs' in 'every wild and irregular direction'. He is delighted by their 'large knots and protuberances' which 'add to the ruggedness of their twisted trunks' and hollow trees whose mosses, and 'decayed substance, afford such variety of tints, of brilliant and mellow lights, with deep and peculiar shades, as the finest timber tree, however beautiful in other respects, with all its health and vigour cannot exhibit' (Price, 1794, Vol. I, pp. 26–27; Daniels and Watkins, 1991, pp. 156–158).

The picturesque soon became the dominant arboreal aesthetic and the ancient Sherwood oaks, however hollow and rotten, found themselves in the forefront of fashion. However, for some, Sherwood lacked a literary and artistic focus. Laird (1810) lamented this state of affairs in his topographical and historical description of Nottinghamshire:

> It is a matter of serious regret, in a picturesque point of view, that none of our landscape painters have ever thought of studying in this forest ... Its style is totally different from the rocks and woods of Claude Lorraine or the savage scenery of Salvator Rosa; but it has a wildness peculiar to itself ... so that there is scarcely a ferny heath, a knoll, or glade, that does not present some novelty to the lover of picturesque beauty
>
> (Laird, 1810, p. 51)

He equates the Sherwood oaks with Gilpin's description of the blasted oak 'ragged, scathed and leafless'. The lack of landscape painters was soon, however, to be replaced by a superabundance of literary tourists in search of the hero of medieval ballads, Robin Hood. He does not figure largely in the late eighteenth and early nineteenth century. The antiquarian Hayman Rooke, for example, does not mention him. Gilpin baldly relates that 'this forest was also the retreat ... of the illustrious Robin Hood ... who ... making the woody scenes of it his asylum, laid the whole country under contribution' (Gilpin, 1791, p. 314). And Laird mentions as an afterthought that 'it only remains for us to notice that famous, but legendary character, ROBIN HOOD whom tradition records as having made this his principal haunt, and of whose popular and interesting story but little is known to any degree of certainty, though his exploits have been celebrated in ballad in every succeeding age' (1810, p. 67).

Things were soon to change. In 1820 Walter Scott's novel *Ivanhoe* was published. This was a runaway success, in England, the United States and on the Continent, and went into many editions. Scott was enormously influential in bringing about a revival of interest in the Middle Ages (Girouard, 1981). A famous episode in *Ivanhoe* tells of Robin Hood and his meeting with King Richard I in Sherwood. Soon, tourists began to flock to Sherwood in order to explore the Forest so vividly described by Scott. The impact of the novel on visitors to Sherwood can be seen through the eyes of Washington Irving the American author who wrote up his visit in his literary miscellany *Abbotsford and Newstead Abbey* (Irving, 1835). Irving writes of his stay at Newstead Abbey, the former home of Lord Byron, and tells of a ride to the Forest. He provides an overblown picturesque description of the forest:

> A ride of a few miles farther brought us at length among the venerable and classic shades of Sherwood. Here I was delighted to find myself in a genuine wild wood, of primitive and natural growth, so rarely to be met with in this thickly peopled and highly cultivated country. It reminded me of the aboriginal forests of my native land. I rode through natural alleys and greenwood glades ... What most interested me, however, was to behold around the mighty trunks of veteran oaks, the patriarchs of Sherwood Forest. They were shattered, hollow and moss-grown, it is true, and their 'leafy honours' were nearly departed; but, like mouldering towers they were noble and picturesque in their decay, and gave evidence, even in their ruins, of their ancient grandeur
> (Irving, 1835, pp. 233–234)

Here Irving uses the full vocabulary of the picturesque sensibility of Gilpin and Price. Irving connotes the ancient oaks with the 'aboriginal forests' of America and is 'delighted' with the 'primitive' and wild wood of 'natural growth'. He sees the riding cut by the Duke of Newcastle in 1709 as a 'natural alley'. However, he is not only beguiled by the contrast between the apparent naturalness of the forest with the surrounding highly cultivated country, but goes on to relish the literary associations:

> As I gazed about me upon these vestiges of once 'merry Sherwood' the picturings of my boyish fancy began to rise in my mind, and Robin Hood and his

men to stand before me... The horn of Robin Hood again seemed to sound through the forest. I saw his sylvan chivalry, half huntsmen, half free-booters, trooping across the distant glades, or feasting and revelling beneath the trees.
(Irving, 1835, pp. 234–235)

Soon tourist guides designed to be purchased by those in search of Robin Hood, such as James Carter's (1850) *A Visit to Sherwood Forest including the Abbeys of Newstead, Rufford, and Welbeck ... With a Critical Essay on the Life and Times of Robin Hood,* began to appear. The great landowners were quick to appropriate this democratic symbol. The Duke of Portland built a new lodge at Clipstone in the forest in 1844. The main room over the arch was 'dedicated by its noble founder to the cause of education, for the benefit of the villagers of Clipstone'. 'The prospects from this room are most beautiful, including Birkland with its thousand aged oaks, the venerable church of Edwinstowe, and a wide expanse of forest scenery' (Carter, 1850, p. 73). The popular author January Searle notes 'on the north side, there are statues of King Richard the lion-hearted, Allan o'Dale and Friar Tuck; on the south side there are similar sculptures of Robin Hood, Little John and Maid Marian' (Searle, 1850, in White 1875, p. 244). Another guide describes three of the sculptures as 'the ancient frequenters of the neighbourhood: one its presiding deity, Robin Hood; the other Little John; and, bearing them pleasing company, as was her wont formerly, Maid Marian' (Eddison, 1854, pp. 194–195).

Later in the century, the new mansion built at Thoresby for Earl Manvers, designed by Salvin and built from 1864–1875 incorporated a vast library fireplace celebrating Sherwood Forest. Its iconography confirms the historical connection between Robin Hood and the ancient oaks for the mid-Victorian mind:

> The first object of striking import is the magnificent chimney piece ... It consists of an elaborately carved representation, in Birkland oak, of a scene in Sherwood Forest, in which are introduced the venerable 'Major' oak, with his knotted and gnarled branches, a foreground of botanical specimens, and a herd of deer – all chiselled with much similitude to Nature. This monument of patience and ability was cut by Mr Robinson, of Newcastle; the wood being from an oak which once flourished in the forest in which the leading feature in the subject forms so proud an ornament. Statuettes of Robin Hood and Little John support each side of the piece
> (Sissons, 1888, p. 58)

The combination of medieval legend and trees old enough to have witnessed scenes depicted by Walter Scott and others was enormously potent. By the mid nineteenth century the ancient oaks of Sherwood had become firmly fixed in the popular imagination as medieval icons. Within a few years individual trees, such as 'Robin Hood's Larder', were imaginatively named and gained credence through being printed on Ordnance Survey maps.

Conclusion

Notwithstanding scholarly research which indicates that the real or mythical figure of Robin Hood had only the loosest of connections with Sherwood

Forest (Holt, 1982), he remains today the main cultural signification of the ancient oaks of Sherwood. But the association with Robin Hood is only one of the layers of meaning which have become attached to the trees. The same individual trees, sometimes alive, sometimes dead, have been ascribed a catalogue of changing values and meanings. They have been prodded and probed, lopped and pollarded, exploited and felled. They have designated status and power and caused legal disputes. They have been the subject of archaeological experiment and aesthetic reflection. They have been categorized as fuel, timber, picturesque, dead and habitat. Yet at their core lies a mystery; their very hollowness ensures that we remain innocent of their ancestry and antiquity.

References

Boulton, H.E. (1959) The forest books of the royal forest of Sherwood. Unpublished MA thesis, University of Nottingham.

Butler, R.M. (1954) Archaeology in Nottinghamshire – achievements and prospects. *Transactions of the Thoroton Society of Nottinghamshire* 54, 1–20.

Carter, J. (1850) *A Visit to Sherwood Forest including the Abbeys of Newstead, Rufford, and Welbeck... With a Critical Essay on the Life and Times of Robin Hood.* Longman, London.

Chambers, J.D. (1932) *Nottinghamshire in the Eighteenth Century.* King, London.

Copley, S. and Garside, P. (eds) (1994) *The Politics of the Picturesque.* Cambridge University Press, Cambridge.

Cowell, B. (1998) Patrician Landscapes, Plebian Culture: Parks and Society in Two English Counties, c. 1750–1850. Unpublished PhD thesis, University of Nottingham.

Cox, J.C. (1905) *The Royal Forests of England.* Methuen, London.

Crook, D. (1979) The struggle over forest boundaries in Nottinghamshire, 1218–1227. *Transactions of the Thoroton Society of Nottinghamshire* 83, 35–45.

Daniels, S. (1988) The political iconography of woodland in later Georgian Britain. In: Cosgrove, D. and Daniels S. (eds) *The Iconography of Landscape.* Cambridge University Press, Cambridge, pp. 43–82.

Daniels, S. and Watkins, C. (1991) Picturesque landscaping and estate management: Uvedale Price at Foxley, 1770–1829. *Rural History* 2, 141–169.

Daniels S. and Watkins, C. (eds) (1994) *The Picturesque Landscape. Visions of Georgian Herefordshire.* University of Nottingham, Nottingham.

Eddison, E. (1854) *History of Worksop; With Historical, Descriptive, and Discursive Sketches of Sherwood Forest.* Longman, London.

Evelyn, J. (1670) *Sylva, or a Discourse of Forest-Trees, and the Propogation of Timber in His Majesties Dominions.* Royal Society, London [2nd Edn; 1st Edn 1667].

Fowkes, D.V. (1977) The breck system of Sherwood forest. *Transactions of the Thoroton Society of Nottinghamshire* 81, 55–61.

Gilpin, W. (1791) *Remarks on Forest Scenery, and Other Woodland Views (Relative Chiefly to Picturesque Beauty)...* R. Blamire, London.

Girouard, M. (1981) *The Return to Camelot. Chivalry and the English Gentleman.* Yale University Press, New Haven and London.

Holt, J.C. (1982) *Robin Hood.* Thames & Hudson, London.

House of Commons (1793) The Fourteenth Report of the Commissioners appointed to enquire into the State and Condition of the Woods, Forests, and Land Revenues of the Crown, and to sell or alienate Fee Farm and other Unimproveable Rents. *House of Commons Journal* 48, 467–511.

Irving, W. (1835) *Abbotsford and Newstead Abbey.* John Murray, London.

James, N.D.G. (1981) *A History of English Forestry.* Basil Blackwell, Oxford.

Laird, F.C. (1810) *A Topographical and Historical Description of Nottinghamshire.* Sherwood, Neely & Jones, London.

Price, U. (1794) *Essay on the Picturesque, as Compared with the Sublime and the Beautiful; and on the Use of Studying Pictures for the Purpose of Improving Real Landscape.* London.

Rodgers, J. (1908) *The Scenery of Sherwood Forest with an Account of Some Eminent People Once Resident There.* Fisher Unwin, London.

Rooke, H. (1790) *Descriptions and Sketches of some Remarkable Oaks, in the Park at Welbeck ..., a Seat of His Grace the Duke of Portland. To Which are Added, Observations on the Age and Durability of that Tree. With Remarks on the Annual Growth of the Acorn.* Nichols, London.

Rooke, H. (1799) *A Sketch of the Ancient and Present State of Sherwood Forest in the County of Nottingham.* Tupman, Nottingham.

Schama, S. (1995) *Landscape and Memory.* Harper Collins, London.

Searle, J. (1850) Leaves from Sherwood Forest. In: White, R. (1875) *Worksop, 'The Dukery', and Sherwood Forest.* Simpkin Marshall, London, pp. 242–244.

Sherrat, A.G. (1965) Hayman Rooke, F.S.A. – An eighteenth century Nottinghamshire antiquary. *Transactions of the Thoroton Society of Nottinghamshire* 69, 4–18.

Sissons, F. (1888) *Beauties of Sherwood Forest. A Guide to the Dukeries and Worksop with Many Maps and Illustrations.* Sissons, Worksop.

Thomas, K. (1984) *Man and the Natural World. Changing Attitudes in England 1500–1800.* Allen Lane, London.

Watkins, C. (1981) An historical introduction to the woodlands of Nottinghamshire. In: Watkins, C. and Wheeler, P.T. (eds) *The Study and Use of British Woodlands.* Department of Geography, University of Nottingham, Nottingham, pp. 1–24.

Notes

1. Nottingham University Manuscripts Department NUMD Ey 510. The date is approximate. I am grateful to Ben Cowell for pointing out this reference.
2. The commissioners report for Sherwood Forest of 1793 is published as 'The Fourteenth Report of the Commissioners appointed to enquire into the State and Condition of the Woods, Forests, and Land Revenues of the Crown, and to sell or alienate Fee Farm and other Unimprovable Rents'. *House of Commons Journal* 48 467–511. This will be referred to in this chapter as HCJ.
3. HCJ, 467.
4. HCJ, 469.
5. HCJ, 469.
6. HCJ, 469.
7. HCJ, 473.
8. HCJ, 473.
9. HCJ, 469.
10. HCJ, 481.
11. HCJ, 481.
12. HCJ, 473.

13. HCJ, 481.
14. HCJ, 473.
15. HCJ, 481.
16. HCJ, 473.
17. HCJ, 473.
18. HCJ, 509.
19. HCJ, 482.
20. HCJ, 504.
21. HCJ, 481.
22. HCJ, 469.
23. HCJ, 473.
24. HCJ, 473.
25. HCJ, 469.
26. HCJ, 473.
27. HCJ, 481.
28. HCJ, 469.
29. HCJ, 481.
30. Edward Harley, the 2nd Earl of Oxford (1689–1741). The *Dictionary of National Biography* (*DNB*) notes that 'habitual indolence, rather than incapacity, prevented him from taking part in public affairs; nor did he care for general society. He preferred to surround himself with the more distinguished poets and men of letters of the day' *DNB*, 8, 1278. 'He had a passion for building, landscape gardening, and for collecting books, manuscripts, pictures, medals and miscellaneous curiosities' *DNB*, 8, 1279.
31. HCJ, 504.
32. HCJ, 472.
33. HCJ, 473.

CHAPTER 9
Landed estates, the 'spirit of planting' and woodland management in later Georgian Britain: a case study from the Dukeries, Nottinghamshire

Susanne Seymour
Department of Geography, University of Nottingham, Nottingham NG7 2RD, UK

> A very laudable Spirit has pervaded the Landowners in this County [Derbyshire], for Improving and Ornamenting their Estates by Plantations, made within the last 50 or 60 years, but principally so in the latter half of that period; and in general, steep, rocky, and barren Lands have been selected for this purpose, which could scarcely be otherwise improved
>
> (Farey, 1813, p. 237)

Introduction: Estates, 'Improvement' and the 'Spirit of Planting'

In the eighteenth century land was the key to social and political status, commanding privileges ranging from the right to vote and be elected to Parliament to the right to hunt. Such was the importance of landed property in these respects that those who made their fortunes from the law, political office, banking or industry generally bought into land to secure their social status. Land also formed a major source of wealth, with returns from farmland, mineral workings, urban property and woodland bringing substantial incomes to their owners. The system of great estates extended through the century as smaller landowners were bought out and larger properties consolidated by the widespread adoption of the practice of primogeniture and strict settlement which ensured that most large properties were passed intact from one generation to another (Beckett, 1986, pp. 43–90).

While land was a central resource for the eighteenth century elite, 'improvement' was a key concept, albeit a much debated one. Condemned by some commentators as a dangerous fad, 'the idol of the age', 'fed by many a victim' in the words of poet William Cowper,[1] for many others it was an important principle for change. For example, the statesman, Edmund Burke, regarded improvement as the treatment of 'the deficient or corrupt parts of an established order with the character of the whole in mind' (quoted in Duckworth, 1971, p. 33) and the term was a principal tenet of a progressive age for many others involved in activities ranging from music to manufacturing. Such a breadth of application of the term encouraged its acquisition of a series of overlapping associations – financial, moral, political and aesthetic.

Much of this concern with 'improvement' was targeted on the landed estate where woodlands were an important focus. Eighteenth century landowners were interested in improving woodlands from a variety of overlapping perspectives: pragmatic, productive, political and pleasurable. A major result of this was an upsurge in the planting of trees in Britain during the century following 1750. Despite a lack of quantitative information, sources, such as Board of Agriculture *General Views*, the Society of Arts' *Transactions*, maps, individual estate records and growing numbers of sylvan publications, confirm this trend, together with an increased interest in planting and woodland management, particularly amongst estate owners (James, 1981, pp. 166–177).

Related to their role as political and social leaders, landowners drew on the qualities of certain trees to represent and naturalize their power and status. By the late eighteenth century the patriotic and patrician associations of trees, particularly the oak, were well established through the use of timber for ships of the Royal Navy and Restoration imagery which associated great trees with great families (Daniels, 1988). Cowley's Restoration contrast of the 'Patrician trees so great and good' with the 'plebeian underwood' (quoted in Daniels, 1988, p. 43) was reiterated in the eighteenth century by Edmund Burke's characterization of aristocratic statesmen as 'the great oaks which shade a nation' (quoted in Sutherland, 1958–1978, p. 377). John Evelyn's *Silva*, the most famous of Restoration calls to landowners to plant trees for the Royal Navy, was republished no less than four times between 1776 and 1812 (James, 1981, p. 318). Landed families readily appropriated old trees, incorporating them as features in landscape designs or using them as central features in family portraiture 'to amplify their pedigree' even if they were not planted by their ancestors (Daniels, 1988, p. 48). Planting trees was firmly established as a patriotic activity whereas felling had radical associations and was seen as divisive. Acorns could take on the symbolic role more usually performed by venison when used as gifts of aristocratic patronage (see below). While oaks and other deciduous forest trees were associated in positive terms with great or established families, quick-growing firs were often associated rather more negatively with the newly rich. For example, the landscape gardener, Humphry Repton, faced with the prospect of such a newcomer as a client, related with disdain how the man mistook for ' "the LARGEST ACORN he had ever seen" ' a stone pine cone which had fallen close to an oak tree, and felt this was 'fit emblem of him ... who had fallen among Gentlemen but could NOT be mistaken for one'.[2] Such associations, however, did not prevent the widespread use of firs as nurse trees in estate plantations, although oaks and other deciduous plantings were those more generally publicly celebrated.

Landowners were equally leaders of taste. During the eighteenth century, interest in gardening blossomed and a new style of design emerged – the landscape gardening movement – in which a taste for more natural looking designs prevailed. Woodland formed one of the key elements of the natural landscaping style, with trees admired for their beauty in groves, clumps and as single specimens, as well as for their ability to mask unwanted features. Much of the planting in landscape gardens and parks took on more than an aesthetic function, while concerns to beautify the landscape extended more widely on to estate lands. Indeed, while predominantly ornamental plantings were most likely in the vicinity of mansions and more commercial plantings on estate lands, many plantations involved a combination of the ornamental, the symbolic and the commercial.

Many landowners at this time were interested in developing the revenues from their estates which often contained mineral, agricultural, urban, cultural and silvicultural resources. The industrial sector of the economy grew rapidly from mid-century, accompanied by increasing urbanization and the emergence of industrial towns such as Birmingham and Manchester. Estate owners benefited from increased urban rents and returns from mineral royalties, some actively exploiting these resources themselves and 'improving' the communications between sites of production and markets (Beckett, 1986, pp. 206–286). Likewise, the agricultural sector grew and became more profitable from the 1740s as population increases and greater demand from the new urban areas led to rises in agricultural prices and rents (Williamson, 1995, p. 14). Many estate owners took the opportunity to exploit these conditions and became actively involved in increasing their rentals through processes such as enclosure, drainage and farm rationalization. Around six million acres were enclosed by Act of Parliament in the period between 1760 and 1830 (Turner, 1986), although these developments had a strong regional focus in the Midlands and north-east (Williamson, 1995, p. 9).

Industrial and agricultural developments fuelled attempts to improve woodlands as products were required for anything from estate repairs, new buildings, new enclosure fences and gates, farm and field rationalization and drainage materials to intensification of hop production or the development of new mineral enterprises such as coal production. Plantation thinnings were a particularly lucrative product since the price of underwood doubled during the eighteenth century (Rackham, 1980, p. 168. The prices analysed are for south Essex). Those of 15 years' growth could be used for pit props, hop poles or poles for enclosure fences.

Many estates were depleted of ancient woodland by the mid eighteenth century and owners would not have been able to implement such structural developments as enclosure or the construction of new farmsteads as easily or as cheaply without their own supplies of timber and poles. Even planting for patriotic purposes (the Royal Navy) was believed to be a profitable as well as a patriotic venture (Thomas, 1984, p. 199). There was good reason for this. In eastern England, prices of oak trees rose by 40% in real terms from 1690 to 1800, and by 1814 stood at 88% above their 1690 value (Rackham, 1980, p. 164).

By the late eighteenth century the links between estate developments, landscape design and patriotic concerns to defend and promote the military and commercial interests of the British establishment at home and abroad were strongly formed. Combinations of 'pleasure, profit and patriotism' thus typically justified calls to plant and celebrations of the 'planting spirit'.

The Dukeries' Estates

By the early eighteenth century, a compact group of parks, forming the basis of more substantial estate lands, had been established in the sparsely populated sand lands of north Nottinghamshire, in the vicinity of Sherwood Forest. These were Clumber (dating from c. 1700), Thoresby (from at least 1589), Welbeck (from at least 1301) and Worksop Manor (from at least 1161) (Eddison, 1854, p. 85; Thompson, 1938, p. 25; Aslet, 1979, p. 2082).[3] The development of these parks and their adjacent estate lands owed a great deal to the former presence of monastic lands and, more particularly during the eighteenth century, to the existence of neglected areas of Crown property and unenclosed lands subject to customary common rights and breck agriculture, which allowed the expansion of these aristocratic interests (Crown Commissioners, 1793, p. 469; Cameron, 1975, p. 58). Since all the parks were owned during the mid eighteenth century by dukes (Newcastle, Kingston, Portland and Norfolk respectively), the area was labelled somewhat disparagingly by Horace Walpole, The Dukeries, and the name has remained (Walpole quoted in Lewis, 1965, pp. 374–375).

Landowners in the vicinity of Sherwood Forest and the Dukeries were highly praised for their 'planting spirit' in the later eighteenth century. Upon his arrival at Clumber in 1760, the 2nd Duke of Newcastle admired developments in the area which achieved a merging of estate and garden, predominantly through tree planting:

> they have cut ridings through the Forest, plowing them up, and then laying them down with Grass Seed, which comes to a fine Verdure, and they have planted Clumps of Trees upon most of the Hills. Mr Mason ... has rowled all his ridings, and pick'd up all the Stones, that [his Estate] is really a perfect Garden ...
> (quoted in Priestley, 1958, p. 203).

A few decades later, the author of the Board of Agriculture report for Nottinghamshire, Robert Lowe, referred to a 'spirit of planting' in the forest sand land area which had 'prevailed much in this district since about forty years' (Lowe, 1798, p. 53). He listed the 'Woods and Plantations in the district of the forest and borders' which in his estimate covered around 7197 acres, plus a number of 'dispersed clumps and plantations of smaller extent' (p. 70). The first substantial plantings in the area, 'chiefly of firs', he traced back to as early as 1750, but in his view the 'greater part' of the plantings were of a mixture of deciduous trees in the last 30 years as it was discovered that all kinds of trees, 'well planted and properly sheltered' would 'succeed very well' on the forest soil (p. 53).

Likewise, Hayman Rooke, in 1799, praised the planting efforts of 'the many respectable Persons, whose Mansions and Parks border on the Forest':

From the laudable exertions of the neighbouring Gentlemen, there is reason to hope, that the uninclosed parts of this extensive Forest of Sherwood will again be imbowered, and, if I might venture to predict future events, I should augur, that Posterity would venerate those majestic Oaks, planted by their ancestors, as Monuments of British Valour.

(Rooke, 1799, p. 23)

Such accounts highlight the multi-functional character of much of the planting in the Dukeries, even of areas of woodland in close proximity to the houses. Likewise, Fowkes (1967) has argued that the parks themselves took on a 'multi-purpose' quality in this period. I will now explore the nature of the 'planting spirit' on three of the Dukeries estates: Welbeck, Thoresby and Clumber.

Welbeck

By 1794, Robert Lowe reported that almost half of Welbeck Park was wooded (although much of this was established woodland) and that the Duke of Portland had planted around 600 of the 2400 acres he had enclosed from the forest within the last few decades (Lowe, 1798, p. 150). When Portland's son, the Marquis of Titchfield, took over Welbeck in 1794, he continued his father's planting programme. Rooke reported that by 1799 Titchfield had made a number of impressive plantations on the old forest, covering 'upwards of a hundred acres' (Rooke, 1799, pp. 31–32; Cox and Whitworth, 1910, p. 378).

Planting schemes at Welbeck from the 1760s to the 1790s were influenced by Portland himself and the professional landscape designer, Humphry Repton, who produced two of his famous Red Books for the Duke in 1790 and 1793 (Repton, 1790, 1793) (Fig. 9.1). However, the majority of the work was carried out under the supervision of Portland's gardener, William Speechley, who undoubtedly influenced planting design and practice. Speechley was an eminent gardener of his day who published widely on a range of topics from kitchen gardening to woodland management and rural economy (Speechley, 1775, 1779, 1790, 1820; Turberville, 1939, pp. 306–321). His *'Account of Plantations at Welbeck'* (1775), submitted to Hunter's 1776 edition of Evelyn's *Silva* (and reproduced by the agricultural writer William Marshall (1785 and 1803) and by Lowe (1798) in the Board of Agriculture report for Nottinghamshire), outlines the objectives of planting and woodland management as practised on the Welbeck estates.

Speechley declared that planting at Welbeck formed 'no more than a part of one great design' (Speechley, 1775, p. 61). Yet it is clear from his account and other estate documents that the plantings were made to serve various ends and were part of a wider campaign of estate improvements: the creation of beautiful landscapes in the irregular picturesque style; the production of commercial timber; the provision of woodland products for use on the estate; the enhancement of the agricultural lands then being developed on the surrounding moorlands; and the encouragement of game.

Speechley had a well-developed appreciation of the picturesque beauty of

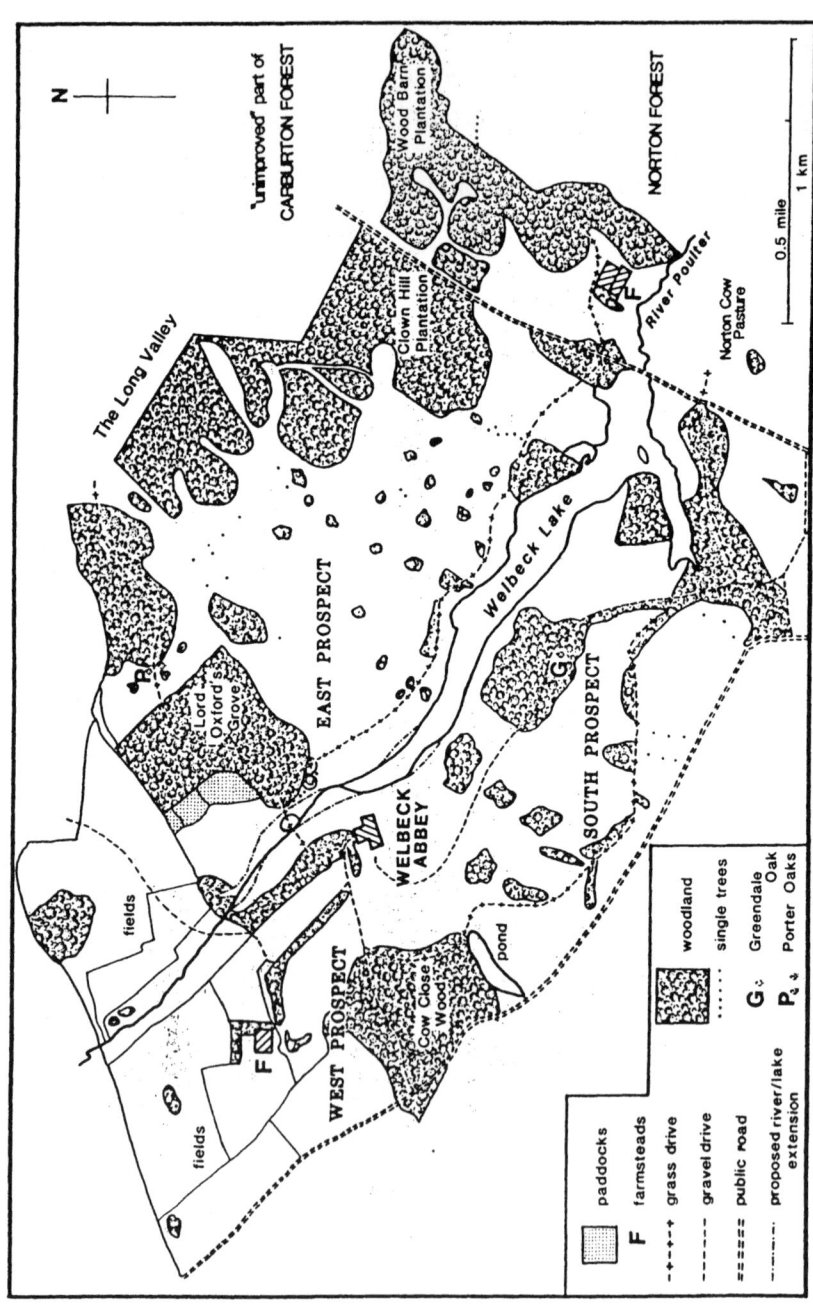

Fig. 9.1. Welbeck Park and its environs in the 1790s, based on Repton's 'Red Books' of 1790 and 1793.

trees and woodland forms. He emphasized the importance of irregularity and variety in his designs and displayed a painter's sensitivity to aspects of light, shade and texture. To this end he advocated the planting of evergreens and exotic trees on the outskirts of woods, to make them 'appear as if scalloped with ever-greens, intermixed sometimes with rare trees, as the *liliodendron tulipifera*, or Virginian tulip trees, &c.' (Speechley, 1775, p. 62). While oaks were the main variety planted at Welbeck, 'irregular patches' of secondary varieties, such as beech, larch and Spanish chestnuts, were scattered 'here and there, throughout the plantations, which, when the trees are in leaf, have the most pleasing effect, on account of the diversity of shades' (Speechley, 1775, p. 62). Birch was also widely planted as the main nurse tree.

Felling, as well as planting, was an important part of woodland management and design at Welbeck. In 1776, Speechley proposed felling 'the greater part of the trees in the wilderness' between the rising ground and the house, leaving only a few single trees 'and here and there a clump for variety'. Speechley's idea, complementing alterations to the Abbey, was to 'catch a view of the Wood on the rising ground which would add to the Beauty of the Scene from the windows of these new rooms when all is compleated'.[4]

Likewise, there were concerns at Welbeck to conserve and celebrate the ancient oaks which grew there. Some of these were of curious shape or immense size and Welbeck was already famous for its oaks by the time Repton visited in the 1790s and was struck by its 'vegetable wonders' (Repton, 1790, p. 19). The Greendale Oak was perhaps the most famous, especially since an archway wide enough to allow a carriage to pass through had been cut into its trunk in 1724. A series of five engravings of the tree were included in Hunter's 1776 edition of Evelyn's *Silva*, patronized by Portland: two landscape views and three plates of its dimensional measurements. The Greendale Oak held considerable symbolic importance for the Duke of Portland as an aristocratic statesman and he distributed gifts of acorns from the tree in the same manner that he made gifts of venison from the Park (Speechley, 1775, p. 68; see also Chapter 8, this volume).

Part of the design of the plantations at Welbeck was also the appropriation of land, to create a more extensive or acceptable view and even to make actual claim to territory on the surrounding moorlands. In his 1790 Red Book, Repton suggested the extension of the kitchen garden plantation, arguing 'the projecting point of a plantation always gives imaginary expanse to the lawn beyond it' (p. 10). The careful siting of new plantations also allowed the appropriation of game from the neighbouring parks of Clumber and Worksop Manor, to enhance the sporting resources at Welbeck. Clumber, especially, Speechley noted, 'does, and for many reasons is likely to continue to abound with the same plenty of game, particularly pheasants'. Speechley's scheme was to develop long, thin plantations running from the borders of these parks into the heart of Welbeck territory, with the purpose 'to make the Plantation a leading string for game from one park to another'. The glades left in the new plantations were also a means of encouraging game and Speechley advised that pheasants were not shot in these newly planted areas.[5]

Planting at Welbeck was also fundamentally related to a rational use of the soil, as advocated by agricultural writers, and to the promotion of agricultural developments on the estate. Speechley's picturesque taste was translated into executable woodland designs through a thorough understanding of the practicalities of woodland management. William Marshall generally approved his planting method, declaring it to be 'in itself a Treatise', although he suggested the improvement of sowing tree seeds directly into the plantation sites rather than replanting from nurseries, to cut down on 'much of the expense' (Marshall, 1803, pp. 172–173).

An adequate supply of trees was essential for the planting campaign and in his *Account*, Speechley paid considerable attention to the creation, siting and management of nurseries. These were generally placed in the centre of the area to be planted and were securely fenced to keep out unwanted animals and people. Internally, they were laid out with a walk down the centre which Speechley reported to be 'exceedingly convenient when we remove the young trees from thence to the plantations' and which enabled Portland to both supervise and display his planting procedures, it being 'wide enough to admit carriages to go through' (Speechley, 1775, p. 58). Speechley was also confident that with a careful thinning procedure, substantial numbers of plants would be available for establishing new plantations.

The planting itself was generally carried out 'in an irregular manner', mainly using oak with birch nurse trees, with 'upwards of 2000 plants upon an acre of land' (Speechley, 1775, p. 65). However, the labour force undertaking planting at Welbeck was organized in a carefully supervised regime. A form of labour specialization was adopted in which the hands were divided into four classes: makers-up, pruners, carriers and planters. Labourers were carefully matched with each task, with 'boys, with some of the worst of the labourers' being used as carriers, leaving the more able workers for the more skilled tasks. In addition, the planters were put to work in pairs, one making the holes, the other setting the tree and treading it firm, in order to make more efficient use of time (Speechley, 1775, pp. 63–64).

Speechley expressed a concern to obtain the commercial goal of a 'length of timber' as well as picturesque effect and his *Account* includes an assessment of growth rates in two oak plantations at Welbeck. It was not just the timber which was regarded as a commercial asset, but also the thinnings. Speechley outlined that 'the first profit of our plantations' came after 4 or 5 years when the lower branches of the larger birch trees were taken off to prevent them from damaging the oaks, 'the birch wood being readily bought up by the broom-makers' (Speechley, 1775, pp. 67–68).

Speechley's scheme was to plant on the 'hilly grounds' of the forest sands, a project aesthetically justifiable and agriculturally beneficial since it allowed the better lands in the valleys to be reserved for farming. Furthermore, Speechley judged that these new plantations would 'in time make the vallies of much greater value, on account of the shelter they will afford' (Speechley, 1775, p. 57). Such planting patterns were very much in line with those, based upon economic criteria, advocated by William Marshall, Arthur Young and other Board of Agriculture reporters. Young made a differentiation, based on rental value, of land likely to be more profitable under agriculture and that likely to yield a better return under trees, in 1799 judging 'Where land can be let at 20s. an acre, it is much more advantageous than what woods would yield in this system' (Young, 1799, p. 222). Much of the land in the Dukeries commanded rentals of less than this figure (Pickersgill, 1979, p. 66).

The plantations established by Speechley also assisted in the enclosure of Carburton, Worksop and Norton Forests where a series of home farms was developed from the 1760s to the 1790s (Fig. 9.1). Only a few accounts remain from the 1770s but these suggest that posts and rails made from the thinnings of Speechley's new plantations played an important role in the enclosure process.[6] Eventually these farms were leased out as large farming units, thereby transforming the traditional forest villages into more conventional enclosed villages and small tenants or owner-occupiers into estate employees or day labourers (Fowkes, 1977; Seymour, 1988, pp. 333–345).

Thoresby

Like Welbeck, Thoresby Park contained some old stands of wood in the late eighteenth century, principally in its southern section which incorporated part of the ancient Royal Forest of Bilhagh. However, both Charles Pierrepont and his uncle the 2nd Duke of Kingston undertook considerable amounts of planting, with Lowe reporting that Thoresby Park and its environs in 1794 contained 981 acres of plantations (see Fig. 9.2). These plantations were mainly of deciduous trees, with smaller areas of fir, birch and wetland coppice. In addition, Pierrepont was preparing to plant a further 306 acres (Lowe, 1798, p. 79; Seymour, 1988, pp. 251–257). A similar combination of beauty, use and patriotic display to that found at Welbeck characterized woodland management at Thoresby.

The Park woodlands included a number of patriotic allusions which highlighted Pierrepont's planting as a public-spirited activity. Some were named after naval heroes of the Napoleonic Wars, for example Howe Plantation and St Vincent Grove (Rooke, 1799, pp. 20–21). Another was called Evelyn's Silva, a reference to the famous Restoration planting tract and its author, to whom Pierrepont traced his own lineage.[7] Yet another of Pierrepont's plantations attained explicit patriotic associations when it was ornamented, in 1799, with a monument called Nelson's Seat, in commemoration of the already famous admiral.[8] Similarly, Pierrepont publicized his oak plantations, although considerable amounts of other trees were planted at Thoresby, and it was a 20-acre oak plantation, guaranteed as well managed, which in 1803 earned him a Gold Medal from the Society of Arts (*Transactions, Society of Arts*, 1803, p. 106).

As part of Pierrepont's planting campaign (and that of his uncle, the 2nd Duke of Kingston), outlier clumps and plantations were established on the moors adjacent to the Park on the east (e.g. Eliza Clump, Augusta Clump and Francis Grove; Fig. 9.2). Such plantings allowed these areas to be incorporated visually into parkland views and also facilitated the further agricultural development of the moorland by setting out a claim to ownership.

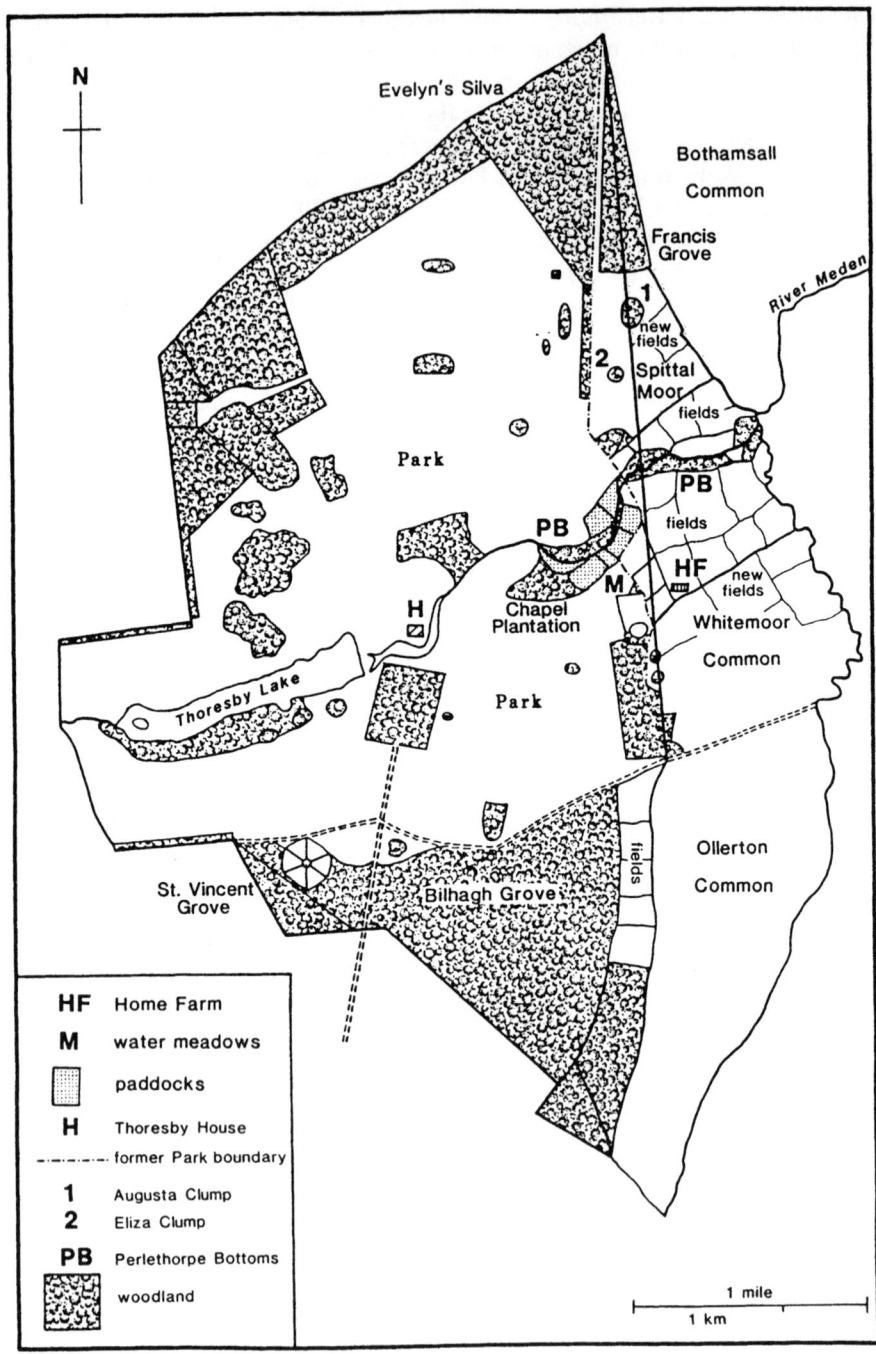

Fig. 9.2. Thoresby Park and its environs, *c.* 1803, based on 'Map of Clumber and Thoresby Parks, 1803' (Welbeck Estate Office) and 'Plan of Thoresby Lands', *c.* 1795.

Subsequently, much of the remaining moor was carved up into farms, including part of the Home Farm. Pierrepont himself was a keen agriculturalist. He was a member of the Board of Agriculture, serving as a Vice President in 1803, and more locally was a founder member of the Retford Agricultural Society and a supporter of the Newark Agricultural Society and its annual show.[9] He built up a 900-head flock of 'improved breeds' of sheep at Thoresby, mainly New Leicester–Forest crosses, and disseminated them to tenants and fellow 'agricultural improvers'. He was also actively involved in promoting enclosures, supporting several enclosure bills for north Nottinghamshire parishes in Parliament (Pickersgill, 1979, pp. 83, 95) and supervising the enclosure of most of the open field and much common land on his own estates (Seymour, 1988, pp. 185–186).

As at Welbeck, Humphry Repton was commissioned to produce a Red Book of designs (Repton, 1791). His suggested 'improvements' were focused on the lake and river area where he set out to compose a serpentine river scene. Trees were essential to the success of this project, with Repton stressing the picturesque possibilities if the river could 'for some distance perhaps be totally concealed from the view, under the shade of the wooded bank' (Repton, 1791, p. 12). Yet even these woodlands were not purely decorative. Repton comments on the strong utilitarian aspect of Pierrepont's interest in picturesque landscape design and entries in the estate accounts indicate that these riverside plantings were a rich source of poles for the local hop industry, located around Retford (Repton, 1791; Lowe, 1798, p. 42; Seymour, 1988, pp. 288–292). Pierrepont's more extensive plantings on the margins of the Park, which formed the backdrop to several parkland views, produced a wider range of products, many for use in estate developments. Products included timber, bark, charcoal, fuel wood, pit props, gates, fencing materials (including sheep hurdles), young trees for planting in other estate woodlands and hop poles (Seymour, 1988, pp. 275–292).[10]

In particular, fencing materials from the Park were used in the enclosures of estate parishes which Pierrepont promoted in both north and south Nottinghamshire. The process was facilitated by the contract made in the 1790s with the woodman, William Wordsworth, to fell trees in Thoresby Park and to convert them into sheep hurdles, posts and rails. Not only large woods but even clumps were a source for such materials, as is revealed by a payment in 1791 to Wordsworth for the conversion of 'Trees in the house & Chappel Plantations & Clumps about the Park'. Apparently these materials for enclosures were provided free of charge, probably constituting part of Pierrepont's investment to secure rent improvements. In 1793, 3660 stakes were sent from the Park to tenants at Kneesall for 'inclosing lands from the fields', while 109 round poles were reserved 'for home use'.[11] A few years later, in 1797, it was reported that Wordsworth and a sawyer at work in Thoresby had prepared a further 2000 rails destined for enclosures on Pierrepont's south Nottinghamshire estates, a debate ensuing about the cheapest method of transporting them.[12] These enclosures proved highly lucrative for Pierrepont. His rentals in the newly enclosed parishes rose significantly, more than doubling between 1789 and 1804 for lands in

Gedling, Holme Pierrepont, Orton, Sneinton and Weston (Purdum, 1978, pp. 313–326).[13]

Clumber

Unlike Welbeck and Thoresby, Clumber was substantially denuded of woodland when the 2nd Duke of Newcastle inherited the property in the 1760s. Yet, by 1771, Arthur Young noted that the Duke was 'planting on so large a scale ... that the place in a few years will not be known' (Young, 1771, p. 423). By his death in 1794, the Duke had planted the greater part of the 1848 acres of trees within the Park and a further 349 acres on his surrounding lands (Lowe, 1798, p. 70) (Fig. 9.3).

Planting formed just one part, albeit a major one, of the programme of 'improvements' instigated by Newcastle from the early 1760s. These included the construction of a new mansion, the digging of a 100-acre serpentine lake, the building of a new classical bridge, temples and lodge gates, the establishment of a 2000 acre Park Farm and extensive enclosures and land developments as well as substantial plantations (Seymour, 1988, pp. 204–210).

Early in his life Newcastle was guided in his landscaping style by Joseph Spence, according to Jacques (1983, p. 72), the 'most respected voice in the north on the laying out of grounds' until the 1760s when 'Capability' Brown began to undertake commissions there. Spence, who had been the young nobleman's tutor on the Grand Tour, encouraged Newcastle in 'an inclination for Planting and Gardening, by pointing out Prospects, & the natural Beauties of places'.[14] Newcastle's interest in planting is apparent from his comments upon arrival at Clumber in the 1760s, when he compared the potential of his property with the already well-wooded neighbouring Park of Thoresby:

> [Clumber] is full as capable of as many bountys at the least [as Thoresby]; the soil in general I shou'd think as good, if not better, and the plantations wou'd thrive to be sure equally as well, The River might be made any thing of, and the Variety of the Hills and Vales are much more considerable and bolder, indeed greater variety of Ground can not be seen anywhere
>
> (quoted in Priestley, 1958, p. 203)

Some people objected to Newcastle's plantings on aesthetic grounds. The Duchess of Northumberland in 1772 found them 'vulgar ... all in square formal figures chiefly pine and larch' (quoted in Clifton, 1979, p. 68) and William Gilpin objected to them on account of their newness in 1776 (quoted in Holland, 1826, p. 195). However, others praised them in terms which linked beauty, usefulness and public service. Viscount Torrington in 1789 described Clumber as 'in wonderful, and hourly improvement' and expressed 'much admiration at the beauty and growth of the plantations, and surprise at the quantity of the pheasants', questioning 'What can be so useful, so noble, or so gratifying as this, forming, from sterility, a charmingly cultivated, wooded

Landed estates in Georgian Britain

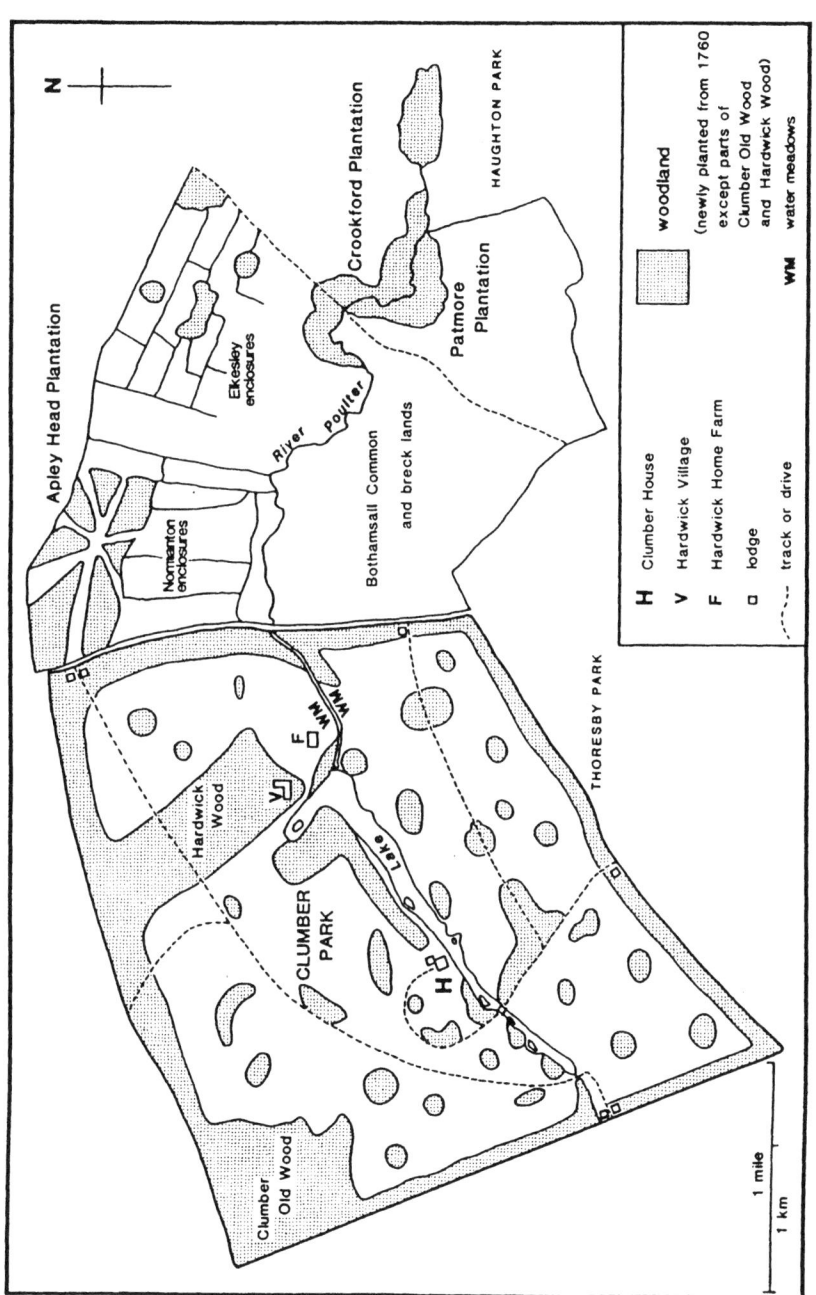

Fig. 9.3. Clumber Park and its environs, 1803, from 'Map of Clumber and Thoresby Parks, 1803' (Welbeck Estate Office).

domain?' (Byng, 1789, p. 9). Agricultural writers were also counted among the admirers of Newcastle's planting. After his visit to Clumber in 1771, Arthur Young commented enthusiastically on the beauty and usefulness of the woodlands:

> the extent of the new plantations is very great, so that they will prove not only an ornament to all the country, but a source of immense profit to the family
> (Young, 1771, p. 423)

A range of concerns contributed to Newcastle's extensive planting schemes: his interests in landscape design and agriculture, his obsession with shooting and a wider concern with the development of the agricultural estate. Mr Marson, the manager of all Newcastle's 'improvements' at Clumber, supervised the tree planting and management (Lowe, 1798, p. 54). He also submitted an account of the woodland regime at Clumber to Lowe's Board of Agriculture Report (Marson, 1798). This indicated that, as at Welbeck, a wide variety of trees were set in the plantations at Clumber, providing a combination of shelter and beauty:

> In general we fill our plantations with a various assortment of American plants; as firs, pines, cedars, &c. besides Scotch firs and birch. These are not only a shelter for the young forest trees, but have a pleasing effect for fifteen or twenty years, for their permanence of verdure, and variety of their foliage
> (Marson, 1798, p. 54)

New plantations were securely enclosed with quick fences which served both aesthetic and practical functions:

> Independently of their ornament, [the quick fences] not only secure the young trees from being injured by cattle, but prevent the sheep from depositing their dung, which in that case, as a manure would be of no use to the farmer
> (Marson, 1798, p. 56)

Newcastle's plantations were in fact vital to the agricultural operation and improvement of his substantial north Nottinghamshire estates. A major concern for Newcastle was the lack of timber on his estate generally, a 1760 survey reporting 'scarce Timber on the Estate to make a Gate Post'.[15] Sheep, and the dung they produced, were a vital part of his improvements and the fences played an important role in ensuring the proper use of this resource. In 1794, 2000 sheep, of an improved New Leicester–Forest cross, were run in the Park, providing fertilizer for the 2000-acre Park Farm Newcastle had developed there over the past 30 or so years. This consisted principally of grass, including 100 acres of water meadows, but about a quarter comprised arable crops: oats (210 acres), turnips (195 acres) barley (76 acres), wheat (67 acres) and rye (65 acres) (Lowe, 1798, pp. 149, 102).[16]

Marson reported that within 15 years of planting, Newcastle's plantations were producing, through thinnings, a wide range of products, including: 'posts, rails, pails; punchwood for the colleries; cordwood, charcoal, hop poles, brush heads, birch brooms, joists [and] rafters' (Marson, 1793, p. 56). The fencing materials derived from the thinnings were of particular importance in the implementation of the large-scale enclosures to create new farms which Newcastle undertook on the former moorland to the east of Clumber. The

Board of Agriculture reporter, Lowe, commented on the critical role of Newcastle's plantations in this process:

> In the extensive enclosures made by his Grace in Elksley, Bothamsell, &c. the quick hedges, which are remarkably fine, were raised with posts and rails, the thinnings of these plantations. I was assured, some years since, that sixty miles running measure had been done in this manner; and by this time it must amount to double that number
>
> (Lowe, 1798, p. 53)

The value of the materials provided by these plantations for the enclosures was not recorded in the estate accounts but may be inferred from the woodland expenses. At least 15% of the total management cost of the Clumber Park woodlands from 1775 to 1793 was directed towards generating products for the enclosures.[17]

There is also evidence that the plantations in and around Clumber provided materials for enclosures on other estate property. When lands were being enclosed in Basford, one of Newcastle's estate parishes in the south of the county, in 1793, £94 was spent transporting rails there from Clumber. As at Thoresby, these fencing products were presumably provided free of charge by the estate owner.[18]

Some of the moorland enclosures were themselves for plantations. A striking example is Apley Head Plantation, a 139-acre design laid out on former moorland adjacent to the north-east corner of Clumber Park in the late 1780s (see Fig. 9.3). This provides an interesting illustration of Newcastle's improving values. In plan, the layout of the plantation is impressive, displaying the wealth and status of Newcastle family (see Fig. 9.4). At the apex a monument to the Duke's late wife, who died at the early age of 33 (Priestley, 1958, pp. 202–203), ringed by cedars of Lebanon, exploited the high ground. Radiating out from this central circle were six views looking towards the seats of local nobles or other nearby places of interest. The grandeur of this design was augmented by the display, within the planted quarters, of over 20 different varieties of trees (see Fig. 9.5).

The plantation also served more pragmatic ends. The many different tree varieties might well have provided supplies of saplings to establish other plantations on the estate as it was the common practice at Clumber to use existing woodlands as nurseries (Marson, 1798). Also, despite the complexity of the planting pattern and range of varieties grown, Apley Head was probably a source of poles for the nearby enclosure fences.

The Apley Head design was more obviously tailored to the requirements of shooting. The sinuous rides within the sections were labelled as 'shooting rides' leaving no doubt as to their major function. The wide range of tree types and the borders of mixed plants would, in addition, have provided the habitat and food supplies to encourage a variety and abundance of game birds. It is also likely that the layout itself, probably a variation on the French goose foot (*patte d'oie*), might have been a specialist sporting design that the Duke had seen during his travels in France. Newcastle's obsession with shooting is well established. He is pictured in Francis Wheatley's 1788 painting, 'The return

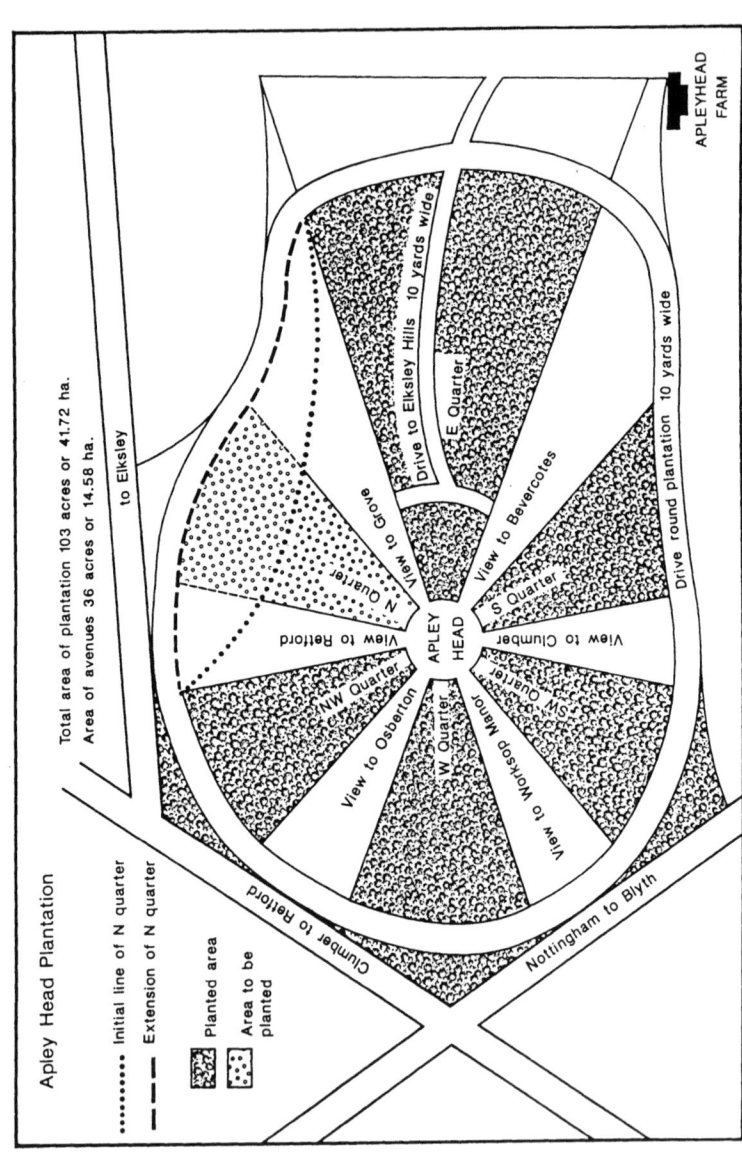

Fig. 9.4. Plan of Apley Head Plantation, near Clumber Park. Source: Sketch of Apley Head Plantation (n.d.), NUMD NeC 4486.

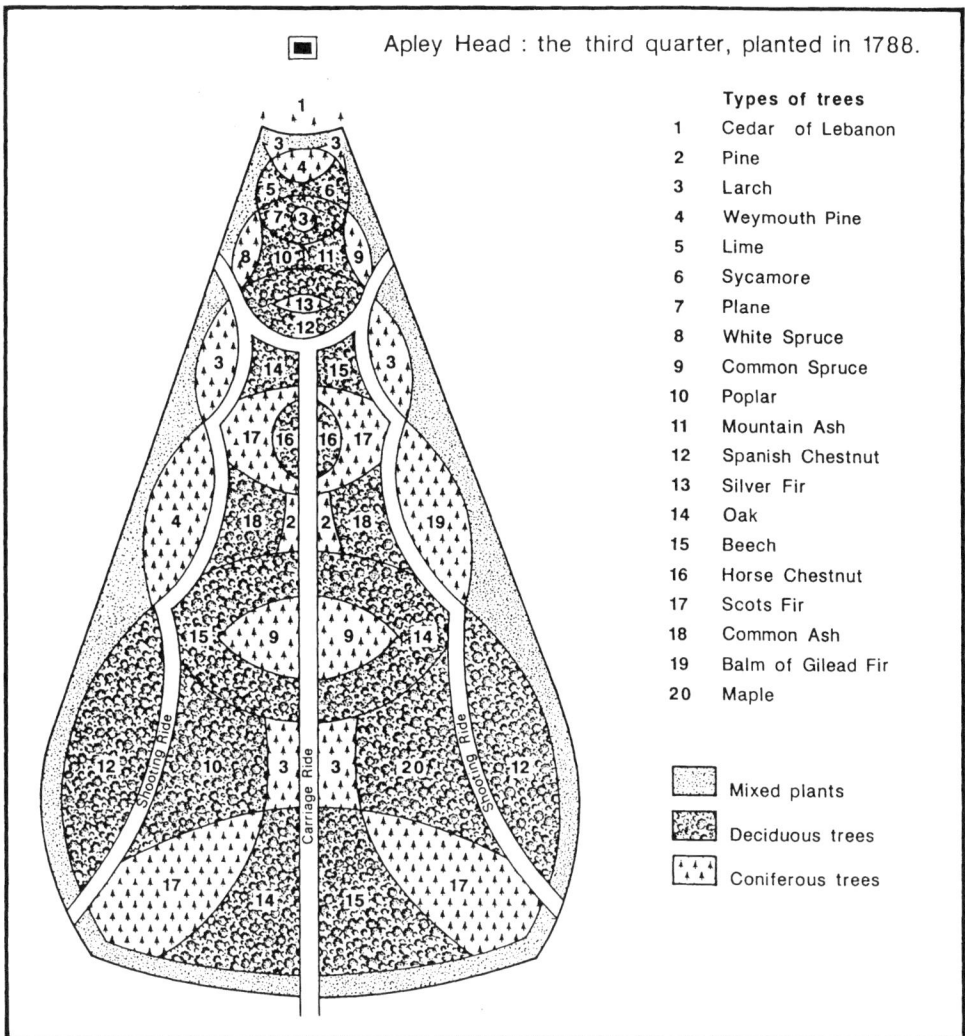

Fig. 9.5. Layout of planted section of Apley Head Plantation (1788). Source: Third Quarter upon Apley Head (1788), NUMD NeC 4343C.

from shooting', among the spoils of the shoot in Clumber Park, sheltered by one of his newly planted groves, his new mansion in the distance.

Conclusions: Estate Woodlands and Estate 'Improvements'

Published sources and field evidence indicate that extensive areas were planted on landed estates in the later eighteenth century. More detailed examinations of estate records indicate that planting of this type was generally undertaken as part of a wider programme of multi-functional estate 'improvements' designed for aesthetic, leisure, economic and propaganda reasons. Locating this 'spirit of planting' within the broader 'spirit of improvement' sweeping through elite landed society provides a new way of understanding tree planting and woodland management during this period.

References

Aslet, C. (1979) Thoresby Hall, Part I. *Country Life,* 165.
Beckett, J.V. (1986) *The Aristocracy in England 1660–1914.* Basil Blackwell, Oxford.
Byng, J. (1789) *The Torrington Diaries. Tour of the Midlands.* Andrews, C.B. (ed.) (1954) London.
Cameron, A. (1975) Some social consequences of the dissolution of the monasteries in Nottinghamshire. *Transactions of the Thoroton Society of Nottinghamshire* 79.
Clifton, J.M. (1979) 'An enchanted palace', Clumber Park, the Newcastle family seat. Unpublished BArch dissertation, University of Nottingham, Nottingham.
Cox, J.C. and Whitworth, R.H. (1910) Forestry. In: Page, W. (ed.) *The Victoria History of the County of Nottinghamshire,* vol. 1. London, pp. 365–381.
Daniels, S. (1988) The political iconography of woodland in later Georgian Britain. In: Cosgrove, D. and Daniels S. (eds) *The Iconography of Landscape.* Cambridge University Press, Cambridge, pp. 43–82.
Duckworth, A.M. (1971) 'Mansfield Park' and estate improvements: Jane Austen's grounds of being. *Nineteenth Century Fiction* 26, pp. 25–48.
Eddison, E. (1854) *History of Worksop; With Historical, Descriptive, and Discursive Sketches of Sherwood Forest.* Longman, London.
Farey, J. (1813) *General View of the Agriculture of Derbyshire, with Observations on the Means of its Improvement.* London.
Fowkes, D.V. (1967) Nottinghamshire parks in the eighteenth and nineteenth centuries. *Transactions of the Thoroton Society of Nottinghamshire* 71, 72–89.
Fowkes, D.V. (1977) The breck system of Sherwood forest. *Transactions of the Thoroton Society of Nottinghamshire* 81, 55–61.
Holland, J. (1828) *The History, Antiquities and Description of the Town and Parish of Worksop.* Sheffield.
House of Commons (1793) The fourteenth report of the Commissioners appointed to enquire into the state and condition of the woods, forests, and land revenues of the Crown, and to sell or alienate fee farm and other unimproveable rents. *House of Commons Journal* 48, pp. 467–561.
Jacques, D. (1983) *Georgian Gardens: The Reign of Nature.* Batsford, London.
James, N.D.G. (1981) *A History of English Forestry.* Basil Blackwell, Oxford.
Lowe, R. (1798) *General View of the Agriculture of the County of Nottingham.* London.
Marshall, W. (1785) *Planting and Ornamental Gardening.*
Marshall, W. (1803) *Planting and Rural Ornament,* 3rd Edn.
Marson, T. (1798) On the method of planting, as practised at Clumber, by the Duke of Newcastle. In: Lowe, R. (1798) *General View of the Agriculture of the County of Nottingham.* London.
Pickersgill, A.C. (1979) The agricultural revolution in Bassetlaw, Nottinghamshire 1750–1873. Unpublished PhD thesis, University of Nottingham, Nottingham.
Priestley, C. (1958) The life and career of Henry Fiennes Pelham-Clinton, 1720–1794. Unpublished MA thesis, University of Nottingham, Nottingham.
Purdum, J. (1978) Profitability and timing of parliamentary land enclosures.

Explorations in Economic History 15, 313–326.
Rackham, O. (1980) *Ancient Woodland: Its History, Vegetation and Uses in England.* Edward Arnold, London.
Repton, H. (1790) *Plans, Hints and Views for the Improvement of Welbeck in Nottinghamshire, a Seat of His Grace the Duke of Portland.* Welbeck Record Office.
Repton, H. (1791) *Thoresby Park in Nottinghamshire, a Seat of Charles Pierrepont, Esq.* NUMD Ma 4P21.
Repton, H. (1793) *Welbeck in Nottinghamshire, a Seat of His Grace the Duke of Portland.* Welbeck Record Office.
Rooke, H. (1799) *A Sketch of the Ancient and Present State of Sherwood Forest in the County of Nottingham.* Tupman, Nottingham.
Seymour, S. (1988) Eighteenth century parkland 'improvement' on the Dukeries estates of north Nottinghamshire. Unpublished PhD thesis, University of Nottingham.
Seymour, S. (1989) The 'Spirit of Planting': eighteenth-century parkland improvement on the Duke of Newcastle's north Nottinghamshire estates. *East Midland Geographer* 12, 5–13.
Seymour, S. (1993) The Dukeries estates: improving land and landscape in the later eighteenth century. *Transactions of the Thoroton Society of Nottinghamshire* 97, 117–128.
Speechley, W. (1775) Account of the plantations upon the estate of his Grace the Duke of Portland, by Mr Speechly, gardener to his Grace. In: Lowe, R. (1798) *General View of the Agriculture of the County of Nottingham.* London, pp. 57–69.
Speechley, W. (1779) *A Treatise on the Culture of the Pineapple and the Management of the Hothouse.*
Speechley, W. (1790) *A Treatise on the Culture of the Vine.*
Speechley, W. (1820) *Practical Hints on Domestic Rural Economy, with an Appendix Containing Several Original Agricultural Essays.*
Sutherland, L.S. (1958–1978) *The Correspondence of Edmund Burke.* Cambridge.
Thomas, K. (1984) *Man and the Natural World. Changing Attitudes in England 1500–1800.* Allen Lane, London.
Thompson, A.H. (1938) *The Premonstratensian Abbey of Welbeck.* Faber and Faber, London.
Turberville, A.S. (1939) *A History of Welbeck Abbey and its Owners,* Vol. 2, 1755–1879. Faber & Faber, London.
Turner, M. (1986) Parliamentary enclosures: gains and costs. *ReFRESH* 3, 5–8.
Walpole, H. (1777) Letter to Lady Ossory, 24 August. In: Lewis, W.S. (ed.) (1965) *Correspondence,* Vol. 32 *The Countess of Upper Ossory,* pp. 374–375.
Williamson, T. (1995) *Polite Landscapes: Gardens and Society in Eighteenth-Century England.* Alan Sutton, Stroud.
Young, A. (1771) *The Farmer's Tour Through the East of England,* Vol. 1. London.
Young, A. (1799) *General View of the Agriculture of the County of Lincoln.* London.

Notes

1. Cowper, W. (1783–1784) *The Task,* Book III, 11.764–765.
2. British Library, Add. Mss 62112. Repton, H. (n.d.) *Memoir,* Pt 2, draft, 170.
3. British Library Add. MS 5750, f. 138.
4. Nottingham University Manuscripts Department (NUMD) PwF 8449. Letter from William Speechley to third Duke of Portland, 25 Jan. 1776; NUMD PwF 8453.

Letter from Speechley to Portland, 17 Feb. 1776.
5. NUMD PwF 8453. Letter from Speechley to Portland, 17 Feb. 1776.
6. Nottinghamshire Archives Office (NAO) DD.5P4/1. Accounts of Joseph Fletcher steward at Welbeck, 1775–1779.
7. NUMD Ma4P22. Plan of Thoresby Lands, c. 1795.
8. NUMD Acc. 680. Thoresby Accounts, Dec. 1799.
9. NUMD Acc. 680. Accounts of the Pierrepont estates, Thoresby disbursements, 1802 and 1804.
10. NUMD Acc. 680. Thoresby Accounts, 1789–1804.
11. NUMD Acc. 680. Thoresby Accounts, 1792 and 1793.
12. NUMD Ma3321/38. Pierrepont estate correspondence, 9 Nov. 1797; NUMD Ma3321/39. Letter from William Sanday [steward] to William Pickin [Pierrepont's land agent], 13 Nov. 1797.
13. NUMD MaS1. General abstract of the estates belonging to Charles Pierrepont in the counties of Nottingham, Derby, York and Lincoln, 1789; NUMD Acc.680. Thoresby Accounts, 1804.
14. NUMD NeC 4140. Letter from Joseph Spence to Lord Lincoln [later second Duke of Newcastle] 28 April 1757.
15. NUMD NeS 105. Clumber estate survey, 1760.
16. NUMD NeI 9, 10. Clumber inventories, 1794.
17. NUMD NeA 248–265. Clumber Accounts, 1775–1793.
18. NUMD NeA 265. Clumber Accounts, 1793.

CHAPTER 10
Need versus greed? Attitudes to woodland management on a central Scottish Highland estate, 1630–1740

Fiona Watson
Department of History, University of Stirling, Stirling FK9 4LA, UK

Introduction

As a general 'environmental' awareness has taken root in the last few decades, so too has public interest in the impact of mankind on the natural resources of our own backyard over a much longer historical period. Though this has undoubtedly the makings of a positive development, forays into the past by groups with a particular contemporary agenda are by no means a guarantee of reliable history. As Oliver Rackham (1993a, p. xviii) most poignantly puts it: 'Why does a pseudo-history grow to accommodate new events that ought to explode it?'

The area of history which has created some of the most deep-rooted mythologies in this respect is the relationship between those who came to 'own' the land and those who worked on it, not least because this issue is still very relevant to contemporary society. At this point, of course, environmental history becomes inextricably linked with economic and social history; for these sub-disciplines, the issue of control — by social and economic forces rather than more overt but often reactive political ones — is vital to establishing 'how' and 'why' certain circumstances arose.

For present purposes, the study of woodland management prior to the Industrial Revolution provides insight into the complicated social and economic networks governing the pattern of rural life; though the specifics of these networks have obviously been superseded, they provide much insight into our own value-systems. In particular, the whole question of who is best placed to manage any natural resource — the local population or a higher, 'central'[1] authority — is often coloured by assumptions of an inherent morality. In the past this morality has been maintained almost exclusively by the landlord, as the quotations immediately below clearly exemplify; in the late twentieth century, however, it is more usually ascribed to, or assumed by, the local community, although the landowning class is still fighting a rearguard action on this issue. This chapter is not concerned, however, with 'right and wrong'; instead, it seeks to illuminate those factors which determine behaviour. The evidence provided by a small central Highland estate in the early modern period allows us to compare the supposed attitudes of landlord and tenants to the surrounding woodlands with what they actually did without the complicating undercurrent of a contemporary environmental debate.

The relationship between tenant and landowner in both England and Scotland over access to, and the treatment of, woodlands was usually a tense one. The following quotations illustrate this basic point, the first referring to conditions in Hatfield Forest in Essex, and the second pertaining to the woods of Urquhart near Inverness.

> The Forest appears irregular ... so as by reason of the small Quantitye of Playnes the multitude of Sheep and other Cattle that depasture there (by pretence of Comonage) ... the Wood are in a great Measure decayed insomuch in case the abuses be not speedily prohibited both Deer and Woods will receive a totall Destruction[2]

> ten years after, we find the laird of Grant again complaining that his woods of Urquhart, which he had been at great pains to preserve, were being wantonly destroyed by the tenants. ... and it seems ... that in this instance the Urquhart people were simply enforcing an old right, including forest pasturage, which had belonged to their ancestors in the loose times in which they lived.[3]

Superficial similarities between the experiences of landowners and tenants north and south of the border should not obscure the fact that there were considerable differences. At a national level, this pertained most obviously to the legal framework under which each set of relationships operated: in England, strong rights of common had developed, ironically enough, out of a wrangle for jurisdiction between Crown and nobility (Rackham, 1993b, pp. 89–90); in Scotland, however, the remnants of effective local power were maintained in the early modern period most particularly by the barony courts. Landowners permitted to hold these courts were given a remarkable degree of control over not only their estates and the activities of those living and working on them, but also the natural resources within them.

The fate of Scottish woodlands was thus – in theory – entirely determined from above; despite a widespread tradition (which was not usually established in law) of allowing tenants to utilize designated woods for building purposes in particular, elite institutions, from the barony court to parliament itself, could, and did, legislate to control all aspects of woodland management. Almost all the evidence relating to Scottish woodlands is to be found, therefore, either in estate papers, or, less commonly, in state archives, providing, naturally enough, a very one-sided picture of management regimes and prevailing attitudes to them.

Parliament had long taken an interest in the preservation of woodlands primarily because timber was such a fast disappearing commodity in the lowlands in the early modern period. Although the first parliament of King James I in 1424 contains the first references to a concern for woodland management, it is highly likely that these were a reiteration of previous acts. A basic fine of 40s. was laid down for those found guilty of stealing greenwood and other destroyers of wood, in addition to the cost of the damage or theft. This parliament underlined the immutable attitude of those in authority with regard to timber and other natural resources, namely that the woods were absolutely the private property of the individual landlord: transgressors were to be punished in the court of the lord from whom the wood had been stolen, whether or not they were caught elsewhere.[4]

Subsequent parliaments expanded on this legislation, encouraging tenants to plant trees particularly. However, most legislative activity, which resulted, apparently, from 'the wood of Scotland being utterly destroyed',[5] tended merely to reiterate previous acts – thereby indicating their lack of success – and increase the penalties for contravention, up to and including the death penalty for persistent offenders. The main impression created by this proliferation of legislation is that, no matter which approach was adopted, the destruction of woodlands continued.

This must have been extremely frustrating for the authorities: from as early as 1535, it is clear that the legislators understood the basics of woodland management, pointing to cutting, peeling and burning, as well as the destruction of enclosures designed to protect woods, as the main problems. Nonetheless, parliamentary statutes, which made no mention of coppicing or pollarding, for example, provided the mere outline of an effective wood management policy; individual landowners were required to promote more detailed legislation themselves in order to provide fully for the needs of their woods.[6]

Such legislation was, in the first instance, hardly worth the paper it was written on if estate owners either could not, or would not, enforce it. It has been asserted that landowners did not take care of their woods if there was little or no commercial profit to be gained from them (Lindsay, 1974, p. 89; Smout, 1993, p. 44); the perceived value of woodland at any given time was therefore crucial to the attention given to its management. This goes a long way towards explaining why a landowner might be tempted to ignore the acts of parliament or even the statutes of his own barony court. However, the equation relating to the 'value' of woods was by no means simple: any attempt at wood management had to take into account the accompanying detraction from the viability of the land to support its tenantry. For example, enclosure, generally regarded as a most desirable form of wood management, was often difficult for the landlord to justify in practice unless there were obvious profits to be made from the woods. In particular, the loss of shelter in winter which enclosure entailed would have required a corresponding reduction in the grazing stock (Lindsay, 1974, pp. 159–160). This was particularly relevant to the Highland economy, due to its reliance on the black cattle trade. It is also difficult to see how Highland semi-natural woods could be enclosed at a purely practical level, since they often encompassed vast tracts of land but with a varying density of tree cover. Also, the local population required a steady supply of timber, though not of particularly high quality, for a variety of uses, from buildings to farm equipment. If alternative materials were unavailable, the landowner would have undoubtedly found it impossible to prevent his tenants from taking it and in such circumstances supervision of cutting became the most practical method of management (Lindsay, 1974, p. 88). However, such supervision was implemented with varying degrees of commitment and success; thus the attitude of the tenants themselves was vital to the future of the woods which formed an integral part of their lives.

Evidence for the activities and attitudes of the Highland tenantry in general is difficult to come by, let alone in relation to woodlands. Not unnaturally, the priorities of those working the land were often restricted to making the best possible living from limited resources, and were thus quite different from those of the landowner. It is also likely that the short-term leases still prevalent at this time gave the lease-holders little loyalty to any particular piece of land and the trees on it, while the landowner, on the other hand, was able to espouse policies which would benefit future generations of his own family. According to one tradition, 'Highlanders never counted it a theft to take a tree from the forest or a fish from the river'. Interestingly, this has been interpreted both as evidence that the local population was profligate with the natural resources at its disposal, and as an indication of an ancient mutually satisfying relationship between Highlanders and their natural environment which was destroyed by commercial landlordism (Nairne, 1890–91, p. 194; Hunter, 1994, pp. 11–12).

No judgement should be pronounced on the relationship between Highlanders and trees until it is established whether timber was used, in whatever quantities, only as a necessary part of the domestic and agricultural circumstances of those living and working on the land and whether consideration was given to the future of the tree being thus utilized. A landlord might resent the pressure which his tenants put on his trees – which were usually worth more if left undisturbed – but this does not indicate that the users were necessarily abusers unless it is considered that their very being there constituted abuse;[7] on the other hand, tenants could do much to alleviate the pressure on woodlands by following management rules which parliament and landowners generally believed would benefit the trees. If tenants did not adhere to these rules, which required care and attention in their execution, then the charge of profligacy stated above can be given some credence. The best evidence for this is to be found in the records of the barony court, which brought to book contraventions by tenants of both parliamentary and baronial statute; some of the best examples of these court books are the six volumes pertaining to the barony of Menzies and Weem in the central Highlands (see Fig. 10.1 for the location of the Menzies' estates).

The Menzies of Weem[8] were renowned in the nineteenth century for their love of woodlands; the fourth baronet, Sir John, was particularly eulogized for having done 'much to cover the barren moorlands'.[9] However, the paucity of references to trees in their own estate papers would suggest, if this was the only evidence available, that the Menzies had no interest in woodlands, either in terms of general management or commercial exploitation.

The barony court records present a rather different picture. A total of 45

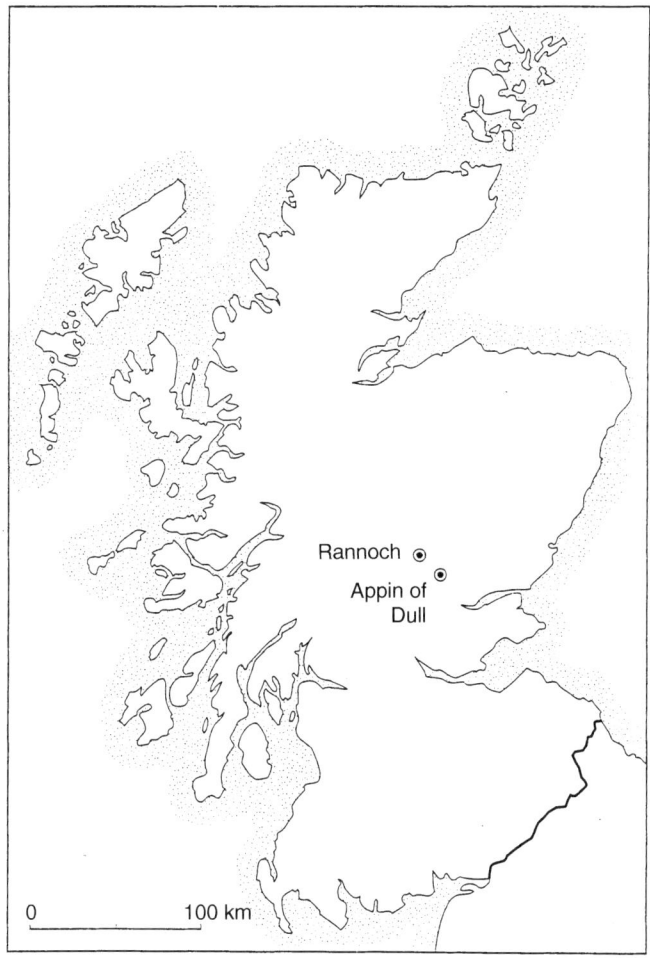

Fig. 10.1. The Menzies' estates.

courts were held in the Appin of Dull (Fig. 10.2) between 1630 and 1738. Not all the records for each court are complete, particularly in the early years, but, nevertheless, their frequency suggests that, in the land surrounding their castle at least, the Menzies thought it necessary to give justice once a year, even if they did not always achieve this. Unfortunately, evidence for only 17 court sessions relating to the Rannoch lands (Fig. 10.3) between 1660 and 1736 remain, although there are references to others which have not survived. This is unfortunate because, as will be shown below, the attitudes of the tenants at Rannoch and their treatment by the laird, as well as the way in which the cases were reported in the records, appear to be somewhat different from those of the Appin of Dull. Nevertheless, a most interesting comparison can be made between the two areas.

From 1677 the laird appears to have seen his duty as the prosecution of

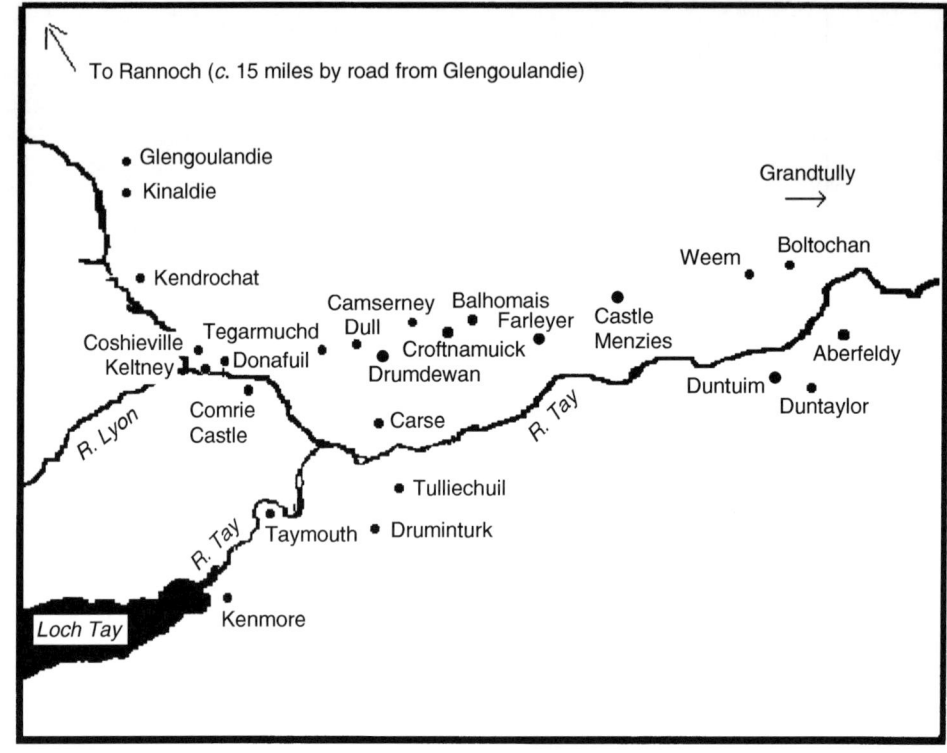

Fig. 10.2. Appin of Dull.

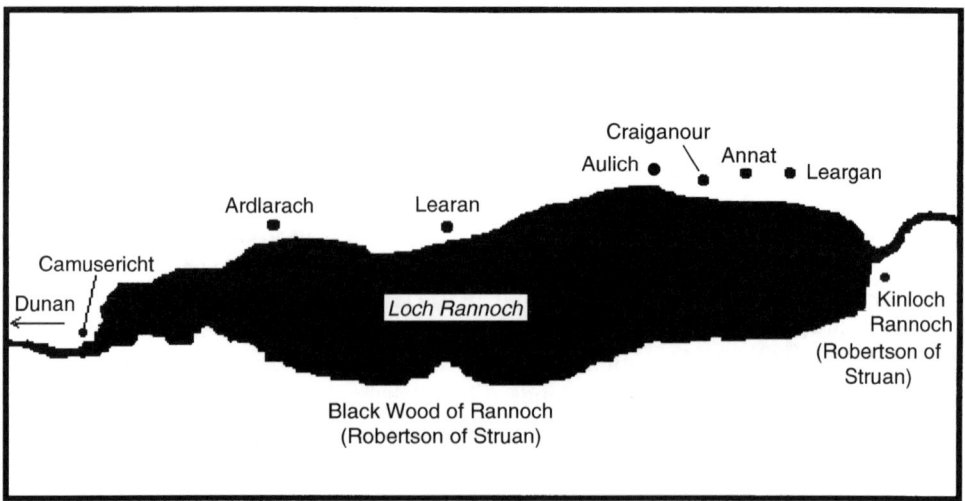

Fig. 10.3. Rannoch.

contraventions of 'the acts of parliament, laws and statutes of this realm', rather than of his own regulations. The subsequent lack of references to specific fines is unfortunate since, without them, it is no longer possible to calculate how much the laird made from the prosecution of woodcutters, or which type of woodcutting, if any, was confessed but not fined. This apparent conformity to national rules regarding wood offences marks a change from the period preceding 1677 on the Menzies' estates at least.

The last court making wholesale references to fines was that dealing with the Appin of Dull[10] in 1673; prior to that date, the highest sum ever demanded of one person by the court was £30[11] for the cutting of a number of types of trees, including the extremely valuable oak. The most usual fine was 40s. Given that the 1639 parliament specified the payment of £20 for a first offence, £40 for a second, and a staggering £100 for a third, it is clear that this barony court, at least, was operating a much more localized and practical system of fines.

The basic 40s. (£2) fine was applied equally to sticks and other insubstantial cuttings of hazel, birch, alder and hained (i.e. enclosed) broom, suggesting that the type of wood, with the exception of oak, was not a factor. The exact fining procedure for more serious wood offences, which included the taking of logs from the above-named tree species, is far from clear; it seems likely that a basic fine was exacted for the crime itself, together with an additional financial penalty calculated from the quality (regardless of species) of the trees cut. In 1632 the laird appears to have almost doubled the basic fine to £3 6s. 8d.; however, the 40s. penalty was re-established the following year, perhaps because there had been resistance to the increase.

As stated above, it is unfortunate that the fines owed by each offender disappeared from the court records, particularly because the amount of money which accrued to the laird from the prosecution of wood cutting may well have played an integral part in management policy. Prior to 1677, there is some evidence to show that the income from fines was substantial and growing (Table 10.1).

Considering that, after 1673, the Appin of Dull court usually secured the convictions of over 100 tenants (and often over 200), which compares well with the 99 who were fined in that year, it is therefore fair to say that the income from wood offences made a significant contribution to the laird's

Table 10.1. Annual income from fines from Appin of Dull.

1630	£157		
1632	£111	1s.	4d.
1633	£76	13s.	4d.
1639	£26		
1650	£76		
1670[12] at least	£283		
1673[13] at least	£485		

Source: SRO GD50/136/1, pp. 23–157.

income. We shall examine, in due course, whether or not this might have had an effect on the overall woodland management policy.

Although the practice after 1673 was to refer vaguely to parliamentary statute as the basis for the punishment of wood offences, the barony court of 1698[14] did in fact specify the laird's own scale of fines: the cutting of oak would incur a penalty of £40, the cutting of ash, £10, and the cutting of alder and birch, £5.[15] Thus, by the late seventeenth century, different species of tree had been accorded varying monetary values but, from this evidence alone, it can only be asserted that these reflect the particular quality of wood on the Menzies' estates – and what was done with it – rather than a national scale. Unfortunately, the records continue to lack references to fines, preventing us from examining the practice in relation to the theory.

This barony court could, and did, make regulations of its own, providing us with an accurate indication of the kind of problems which the laird felt should be tackled. Some of these regulations can be construed as a genuine attempt by the Menzies to provide for the welfare of their woods; others should be regarded more as a reiteration of the laird's prerogative rights over his estate. It is also clear from the records that they formulated policies of wood management in response to the needs of the particular woodlands within their domains, and also with reference to the attitudes of the communities living alongside them. There is thus a clear distinction made between the woods of north Rannoch and those of the Appin of Dull.

The bulk of the records of the barony court refer to the area of which Menzies Castle was the centre, known from the Middle Ages as Apnadull or the Appin of Dull. This is not particularly surprising: the area in which the lord was most obviously present is the most likely to be well regulated and controlled. The tone of such control was set in 1630 by the first recorded statute of the Menzies' court books: no-one was to gather wood – dry or otherwise – within the laird's woods under the threat of a fine of 40s.[16] This statute was thus the basis for the large-scale convictions which followed, since the collection of the kind of loose-lying, low-quality timber implied by the above is the least that could be expected from a community dependent on wood for so many aspects of its life and livelihood. The laird had thus stated that his right over all wood was unequivocal, irrespective of the needs of his tenants.

Other statutes were more specific: goats were ordered to be kept out of the woods in 1634. Most landowners made such a regulation, given the devastation to young trees which these animals were known to cause.[17] In the following years, although only one person was prosecuted for keeping goats (and presumably letting them wander into the woods), reference was certainly made to trees whose bark had been peeled by these animals.[18] However, they were by no means the only guilty party: 30 trees in an enclosed park on the Craig (Rock) of Dull were found destroyed in 1722 and the culprit was deemed to be a ram caught red-handed. Its owner claimed that his ram was not the only beast to have entered the park, but he was nonetheless made responsible for the entire damage and a hefty fine of £300.[19] This clearly suggests that trees were considered to be under threat from all grazing animals; equally, this is strong evidence for a distinction in treatment between the parked woodlands, surrounded by dykes, where the laird was intent on cultivating young trees of a variety of species, and the less valuable, and thus less protected, unparked woodlands from which tenants took their wood supplies.

In 1665, the laird of Weem took up the example of the parliament of 1661 by encouraging tenants to plant trees annually 'at their several dwelling places'.[20] These trees were to be got from the Wood of Weem, implying that the laird had already set up a nursery in the plantations behind his castle.[21] However, the Menzies of Weem do not seem to have fully understood the practicalities of this otherwise admirable policy, certainly in comparison with Duncan Campbell of Glenorchy. As early as 1615, Campbell, whose family was developing a tradition of active woodland management, had also ordered his tenants, via the barony court, to plant trees in their kailyards (kitchen or cabbage gardens situated next to the house). However, it was further stipulated that these trees were later to 'be sett in the maist comodious pairtes of their saidis occupatiounes' – a far more sensible arrangement which would permit these new trees to flourish in an appropriate soil at the most accommodating location.[22]

There is some evidence to suggest that the tenants of the Appin of Dull did engage in planting in accordance with the 1665 statute. Unfortunately, this evidence is entirely negative, coming to light when tenants were prosecuted for having cut down the handy planted trees in their kailyards. Fines of over £20 for such counter-productive activities are recorded, illustrating how seriously they were taken.[23]

There was no aspect of estate life, however particular, in which the laird would not intervene in the interests of his woods. In 1698 he ordered that the proliferating number of cottar houses which had grown up in the large township of Camserney – 'whereby stealing and other vices are encouraged and that the woods are destroyed [by] these cottars (tenants occupying a cottage with or without land attached to it)' – should be reduced to six.[24] In 1717 tenants were ordered to use stone or turf dykes instead of hazel or any other timber as the fencing surrounding their corn and kail.[25] The main thrust of this statute was undoubtedly to make farm enclosures more secure. Such ostensibly sensible agricultural practices, which were clearly in the interest of the tenants, were nonetheless imposed by the laird, indicating a conservatism on the part of the former which was in keeping with their suspicion of improvement in general, and enclosure in particular. Of course, the key-word here may well be *imposed*; tenants were not necessarily unhappy with the measures themselves but resented the method of their introduction.

Nevertheless, the 1717 act appears to have been largely successful, as confessions of the use of hazel for fencing virtually disappear from the court records thereafter. However, one incident from 1718 illustrates poignantly the differing priorities of laird and tenant. Ewen McInskellich, servitor to the minister at Weem, admitted to contravening the act by cutting new timber, hazel and others, 'to be pealing [fencing] put about the corn yard'. McInskellich was asked if the minister had ordered him to do this, to which he replied that 'he had no such orders but that his master ordered him to take care of the corn yard'.[26] For those who worked the land, the responsibilities of farming often overrode a concern for woodlands.

Though part of the Menzies' estates from the Middle Ages, the barony of Rannoch suffered from, or rejoiced in, depending on your point of view, not only its distance from the controlling eye of the laird, but also its remoteness *per se*. There are thus no records of a barony court held for the area until 1660, although a document from the Menzies' correspondence notes 'the names of those who doeth destroy the woods of Rannoch' in December 1632. However, none of the accused were Menzies' tenants, but rather had come to the north Rannoch woods from neighbouring lands.[27]

The Privy Council records for the seventeenth century also create an impression of a difficult and lawless area over which the Menzies of Weem exerted little control at best, and were openly flouted at worst. In 1642, during one of the many lulls in the Covenanting wars against King Charles I, Sir Alexander Menzies petitioned the Scottish Privy Council, claiming that Patrick Murray (McGregor) of Glenstrae was 'uplifting his [Menzies'] rents and sorning (foraging) upon and oppressing tenants *as if the arm of justice was not able to reach him*'.[28] Even as late as 1661, the laird, grandson of the above, sought to be excused from responsibility for the good behaviour of the McGregors in Rannoch, stating that they were still 'so far from being obedient that they will not pay their duties'. However, the restoration of Charles II in 1660 brought about a desire on the part of the Crown and – to an extent – the means at its disposal to impose its will throughout Scotland: a party of foot guards, which was already in the area collecting taxes, was duly despatched to repossess the barony for the laird of Weem.[29]

The business transacted by the first recorded court dealing with the barony of Rannoch – that of 5 July 1660[30] – indicates nonetheless that the laird at last believed he could educate his tenants there and make them accountable for those actions which were outlawed by both national and local statutes. In the first instance, a total of 71 tenants from nine farms – surely the full complement of adult males – were brought before the court and 'every one of them for their own parts' was fined for 'cutting, burning and destroying of woods, pasturing in the forests, killing of deer, roe and muirfowl within the bounds and barony of Rannoch'. There was thus a great deal of work to be done even to bring them up to the level of obedience demonstrated by the far-from-model tenants in the Appin of Dull.

To this end the court promulgated 15 statutes encapsulating the basic rules of Menzies' estate management.[31] Five of these related to woods. The peeling of bark (used in tanning) from any tree over a foot high was outlawed, under penalty of a £20 fine per tree to discourage this particularly destructive practice. A fine of £20 was specified for the cutting of timber 'a foot high above the ground'; this was presumably intended to encourage a rudimentary form of coppicing, to stimulate the growth of shoots. As well as the usual fines for cutting without permission, tenants or cottars were prohibited from allowing goats or swine to pasture within the laird's woods. A further two rules forbade the cutting of broom for burning or thatching and the gathering of dry sticks without permission. However, in 1702 it was again made clear that the use of wood for buildings would only be permitted if 'it was cut close over alike with the ground', suggesting that the Rannoch tenants had still not taken on board the sensible tree-management techniques advocated by their master.[32] All in all, the history of Rannoch suggests that it was an area which had taken a long time to submit to the jurisdiction of both the local landowner and central authority and which, despite increased control in the period after 1600, still proved difficult to administer.

The regulations laid down in the baronial courts of the Menzies of Weem, together with the use of parliamentary statute, indicate a very hands-on approach to woodland management, which combined both a concern for their rights as owners of the woods and a care for, and degree of understanding of, the needs of the trees themselves. Control of the use of woodlands on the Menzies' estate was thus firmly regulated – in theory at least.

One of the most interesting aspects of the information provided by the barony court books is the clear evidence for a near total reliance of the local population on wood for almost all aspects of their life and work. Timber was the main component of all constructions: the houses themselves; buildings used for agricultural purposes such as barns, byres, sheep cots, pens, bothies and soot houses; and those which serviced the needs of the community as a whole, such as mills and kilns. These wooden constructions, though serviceable, were not always equal to the task, however: John Reoch in Kirktoun of Weem, for example, confessed the use of two ash forks to support his house which 'fell in the night time' in the winter of 1699.[33] Unfortunately, there were others whose houses suffered a similarly traumatic fate. Timber was also a prominent feature of the inside of a house: alder was the wood most frequently employed therein, proving ideal for weaving into partitions and wattles. Agricultural implements also relied heavily on all types of wood – ash, alder, birch, oak, willow – for their manufacture. Timber of all types and sizes was thus regularly required by the Menzies' tenants; however, they were by no means automatically allowed to use it.

The lairds of Weem were sometimes explicit in their exasperation at their tenants' insensitivity to their attempts to encourage new growth: in 1691 the court recorded plaintively that the inhabitants of the Appin of Dull, despite the various parliamentary and baronial statutes to the contrary, 'yet nevertheless they frequently cut green oak, ash and other green timber'; in 1700 the Appin of Dull court was held specifically to deal with 'the cutting of woods and destroying the young growth of timber growing on the said Sir Alexander's lands'.[34]

The tenants in the Appin of Dull, according to the laird, should have known better. This was not the case in Rannoch, where the courts were still concerned with contraventions of rules for management, such as cutting high (above a foot from the ground), peeling and burning; there was still a long way to go until these tenants were trained to act responsibly when they cut timber. Even by 1717, some 40 years after the first recorded Rannoch court, Donald Bane, a tenant in Dunan at the west end of Loch Rannoch, noted that he had seen more than 100 trees peeled.[35]

Getting their tenants to cut close to the root was one of the main objectives of the Menzies lairds. References to cutting from the root certainly increase between 1660 and the last recorded court in 1736; however, in this last year 25 tenants also confessed to cutting high. It cannot therefore be said that the Menzies lairds had managed to change significantly the habits of their Rannoch tenants; however, at least they were now able to enjoy the profits of the latter's non-cooperation.

In the Appin of Dull, timber was sometimes given free, particularly for construction work, but this was more often not the case. The evidence suggests that the tenants felt that they should have been allowed to take wood freely for the use of their buildings: in 1686 John Calcater in the Mains of Weem admitted that he cut timber yearly for building his houses, but there was no mention of absolution; on the other hand, when Patrick Ferguson in Drumdewan confessed all wood except oak for the use of his buildings in 1701, he was 'absolved accordingly'.[36] This conflicting evidence may indicate that the laird operated a quota system in order to ease the pressure placed on his woods by the needs of his tenants; in other words, a grant of wood was intended to last a certain period of time. Unfortunately, because fines are not recorded, it is not possible to ascertain for certain what was and what was not allowed.

Tenants could petition the laird for liberty to take timber and this was sometimes granted. However, incidences are too infrequent to suggest that this was usual. Nor did it necessarily make much difference: there is certainly at least one instance where a tenant cut wood despite being refused permission.[37] There are two explanations for this apparent recalcitrance: either the tenants were refusing to adhere to a responsible system of wood management and were determined to use as much timber as often as they wished without any concern for the future of the woods; or, the tenants were reacting against increasing restrictions applied to a resource which they had little choice but to use and to which they felt that they should have access. Such explanations do not alter the level of pressure placed on the woods but they do have a bearing on what lessons might be learned from this situation.

The tenants seem to have understood the overall intention behind the regulations since they sometimes claimed, in their defence, that the timber which they took was 'useless', that they did no 'skaith' (damage) to the woods, or that they only took wood which someone else had already ruined. Such excuses must doubtless be taken with considerable amounts of salt but may also on occasions indicate the truth. Some tenants claimed a more active involvement in tree management: Patrick McPhettie in Tullichuil, for example, although not exempt from prosecution himself, declared that 'when he did see any in the wood, he put them off cutting'. He may well have landed himself with a greater responsibility than he realized since the wood of Tullichuil was one of the most popular with the Appin of Dull tenants, especially 'under cloud of night'.[38] Nevertheless, the cutting of green wood continued, presumably because it was either too desirable or too essential.

Although the tenants of the Appin of Dull were not usually explicitly accused of the kind of offences commonly prosecuted in Rannoch, this does not mean that wood was never cut high, peeled and burned by them: the court of 1711 was forced to deal with exactly these crimes.[39] Thus, although the inhabitants of the Appin of Dull were better educated in the rules of wood management and thus generally better behaved, many still contravened the basic regulations; once the Rannoch tenants were being brought regularly to court (c. 1700), the control exerted by the laird of Weem over all his tenants varied only in the extent to which they conformed to his regulations. The methods of control were the same on both estates.

One of the most fascinating aspects of these court procedures is the picture which they convey of an intricate social and moral network which caused the inhabitants of these lands both to confess their own wrongdoing and to inform the authorities of those of their neighbours. The oath of a defendant was accepted without proof in these more god-fearing times; however, if a man claimed not to have contravened the statutes and someone else asserted that he had been seen doing so – as happened on a number of occasions[40] – the penalty for such perjury was high. Tenants had to be exact in their accusations, however, as Patrick Cameron in Carse found out in 1711: he was ordered to establish the identity of a man he had seen at a distance carrying an alder tree, or else himself pay a fine of £20. Fortunately his memory came to the rescue and he was able to inform on Duncan Deor in Dull, who confessed to the same.[41]

The regularity with which tenants informed even on those closest to them, together with the fullness of their own confessions, suggests that this system of control was largely successful. However, as noted above, some went to great lengths to avoid detection, particularly by cutting at night. The high numbers of those brought to court – which on a few occasions appears to have corresponded more or less to the total adult male population[42] – suggests that the authorities were keeping tabs on the wood-cutting activities of every single adult male in the area, even if some could, and did, prove their innocence in court. In other words, an individual tenant or cottar escaped a summons to court only if the laird and his woodkeepers were absolutely sure that no infraction had been committed.

For those convicted of offences, a fine was the usual penalty exacted; however, if wood was taken from one of the laird's parks (usually planted with pine, which was not found in – or at least taken from – the unenclosed woods), a much more severe punishment was exacted. In 1725 Alexander McOmie, a mason in the kirktoun of Weem, and his brother James were accused of 'cutting a long planted fir (pine) wood above the nurseries in the east park (above the castle) about Martinmas last'. James, as the one caught in the act, was also deemed responsible for the cutting of a further 12 planted firs (pines), unless he could find alternative guilty parties; he admitted cutting the tree for his brother's use but had not actually carried it off. Nevertheless, the baillie insisted on punishing both brothers for all the pines found cut, which amounted to a joint fine of £240 scots, with the proviso that some of this could be recouped if the real cutters were found within 6 months. However, a mere financial exaction was not deemed sufficient for the man who had actually committed the crime. The court further ordained that James was to stand in the jougs[43] in Weem kirkyard on three consecutive Sundays while the parishioners trooped into church 'with a paper on his breast bearing this inscription "I stand here for cutting of Sir Robert Menzies of that Ilk his planting"'.

Such a public humiliation would no doubt have served as 'the example and terror of others to commit the like in time coming' and underlined the seriousness with which the laird took his planting.[44] The significance of this form of punishment is further underlined by the fact that although large fines were ordered by the court for significant offences, the actual fine was often much less; instead, the remainder was – sensibly – held hostage for the future good behaviour of the offender. In 1726, for example, Alexander Wilson in Aulich in Rannoch was fined £100 for audaciously selling timber in Perth; £12 of this was to be paid by the following Martinmas (11 November) but the remainder was to be demanded 'upon his next transgression or when Sir Robert shall order it to be taken up by a writ under his hand'.[45]

Indeed, despite occasional examples of apparent severity, any discussion of control must address the general question of enforcement. Certainly there is some evidence to suggest that many of these strict regulations remained theory rather than practice. Non-attendance at court was a problem: in 1698 20 people were recorded as being absent from the Appin of Dull court for the second time in a row; in 1701, Donald McInnes in Kendrochat finally appeared before the baillies, having been absent from former courts; in 1711 John Dow, the miller in Keltney, was fined £10 in his absence 'in respect he never answers to court'.[46] The evidence is also not available to ascertain how promptly those convicted of wood offences actually paid their fines, or even if they paid them at all. Certainly Donald Murray in Craganour in Rannoch managed to accumulate a £120 fine between 1706 and 1717, when the sum was recorded as still outstanding.[47]

The extent of control exerted by any authority rests, to a degree, on the severity of the sanctions which it can impose to enforce its dictates. There is no evidence to suggest that the lairds of Weem would evict a tenant for wood offences alone, although there is an example of such behaviour being included among a number of points in a case for eviction. Imprisonment was another option and, again, there is one example of a cottar being warned that if he could not find someone to stand surety for his fine, he would go to prison. Public humiliation, in the stocks or the jougs, as above, was also an option.[48]

Nevertheless, as the graphs in Figs 10.4 and 10.5 illustrate, the Menzies' tenants did not appear to regard the threat of such punishments as a deterrent: the percentage of the population believed by the laird and his officers to have cut wood and who either confessed their offences or did not attend the court fluctuated between 70% and 100% for most of the period. Or, to put it more usefully, the numbers of those absolved from cutting wood[49] did not exceed 30%. However, the graphs also show that the years from 1720 to 1727 for the Appin of Dull and from 1723 to 1732 for Rannoch were exceptional.

The increase in the efficiency of the courts in preventing wood-cutting was almost certainly connected with the payments made to John McInnes,

Woodland management on a Scottish estate

Fig. 10.4. Incidence of wood-cutting in the Appin of Dull, expressed as a percentage of tenants summoned to court 1650–1738.

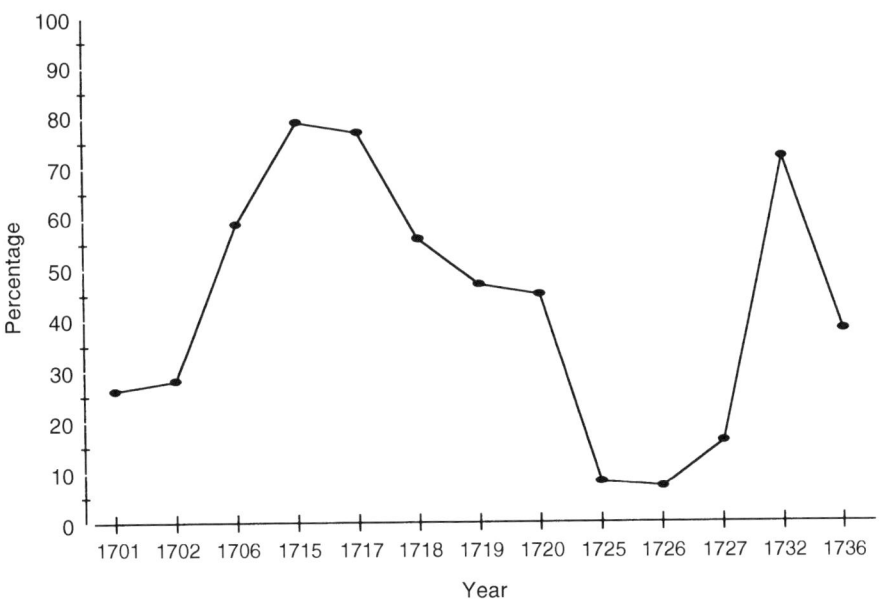

Fig. 10.5. Incidence of wood-cutting in Rannoch, expressed as a percentage of tenants summoned to court, 1701–1736.

park and woodkeeper of Dull, in 1725, 1726 and 1727.[50] The estate had employed men to oversee the woods before: in 1634 Dougal Stewart in Glengoulandie was appointed forester of the woods of Glengoulandie and James Menzies in Kendrochat forester of the woods beneath Glengoulandie to the foot of the Keltney burn. They were both to receive for their fees half of the wood confiscated, together with the axes of the cutters, a considerable incentive for devotion to duty. Unfortunately, we can draw no conclusions about their effectiveness. In 1682, Duncan Roy from Rannoch was described as forester, but considering the trade in illegal venison in which he appeared to be engaged, it is unlikely that he proved much use in keeping his neighbours on the straight and narrow. In 1701 an unidentified Alexander Menzies was named as forester in the Appin of Dull, supervising the cutting of a couple of ash required by a tenant in that year.[51]

None of the above woodkeepers were named in the records as an estate officer and this may hold the key to the efficiency of John McInnes. His salary, which averaged around £13 a year, indicates that he was expected to take his job very seriously, and there is considerable evidence to suggest that he did just that. As a tenant in the township of Dull itself, he was centrally placed to keep an eye on the barony and there is no doubt that he was active in his duties the length and breadth of it, apprehending tenants in the act of cutting and forcing them to appear before the court.

One of McInnes's responsibilities appears to have been the sale of timber to the Menzies' tenants: John Menzies in Balemenoch, for example, was absolved from cutting in 1732 because he had purchased sufficient wood from the forester 'for all uses he had to do with'.[52] This system of buying timber – or at least the enforcement of that system – may have contributed to the increased numbers of absolutions for cutting in these years. However, if the timber still came from the Menzies' woods, the pressure placed on them was not necessarily reduced.

MacInnes was also active in attempting to force the payment of fines, complaining in 1727 against Alexander Boyd, the late tenant in Tomintogle, who had been fined £50 for taking bark from an oak tree 3 years previously. Boyd was only required to part with 1 merk, but he still refused to comply, providing evidence of continued resistance to the management policy and the limits of McInnes's effectiveness.[53]

Nevertheless, the inhabitants must have watched the approach of the forester with trepidation – he was quite capable of conducting a search of their dwelling places: in 1732 Alexander Bigom in Duntaylor, although confessing only a stick of ash, was called to account for 'birch and ash split' which John McInnes found in the east end of his house.[54] Not surprisingly, the latter's activities aroused considerable resentment which, on one occasion, descended into physical violence. Thomas Menzies of Tegarmuchd was brought to court by the forester in 1728, accused of beating McInnes 'without any just provocation' when the latter had commented on the roots of some timber that had been cut as they happened to be passing.

In fact, McInnes was not as blameless as he wished to make out, appearing in this, and other examples of his activities, as aggressive in the pursuit of his duties. The baillie, having heard all the witnesses, decided judiciously that there had been a 'mutual battery', for which both were charged £5. The ultimate winner of this fight, was, of course, the laird, who picked up the £10 fine.[55] McInnes, although clearly a conscientious woodkeeper, was also not above trying to work in a few perks of the job: in 1731 Sir Robert Menzies himself complained that his forester had cut several alder trees more than he had been given permission to take, for which he was fined £20.[56]

John McInnes was the forester and park keeper of the woods of the Appin of Dull; James Menzies in Annat was mentioned as wood officer in Rannoch at around the same time (1727). Despite no record of a salary similar to that paid to John McInnes, Menzies appears to have been effective, seizing the bark taken from trees by tenants in neighbouring Aulich.[57] This kind of direct action might also, therefore, have been responsible for the clear increase in the incidence of absolution from wood-cutting which occurred in the late 1720s in Rannoch (see Fig. 10.4). Despite the dangers of crediting individuals with making a significant difference in any situation, especially when those individuals were not in a position of real power, this evidence suggests that if the lairds of Weem seriously wished to alter the habits of their tenants, then they had to employ a man committed to the task and pay him a decent salary. Unfortunately James Menzies appears in the records only in 1727 and, after 1732, there is no further mention of John McInnes, either as forester or even as a tenant. More importantly, if the latter, in particular, had either left or died, there is no evidence of any successor as woodkeeper. Perhaps the ground officer added the duties of forester to his own; this had certainly been the case in 1698.[58] If no woodkeeper was appointed after McInnes, the priorities of the Menzies' wood management policy begin to look less altruistic.

The exemplary punishments inflicted on those who interfered, directly or indirectly, with the laird's planting, indicate that, while great efforts were made to protect and preserve the plantations of pine, elm and sycamore, the care of the semi-natural woodlands does not, with the exception of a brief period, seem to have been considered important enough to expend the kind of resources, such as the salary of a woodkeeper, which would have made the regulations effective. It should also be noted that the few references to any form of enclosure – such as dykes or fences – apply only to the plantations, not the semi-natural woodlands. Enclosure was not always effective in excluding sheep or goats, as we have seen; however, since there is no evidence to suggest that the semi-natural woodlands were at all protected, they were extremely vulnerable to the effects of grazing animals.

Conclusion

The lairds of Weem were clearly aware of the kind of regulations which would help to preserve their woods. The statutes promulgated in their courts were sensible of both contemporary ideas regarding woodland care and management and the particular problems pertaining to Rannoch and the Appin of Dull.

However, there can also be no doubt that considerable pressure was placed on the Menzies' woods by the needs of the local population. These needs should not be regarded as profligate in themselves, but there is evidence to suggest that many tenants ignored the court regulations and inflicted unnecessary injury on the trees. It certainly cannot, therefore, be argued that those working closest to the land had any intrinsic regard for the preservation of the natural environment in which they lived; indeed, why do we presume that they should have done when they were categorically excluded from involving themselves directly in managing the woodlands?

The lairds of Weem, as with most landlords, bemoaned their tenants' insensitive treatment of their woodlands in the official records. However, there would appear to be a degree of hypocrisy in such an attitude. One of the main incentives behind the commitment to the prosecution of wood offences appears to have been the lucrative income from fines. Although the evidence after 1670 does not allow us to assess exactly how much the court brought in, the amounts recorded before that date suggest that the total could be as high as £500 per annum. Obviously a degree of non-payment and the system of remitting the greater part of the highest fines as surety for future good behaviour must be offset against any potential total figure; equally the alternative policy of selling timber to tenants would have brought in some income, though presumably a straightforward sale would have cost the tenant less than a fine. Nevertheless, would it be too cynical to postulate that the reason that the laird of Weem did not apparently employ a forester on the same terms as John McInnes was precisely because the latter's success in cutting down the numbers found guilty of wood offences brought about a sharp decrease in the income from fines?

Certainly the judicial process as a whole, except under McInnes in the 1720s, did not act as a deterrent. The overall impression conveyed by this judicial hyperactivity is one of a tacit agreement that the woods could be used so long as an attempt at good practice was made and a full account given in court; in other words, the laird was content to allow his tenants access to wood in a reasonably controlled manner on condition that he received payment for it in fines. By the early eighteenth century, explicit mention is made of the sale of wood by the woodkeepers to the tenantry; prosecution then followed if more wood was cut. Unfortunately, this sensible arrangement does not appear to have been particularly widespread even at this late date.

Despite their ostensible cooperation, many tenants were clearly unhappy with these arrangements. Some voted with their feet and continually refused to pay their fines. The evidence suggests that, if the tenant was well behaved in other areas, little was done to force such payment. Others took evasive action, making the 15-odd mile journey to the Rannoch woods, or else waiting until darkness to cut timber closer to home, in the vague hope, and sometimes the deliberate intention, of evading detection because they could not be identified. Still more may have felt that they just could not afford to pay for all their necessary timber, particularly in the light of the fact that the commodity was so readily available. The peeling of bark, which could cause such damage to the tree and therefore provoked one of the most expensive fines, may well have been pushed 'underground' for that very reason.

The bottom line remained that wood was cut so often and in such large quantities because it was essential to the lives and livelihoods of the local tenantry. The records hint at a sense of frustration provoked by a small-scale financially orientated wood policy which forced them to own up, for example, to an act of kindness in giving a parcel of wood free to the boy servant of a blind violer.[59] Ultimately, as Ewen McInskellich, the servant who fell foul of the court because he had done his best to take care of his master's corn yard, serves to remind us, neither the laird, nor late twentieth century observers, can really expect a population often living between poverty and bare subsistence to be the responsible party. Not all the Menzies' tenants were poor, certainly, but perhaps few could afford to put environmental considerations before the requirements of their farmland.

Acknowledgements

This paper was written as a result of research forming part of a 3-year project entitled 'Sustainability in the management of Scottish woodlands, 1600–1900', funded by the ESRC's Global Environmental Change Programme. I owe an immeasurable debt of gratitude to Professor Chris Smout, both for helpfully commenting on this particular paper and for his inspiration and courteous direction throughout our work on this project. I would also like to thank Professor George Peden and Dr Ronnie Lee for helping me to refine earlier drafts; this final version is, of course, entirely my own responsibility.

References

Hunter, J. (1994) *Everyone Who Ever Mattered is Dead and Gone*. Institute of Environmental History, University of St Andrews (Unpublished paper from conference on Cultural Environments 2), St Andrews.

Lindsay, J. (1974) The use of woodland in Argyllshire and Perthshire between 1650 and 1850. Unpublished PhD thesis, University of Edinburgh.

Nairne, D. (1890–1891) Notes on Highland woods, ancient and modern. *Transactions of the Gaelic Society of Inverness* 17, Inverness.

Rackham, O. (1993a) *Trees and Woodland in the British Landscape*. Dent, London.

Rackham, O. (1993b) *The Last Forest*. Dent, London.

Smout, T.C. (1993) Woodland history before 1850. In: Smout, T.C. (ed.) *Scotland since Prehistory: Natural Change and Human Impact*. Scottish Cultural Press, Aberdeen, pp. 40–49.

Notes

1. A landlord is considered here to be a more 'central' authority, because he often has decision-making control over parcels of land spread over a wide area.
2. Report of [corrupt] Royal Commission, 1639, quoted in O. Rackham, *The Last Forest* (London, 1993), 93. The forest referred to was Hatfield in Essex.

3. D. Nairne, 'Notes on Highland woods, ancient and modern,' in *TGSI*, vol. XVII, (1890–91), 194.
4. *Acts of Parliament of Scotland* [*APS*] c. 10, II.7; *APS* c.11, II.7.
5. *APS*, c.27, II.7; c.7, II.343; c.284, VII.263. This destruction had apparently taken place by 1503 but really only refers to the accessible woods of the Lowlands.
6. *APS*, c.15, 16, II.242, 251; c.8, II.343; c.22, III.145; c.48, III.460; V.605a; c.284, VII.263.
7. This attitude, which certainly underpinned the nineteenth century clearances, cannot be said to be prevalent in the seventeenth and early eighteenth centuries.
8. The owners of this estate will be described variously as the Menzies of that Ilk, the Menzies of Weem or the lairds of Weem. Their main estate was situated in the Appin of Dull, centred on Weem and a few miles west of Aberfeldy; they also held land along the north shore of Loch Rannoch. See Figs 10.1–10.3.
9. D.P. Menzies, *The Red and White Book of Menzies* (Glasgow, 1894), 405.
10. Fines are never recorded for Rannoch.
11. All monetary values referred to in this article are in scots, i.e. one-twelfth of sterling values.
12. Records for the court of 1670 have been entered in a singularly curious manner, in that an incomplete list of those convicted of wood crimes, including their fines, is given, followed by a fuller list of all tenants summoned to appear at the court, including those who had not offended, and further details of the offences of those found guilty. If the records had continued in this vein, our understanding of the court's proceedings would have been greatly enhanced.
13. The fines of those who had committed the most serious offences, such as the cutting of 60 birches and 20 willows, have been left blank, presumably so that a higher authority, perhaps the laird himself, could pronounce the judgement.
14. The year of the last Scottish parliament to mention woodlands, although it merely ratified previous acts (*APS*, c.35, X.175).
15. Scottish Record Office (hereafter SRO) GD50, section 136/2, 452. No mention is made of hazel, suggesting that its use was so widespread and regarded as unworthy of punishment.
16. SRO GD50/136/1, 27–28.
17. SRO GD50/136/1, 45; see *British Wildlife*, vol. 5, no. 2, Dec. 1993, p. 78, for the particularly destructive nature of goats' grazing.
18. SRO GD50/136/1, 110; GD50/136/3, 976.
19. SRO GD50/136/4, 1392–1397.
20. *APS*, c.284, VII.263.
21. SRO GD50/136/1, 107.
22. SRO GD112/17/4, 1.
23. For example, SRO GD50/136/2, 667; GD50/136/3, 1258; GD50/ 136/4, 1444.
24. SRO GD50/136/2, 432.
25. SRO GD50/136/3, 1013–1014.
26. SRO GD50/136/3, 1086.
27. In 1632 they came from Kinloch (Rannoch) [NN662 587], Drumchastle [NN684 588], Drumchin, Trinafour [NN725 646] and Dalchalloch [NN729 647]. SRO GD50/131.
28. *Register of the Privy Council of Scotland* (hereafter *RPCS*), 2nd Series, Vol. VII, 1638–1643, 177–178.
29. *RPCS*, 3rd Series, Vol. 1, 1661–1664, 118–119, 370–371.
30. The laird of Weem thus held his first post-Restoration barony court less than 2 months after the return of Charles II to Britain and 10 years after the previous court.

31. Although not all of these regulations have survived in the court records for the Appin of Dull, it is almost certain that they had applied there since the 1630s at least.
32. SRO GD50/136/1, 58–73; GD50/136/2, 551–552. The tenants in 1702 were clearly being urged to cut from the root, not the branches, of the tree.
33. SRO GD50/136/2, 520.
34. SRO GD50/136/1, 354; GD50/136/2, 476.
35. SRO GD50/136/3, 1008.
36. SRO GD50/136/1, 317; GD50/136/2, 512.
37. SRO GD50/136/3, 894.
38. SRO GD50/136/3, 1162; see, for example, GD50/136/2, 430–431; GD50/136/3, 1162–1163.
39. SRO GD50/136/3, 809.
40. See, for example, SRO GD50/136/2, 439.
41. SRO GD50/136/2, 821.
42. In each of the years 1711, 1718, 1719, 1724 and 1738 the court summoned at least 250 men from the Appin of Dull to answer wood-cutting charges.
43. The jougs were an instrument of public punishment consisting of a hinged iron collar attached by a chain to a wall or post and locked around the offender's neck (*The Pocket Scots Dictionary*, p. 133).
44. SRO GD50/136/4, 2621–2625.
45. SRO GD50/136/4, 1548; GD50/136/6, 2671–2672.
46. SRO GD50/136/2, 435–456; GD50/136/2, 506; GD50/136/3, 826.
47. SRO GD50/136/3, 1058.
48. SRO GD50/131; GD50/136/2, 453.
49. Those who cut wood but were absolved from the offence because they had liberty are still counted as wood-cutters.
50. SRO GD50/138/2/13. Although further wage payments have not been found, McInnes was still in the job as late as 1732.
51. SRO GD50/136/1, 45–46 GD50/138/2/71 GD50/136/2, 524.
52. SRO GD50/136/4, 1711.
53. SRO GD50/136/4, 1446; GD50/136/6, 2734.
54. SRO GD50/136/4, 1712.
55. SRO GD50/136/6, 2899–2901.
56. SRO GD50/136/6, 3033.
57. SRO GD50/136/3, 1609.
58. SRO GD50/136/2, 431.
59. SRO GD50/136/4, 1592.

CHAPTER 11
Woodland management and timber supply for ship masts in eighteenth century western Liguria (Italy)

Gaudenzio Paola[1] and Furio Ciciliot[2]

[1]Istituto Botanico Università, C.so Dogali, 16136 Genoa, Italy; [2]Centro Studi Attività Marinare, via alla Costa, 17047 Vado Ligure (SV), Italy

In 1709 a sale contract was made between a timber merchant from Sanremo and the Municipality of Pigna for the purchase of 3000 fir trees. This contract, together with a report prepared for the Princes of Piedmont by an official who surveyed the Pigna woods 1 year later in order to verify the progress of the sale and the availability of trees in the area, allows us to consider the type and management of woods in western Liguria in the early eighteenth century.

The Gulf of Genoa is, together with the Gulf of Venice, the area of the Mediterranean basin at the highest latitude. The main ridge of the Ligurian Alps and Apennines, developing parallel and generally very close to the coastal line, clearly separates the sea-facing slopes of the region from inland areas. Moreover, the coastal slopes are for the most part southerly exposed and are therefore protected from northern winds. As a result, coastal Liguria benefits from a milder climate than would be expected given its latitude (Paola et al., 1991; Barberis et al., 1992).

This coastal portion of Liguria is usually considered Mediterranean, and it can be delimited in a broad sense following the distribution range of *Olea europaea* (olive tree) (Paola and Minuto, 1996). The width of this Mediterranean belt is not uniform in the different parts of the region and, except for some special locations, it could not supply timber suitable for shipbuilding. Those woodland areas which could supply the shipbuilding timber (deciduous oak for the hull, fir for masts, beech for oars) could be found in the higher areas, where the influence of a central European climate is present.

The woods investigated in this chapter are on the southern slopes of the Maritime Alps, not very far from the Mediterranean Sea (Fig. 11.1). Their exact position is not clearly indicated in the ancient documents, but a critical reading of these, together with information derived from nineteenth century maps, enables us to place them on the slopes of Mt Alto (1269 m above sea level). Although not very high, these woods were situated in a mountain climatic belt and in the past beech (*Fagus sylvatica*) and fir (*Abies alba*) were probably the dominant trees. However, it is necessary to realize that in past centuries large

© CAB INTERNATIONAL 1998. *European Woods and Forests: Studies in Cultural History* (ed. C. Watkins)

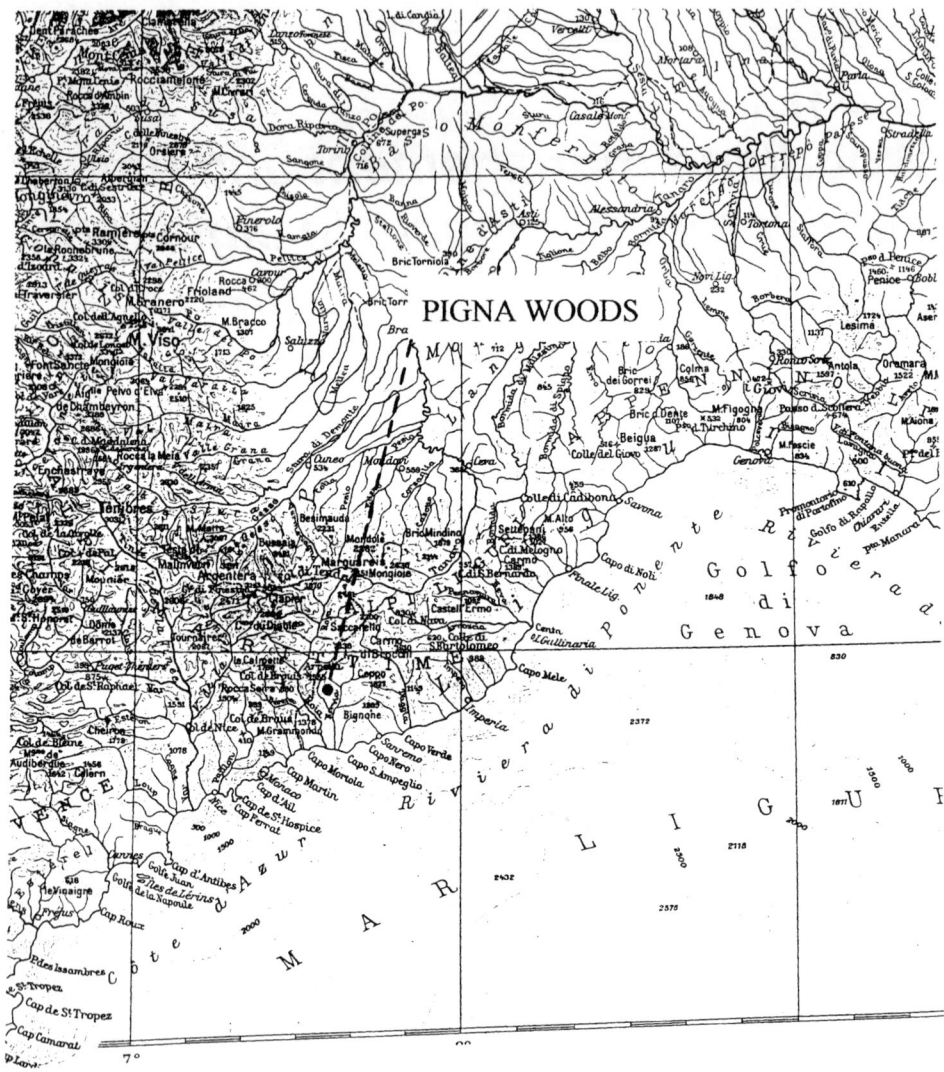

Fig. 11.1. Pigna woods on the southern slopes of the Maritime Alps.

areas of the Ligurian mountains were heavily exploited for pasture and livestock rearing was the most important economic support for the local population. Consequently, we can assume that first, the area of 'true woods' which could supply shipbuilding timber was probably not very large, and, second, human intervention over extensive areas would have reduced the number of trees per hectare and many sites would be similar to pasturelands with scattered trees.

The woodlands studied were the property of the Municipality of Pigna, the last important inland village in the valley of the River Nervia. The woodlands were far from the village, but in the past all inland villages of the valley, whose economy was based mainly on livestock breeding, held large areas on the slopes of Maritime Alps in order to obtain summer grazing. At the beginning of the eighteenth century, Pigna and all its property was located in the State of Piedmont. Consequently the woods from which conifer trees suitable for masts could be obtained were outside the territory of the Republic of Genoa, and under the jurisdiction of the Savoia, the Princes of Piedmont.

The mast is probably the most important single piece of wood for a sailing ship. The stresses which it has to support are enormous and only a few kinds of wood are suitable. Wood from coniferous trees, especially the fir, is the most commonly used for masts and lateen yards. In 1709 a timber trader from Sanremo signed a contract with the Municipality of Pigna for the purchase of the very large number of 3000 fir trees to prepare masts and lateen yards for ships (Vicario Di Tenda, 1710). As only two masts and some lateen yards were usually necessary for a medium-sized vessel, 3000 tree trunks would be sufficient for several hundred large vessels.

He bought the 3000 trees in bulk, with the freedom to cut down those which he chose. The fixed price included the construction of the roads needed to carry out the logs from the wood to a fixed site down in the valley, named Barbaira Bridge (Fig. 11.2). He paid 3610 lire; consequently, on average and irrespective of size, each tree was bought for 1 lira, 4 soldi and 1 denari.

About a year after this contract was signed a government official of the Princes of Piedmont was sent to Pigna woods to take a census of the trees suitable for shipbuilding. He surveyed Pigna woods on 10 October 1710. At this date only 300 trees for lateen yards had been cut down and carried away; another 350 trees had been cut down but were still lying in the woods. One year after the contract of sale, only 20% of the stated number of trees had been cut down and only 10% had been removed.

Of course, the entire operation was a long-term business for the timber merchant, from which he expected a very high gain, and the sale contract specified a period of 6 years in which to complete the job. Consequently, the felling and sale of the trees proceeded slowly, probably because the merchant wished to keep the prices high. So, more than 1 year after the date of the contract (26 July 1709), most of the trees were still growing in the woods.

The official of the Princes of Piedmont was interested both in the state of the contract signed by the Municipality of Pigna and in the number of trees suitable for shipbuilding in the area. As far as the contract is concerned, he pried into the timber merchant's financial affairs and he assessed the selling price of each kind of tree (see Table 11.1; the buying price and the approximate weight of each kind of tree has been added for comparison). He also calculated the carriage costs from the wood to Barbaira Bridge, and from there to Sanremo (see the first and the second values in brackets in the column 'carriage price' in Table 11.1).

The carriage price is more than the buying price, but the final selling price allows a very high gain. We can note the low price of a tree in the wood and how its dimensions were of no importance in the agreement of sale, although the difference between the smallest and the biggest trees is very high with the height varying from 12.5 m to 25 m and circumference from 25 cm to 375 cm,

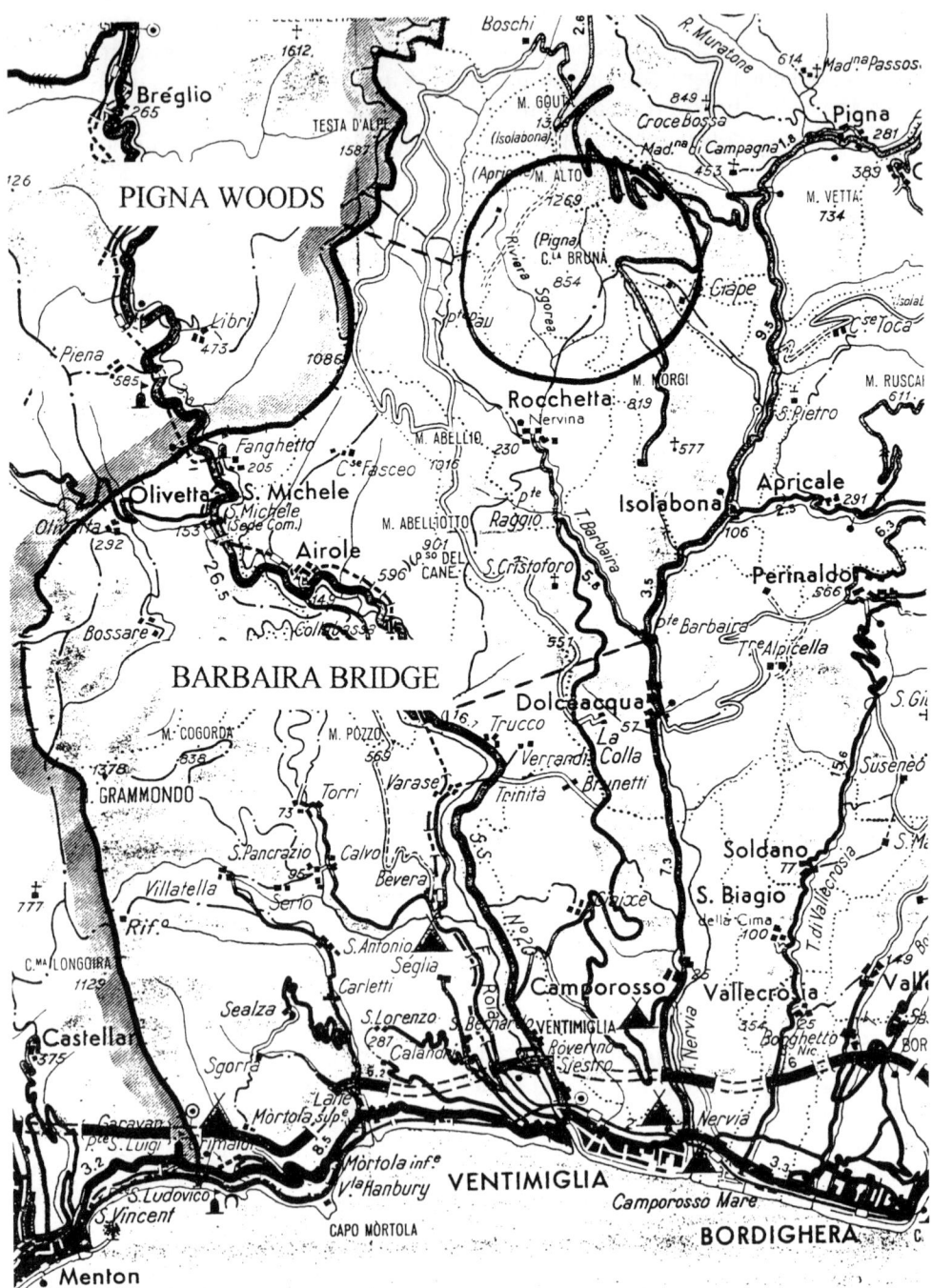

Fig. 11.2. Pigna woods (circle) on the southern slopes of Mount Alto, west of Pigna village.

Table 11.1. Measures, costs and weights of the trees from Pigna woods suitable for masts and lateen yards.

Height (m)	Circumference (cm)	Buying price (lire.soldi. denari)	Carriage price (lire.soldi)	Selling price (lire)	Approximate weight (quintals)
12.5–15.0	25–50	1.4.1	8.10 (6+2.10)	40	0.3
20.0–22.5	75	1.4.1	12.10 (10+2.10)	80–90	4
15.0	100–125	1.4.1	18 (15+3)	200	5
25.0	350–375	1.4.1	70 (60+10)	2000	108

with consequent variations in weight. It is interesting to see that the price of 2000 lire for the biggest mast is about the price of the hull of a 12 m long trade vessel of the tartana type.

Table 11.2 shows the results of the survey of the Pigna woods by the official of the Princes of Piedmont. According to this survey, the Pigna woods were fairly rich in timber and the sale of 3000 trees did not affect significantly their richness. Even assuming that the timber merchant probably cut down all the biggest trees, the data suggest that the amount of timber removed was not excessive from the point of view of sustainable forestry.

Today the vegetation of the area is a mosaic of pastureland and beech forest, pure or mixed with fir or, in certain areas, pure fir or Scots pine forest. It is likely that at the beginning of the eighteenth century the situation was very similar: although beech is not mentioned in the document this species was of little interest because large quantities were available much closer to the shipyards. Considering that the area concerned at Pigna is about 900 ha, the number of fir trees counted is very low, and even if we double the number of trees of different species it results in a very low density of trees per hectare. However, it is necessary to keep in mind that the area, even if known as Pigna woods, was devoted mainly to livestock rearing, and consequently it was probably more like a pastureland with scattered trees than a dense stand of trees. It can be assumed that the trees, scattered or in little woods, were maintained in order to obtain income from time to time by felling fir trees.

It is interesting that some circumferences are absent from the two sets of data reported by the official of the Princes of Piedmont. We think that as far as carriage and selling prices are concerned (Table 11.1), the official intended

Table 11.2. Availability of trees suitable for masts and lateen yards in Pigna woods, 1710.

Use	Circumference (cm)	Height (m)	Number of trees
Mast for ship (vascello)	300–375	25	325
Mast for galley (galea)	225–250	25	500
Mast for barche	150–200	17.5–20	7,500
Lateen yard for galley	50–100	15–17.5	10,000
Lateen yard for barche	25–50	10–12.5	2,000
Total			20,325

just to give general indications about the subject. The absence of some circumferences from the stock-taking related to the richness in trees of Pigna woods could be more interesting (Table 11.2). A rough mistake does not appear plausible in an otherwise extremely precise report. The second possibility is that no fir trees with circumferences of between 250 cm and 300 cm, or between 200 cm and 225 cm, or between 100 cm and 150 cm were present in the woods, but this hypothesis also appears to be hard to sustain. Another possible conclusion is that only trees with circumferences suitable for different types of masts and lateen yards were taken into consideration and that other trees were not included. Reports made by the same official for other woodlands are prepared in the same way listing only the trees with diameters suitable for masts and lateen yards.

One possible ecological consequence of this hypothesis is that many large trees would not have been felled. As the fir reaches its maximum commercial height and diameter at about 100–120 years (suitable for ship masts; see Table 11.2), a selected felling allows a repeat sale to be held every 30–40 years. This is a similar time interval between fellings as that used for the sustainable management of oak. Another consequence is that the density of trees in the area is rather greater than a superficial examination of the document might suggest.

In Liguria, as in much of Italy, the woodlands were the only suppliers of fuel and construction material and consequently their exploitation could not result in their destruction over the entire territory. Woodlands had to be managed, at least in some areas, in order to maintain a supply of energy. The documents we have considered suggest that the management of the Pigna woods left a lot of trees of various diameters standing. So we can see that in a mountainous area, formerly all covered with woods and later largely deprived of most woodland and changed into pasture, the woodland is managed in the eighteenth century in a shrewd way. We know nothing about the ecological sensitivity of Pigna people, but undoubtedly a selected cutting allows for the sustainable production of timber.

The Pigna woods, located in a mountainous area and far from the coast or any town of importance, could supply timber only for the masts of vessels, and the local laws and traditions were probably sufficient to preserve them. Many other Ligurian woods, in contrast, located near the towns and many industries such as potteries, iron-foundries and glassworks, needed to be preserved with the help of strong laws and the employment of special officials of the Republic, to avoid over-exploitation.

References

Barberis, G., Peccenini, S. and Paola G. (1992) Notes on *Quercus ilex* L. in Liguria (NW Italy). *Vegetatio* 99–100, 35–50.
Paola, G., Barberis, G. and Peccenini S. (1991) *Pinus halepensis* formations in Liguria (NW Italy). *Botanika Chronika* 10, 609–615.
Paola, G. and Minuto, L. (1996) Indigenous and exotic species as markers of the climatic

limits of the Mediterranean Region in Liguria (North Western Italy). *Proceedings of the International Colloquium. Mediterranean: Climatic Variability, Environment and Biodiversity,* Montpellier 6–7 April 1995, Maison de l'Environnement, Montpellier, pp. 180–184.

Vicario Di Tenda (1710) Lettera del Vicario di Tenda a Sua Altezza Reale concernente la vendita fatta dalla Communità di Pigna all'Abbate Sardi di San Remo di tre milla alberi per uso di barche et altri navigli da mare. Con uno stato delle qualità e prezzi di detti alberi esistenti nei boschi di Pigna. Archivio di Stato di Torino, materie economiche, caccie e boschi, 77, mazzo I.

CHAPTER 12
Afforestation policy of the Zionist Movement in Palestine 1895–1948

Nili Liphschitz[1] and Gideon Biger[2]
[1]Institute of Archaeology, [2]Department of Geography, Tel Aviv University, Ramat Aviv, 69978 Tel Aviv, Israel

Introduction

Afforestation activities are considered among the main geographical processes which have changed the landscape of Israel during the last 100 years. Several hundred years of neglect and destruction resulted in deforestation of the natural forests of Palestine. Written memories of pilgrims and travellers, who visited the country during the nineteenth century, describe the total desolation at the time and the absence of maquis and forest trees (Tristram, 1977). The renewed settlement of Palestine, which started with the immigration of the first German Templars to the Holy Land during the 1860s, and continued with the establishment of the Jewish settlements from the 1880s onwards, led to a prominent change in the attitude towards forest and afforestation concepts, and, consequently, to the planting of forest trees in various regions of the country. The main afforestation activities took place during the British Mandate period, between 1920 and 1948. During that time two main agents were involved in the afforestation activities of the country: the department of forestry of the British regime and the Jewish National Fund (JNF). British afforestation policy and activities have already been discussed by us (Cohen and Biger, 1987; Liphschitz and Biger, 1994a). Much of the forestry literature dealing with afforestation by the Jewish population in Palestine has been written for propaganda reasons. Several scientific publications have been published, most famous among them is a book written by J. Weitz (1970), the head of the forest department of the JNF. The first plantings undertaken by the Zionist Movement up until the First World War have been described in detail by Shilony (1990). Other research, concerned with afforestation activities and policy of the JNF after the establishment of the State of Israel, has also been published (Kliot, 1993). This chapter will focus on the afforestation policy of the Zionist Movement in Palestine up to the establishment of the State of Israel, especially during the British Mandate period. The aim is to present the reasons and motives behind a very important geographical–historical process which has changed the landscape of Israel.

First Ideas: the Afforestation Process During 1895–1907

The first forest trees were planted by the Jewish settlers with the establishment of the first settlements. Gardens and small groves were planted according to plans (Ben Artzi, 1988), but the main activity which led to planting of a true man-made forest was carried out at the settlement of Hadera. Hundreds of thousands of eucalypts were planted, in order to dry the Hadera swamps (Biger and Liphschitz, 1994). Although the planting was aimed to enable a rapid settlement in the area, it eventually resulted in a true man-made forest.

The idea to plant forest trees in Palestine by Jews was initially raised by the headmaster of the first agricultural school in Palestine, Miqwe Yisrael, in a conference organized at Jaffa in 1894. He thought trees should be planted to reduce the high costs of timber and improve the bad air, but his plans were not realized (Myoraq, 1992).

The leaders of the Zionist Movement, founded at 1897, who were not familiar with the situation in Palestine, were not convinced in the necessity of afforestation (Shilony, 1990, pp. 71–72). After becoming aware of the poor conditions of the country, J. Kremnizki, the future first director of the JNF, realized the urgency of its reafforestation. T. Herzel, the leader of the Zionist Movement, suggested the establishment of an organization that would raise money from donations and be involved in the afforestation process (Herzel, 1896).

In 1898 T. Herzel visited Palestine for the first time, and saw for himself the desolation of the country. After this visit Herzel asked Professor Otto Warburg, a famous botanist and an expert on settlement in tropical regions, to prepare a detailed survey on the plants of Palestine and offer settlement possibilities. One year later Warburg gave Herzel his survey (Warburg, 1948, pp. 77–79), which included data on the different climatic regions of Palestine, and details of the essential conditions demanded for tree growth in those regions. Warburg was convinced that afforestation would improve both the climate and the economic conditions of the country. He suggested various forest trees:

> the afforestation is very important; Primarily – *Pinus maritima* that grows near the Mediterranean coast (up to 400 m elevation), and also *Pinus halepensis*; It is worth trying also *Quercus ilex* and *Quercus suber* probably for the last species it would be too dry; It is also worth to try *Pistacia lentiscus, Liquidamber orientalis, Pistacia terebintus, Pistacia vera*, several plants of the *Labiatae* family that grow on the mountains and can be considered as perfume plants, and also various species of *Astragalus* that can supply traganth or laudan.

He suggested bringing to Palestine numerous species of fruit trees, forest trees and garden trees from Africa, Asia and Australia, in order to acclimatize them, and – finally – added a long list of other recommended species like palms, coconut, *Cedrus* and various decorative bushes. Zelig Soskin and Aharon Aaronson, agronomists who lived in Palestine, supported his ideas (Livne, 1969).

At that time the Zionist Organization activities in Palestine did not accept the concept of afforestation with non-fruiting forest trees. In August 1903, Otto Warburg suggested the making of terraces on the mountain slopes in order to plant Mediterranean fruit trees – grape vines, olives and carobs (Warburg, 1948, pp. 158–159). The olive, the most typical Mediterranean fruit tree, seemed the most suitable for afforestation, due to its wide distribution in the country, its adaptation to the local conditions, and the possibility of exporting its products. An organization, 'Olive Tree Donations', was founded to raise money and plant olives on lands purchased by the JNF but unsuitable for agriculture.

'Herzel Forest'

In order to encourage people to donate money for afforestation it was decided to organize an official committee under the name of T. Herzel, who died in 1904, and plant a forest of 10,000 olive trees in his memory. The organization looked for a suitable area for the 'Herzel Forest'. M. Berman was nominated as the director of a farm which would be established on the JNF lands. From the very beginning the Jewish experts who lived in Palestine, who were familiar with the problems of the area, the taxes, and the growth conditions necessary for olive cultivation, were against this method of afforestation. Criticisms of the enormous expenses involved in the establishment of the administration, and against the preferred conditions offered to Berman, were raised within various authoritative circles. On 15 December 1905 the head committee of the Zionist Organizations in Palestine sent a letter to the executive committee in Wien, protesting against the whole idea (Shilony, 1990, p. 76). These claims, based on aquaintance with the local conditions, were not accepted by the management that stayed abroad. On November 1906 the Palestine Committee decided to allocate 200 dunams near Hadera for olive plantations. This land had been purchased by the JNF 3 years previously. It was also decided to prepare a nursery for expanding the plantations in the future. Berman was nominated as responsible for the operation of these decisions. The first 18 olive trees were planted in February 1907 by David Wolfson, the head of the Zionist Organization, in the first grove of 'Herzel Forest', and the first tree was named 'Herzel Tree'.[1] A nursery of 2000 tree seedlings was founded at the site.

Later on, when it became obvious that the soil in the Hadera area was not suitable for olive growth, and the farmers as well as some of the leaders were against the idea of planting trees at a far distance from the settlements, Beit Arif (Ben Shemen), which was previously purchased by the JNF, was considered as a possible alternative for plantations. Since the area was designated for the graduates of the Qiriat-Sefer high school, it was decided to build a temporary nursery there. Thus, it would be possible to start the afforestation at any moment, and evacuate the area if and when it was demanded for another purpose.[2] The constitution of the 'Olive Tree Donations' was completed in January 1908. The propaganda for the promotion of money-raising was undertaken, aimed to enable the planting of 100,000 trees. On May 1908 Berman started to build an olive nursery for 'Herzel Forest' on the land of Beit Arif (Ben Shemen). The nursery was ready after 1 month, and 12,000 seedlings were planted there. Meanwhile the work in another site, Hulda, had started.

In view of the failures to settle Ben Shemen, and the danger of losing the ownership on the land, following its neglect, the JNF decided to use the stony grounds for the 'Herzel Forest'. The forests in Hulda and Ben Shemen were planned to carry 15,000 trees each, and concomitantly it was planned to plant a third forest in the Galilee (Warburg, 1948, pp. 132–133). On 11 April 1910 the final version of the contract between the JNF and 'Olive Tree Donations' committee was accepted, transferring the responsibility for the olive plantations to the JNF. Dr Arthur Rupin, the director of the Palestine Office of the Zionist Managenent, initiated and actually executed the plantations at Hulda and Ben Shemen (Rupin, 1937), with the money raised by the 'Olive Tree Donations' committee. Donations raised by the Jewish community of Dasau, Germany, enabled the planting of another area – the mountain slopes above Kinneret Farm[3] – called 'Dasau Forest', to start in spring 1911.

Change in the Afforestation Policy

The great hopes and expectation concerning the afforestation enterprise were not fulfilled. Already during summer 1911 suspicions were raised that the actual costs of the planting of each seedling would be much higher than estimated, and that the 'Olive Tree Donations' committee would not be able to plant the number of trees promised to the donors. In April 1911 Dr M. Bodenheimer, one of the leaders of the JNF, sent a letter to the Executive Committee, in which he claimed that the calculated budget for planting per single tree was not realistic. He also attacked the high expenses involved in the building of a management-house in Hulda, and the extremely high costs involved in the afforestation activities.[4] At that time a criticism against the conditions of the plantations was raised. In a report prepared by A. Aaronson, at the end of May 1911, concerning the activities undertaken at Hulda and Ben Shemen, he criticized the poor conditions of the olive plantations and nurseries at both places, and claimed that the costs per olive tree were higher by 50% than in the original budget. Another criticism was the use of Arab workers in the afforestation activities in 'Herzel Forest'.[5] On 7 August 1911, in a meeting of the General Assembly of the JNF in Basel, Bodenheimer again attacked the olive plantations. David Wolfson, the president of the Zionist Organization, supported Bodenheimer, and criticized the policy of Warburg, who ignored previous suggestions and the comments (Levontin, 1925). It was decided to pass over the management of the afforestations to the JNF. Consequently, Bodenheimer, who was the head of the JNF, became the manager of the 'Herzel Forest' enterprise. After he was convinced that the profits from olive trees would stay low even after maturity, in spite of their fruit production, he decided on a basic change in the afforestation policy. After his visit to Palestine, in 1911, he wrote:

> This visit of mine to Palestine was most important for me for several reasons. While staying in the country I became convinced that the olive plantation, which was so highly recommended, was a mere failure. Only in a mixed economy has it some value. Nevertheless, for such an economy we must offer outstanding conditions, and a lot of money, which makes olive production unprofitable. Moreover, one cannot speak on forests in Palestine in the concept of a European forest. After Dr. Rupin's advice I decided to make a drastic and a principal change. According to my advice only eucalypts, cypress, and gopher [probably *Cypress* spp.] trees would be planted, trees which can improve the landscape and the climate. Such forests demand investment of money for the long run, but are blessed. Such an enterprise can be undertaken only by a public organisation. Therefore I see in it a must-mission of the JNF.[6]

A real change in policy, due to the economic failure of the olive plantations, had taken place. From that period onwards plantations of forest trees replaced the plantations of olives and other fruit trees. This prominent change had an economic and a national meaning. Future economic profits were given up, and replaced by fulfilment of national targets, concerning improvement of the landscape and climate: money was invested in afforestation projects, planned for large areas, and missions were undertaken by a public organization, whose aims were not necessarily economic, but national (Shilony, 1990, p. 94). This change took place after the nomination of Warburg to the president of the Zionist Movement in the summer of 1911. Although at first he was in favour of olive planting, Warburg accepted the change in the afforestation policy due to circumstantial reasons, and later supported the new policy in public. Already before the outbreak of the First World War, about 250 dunams of pine had been planted at Hulda, and hundreds of pine and cypress trees were planted at Ben Shemen. Numerous eucalypt trees were planted near Kinneret and in the environs of Merhavia and Degania, in order to dry flooded areas adjacent to those settlements.[7]

The first experimental research works had started at the agricultural research station. Experiments to acclimatize various exotic forest trees from abroad were made during the period 1911–1914 by Aharon Aaronson in his station at Hadera. He also looked for potential forest trees in neighbouring countries, and carried out research projects on forest trees that were locally extinct. He laid the foundations of an afforestation policy which aimed to avoid extinction of existing species and to protect the soil. He presented his ideas in a publication entitled 'On the Deforestation and Afforestation of Palestine' which he gave to the Ottoman Governor of Beirut on his visit to the station on 1913 (Aaronson, 1913).

By the eve of the First World War an afforestation policy had been accepted by the Zionist Movement. This policy is still in place. Afforestation was considered as a positive way to fulfil the Zionist plan to change the landscape of the country and settle it with Jews. The JNF, which was previously involved only in land purchase for settlement purposes, became actively involved also in afforestation of land unsuitable for agriculture. Afforestation was aimed to keep the ownership of these lands, and to supply work for unemployed Jewish immigrants. The change from planting of fruit trees to forest trees resulted in a shift of the afforestated areas to the mountain regions (and later to the northern Negev desert), where agriculture could not succeed. The main afforestation project of the Jewish population in Palestine at that time, afforestation of the Hadera region, was not continued by the Zionist Movement, because it was finally understood that only drainage of the area, and not eucalyptus plantations, would dry the swamps (Biger and Liphschitz, 1994).

Afforestation Policy at the Beginning of the British Mandate

The First World War prevented proper care of the fruit plantations in Palestine, and the trees did not survive, whereas the forest trees did not suffer from the neglect and absence of treatment, and did survive. When the war ended the agronomist Aqiva Etinger was nominated as the acting manager of the JNF and the Settlement Department of the Zionist Organization in Palestine.[8] Etinger accepted the viewpoint of Bodenheimer and tried to compensate the tree-donors, and declared that afforestation would be made

> Not by fruit trees, which are not forest trees, but by barren trees which are by their nature forest trees; barren trees would be planted not in valley soils, which are suitable for growth of intensive agricultural crops, but only on gravel hills and stoney mountains
>
> (Weitz, 1970, p. 186)

This concept was a guideline for the afforestation activity of the JNF. The idea was to plant 'numerous forest trees, which grow rapidly, and do not demand a long treatment; nevertheless these trees would be planted in soils which are not suitable for agricultural purposes, except for afforestation, i.e. stony soils, swamps and moving sand dunes' (Weitz, 1970, p. 99).

Immediately after his appointment, Etinger laid the foundations of a wide-scale afforestation project by the establishment of nurseries. Already by 1919 about half a million seedlings for plantations were raised, in different nurseries, in various places of the country. Experiments of direct sowing with *Pinus pinea* seeds were carried out during the same period, without using seedlings. This method had previously been tried by the German Templars on Mount Carmel. The afforestation project was concentrated in three regions: the mountains, the swamps and the sands. Afforestation methods included planting seedlings in swamps and sands, as well as sowing combined with planting in the mountains. In 1920 the afforested areas covered 1800 dunams in the mountains, swamps and sands. About 275,000 different trees were planted, and 350,000 *Pinus pinea* seeds were sown, all in all 625,000 trees. During that year above a million seedlings were raised for the afforestations of 1921.[9] In July 1919 Josef Weitz was invited by Etinger to join him, in the development of fruit-tree plantations in the agricultural settlements, and in the afforestation project for the JNF (Weitz, 1970, p. 99). Until his retirement 45 years later in 1966, Weitz had a great impact on afforestation.

First Steps: 1918–1928

During the period under discussion special attention was paid to afforestation policy and planting. At the beginning, the concept of planting fruit trees was still accepted as one of the aims of the comprehensive afforestation enterprise.[10] The afforestation project was intended to renovate the landscape of Palestine by planting on mountain-slopes on the one hand, and to dry the swamps and improve the air of the valleys on the other. Afforestation was also considered as an economic project, intended to develop fruit-tree plantations for export and supply work for the unemployed Jewish immigrants. The forests were looked upon as a future source for construction timber, instead of importing it from abroad. Moreover, the afforestation enterprise was also regarded as an educational mission, through which Jews would be trained for agricultural work, while their income would contribute to the development of education and cultural life in the future.

Some of those targets were achieved during the period under discussion, while others were never fulfilled. Nevertheless, long-term planning greatly influenced the character and shape of the plantations. At the beginning of 1921 the special importance of the afforestation for the country was recognized, and the determination of new afforestation policy as well as the methods of performance were promoted.[11]

> The forests have an important function in the human history, from the very beginning till now. Country with many forests is blessed, becoming attractive to people to come and settle, since forests have three main conditions essential for settlement: soil, air and water. Miserable is the country that had become deforestated and her trees cut down. The absence of forests disturbs the harmony of natural creation and offers room for evil powers, which corrupt the air against the local inhabitants. When Israel came to inherit the land they were ordered by God on planting ... and our predecessors fulfilled this mission very intensively ... However, since the exile of Israel, the country became desolated; Her enemies eroded her from her tree-garment, until she turned to be more and more devastated. For hundreds of years till today, the country is totally naked.

This romantic point of view is more imaginary than realistic. Recent investigations have shown that Israel was never covered by true forests, similar to the European ones, due to its climate, which is not conducive for the growth of dense tall trees (Liphschitz, 1986, 1988; Liphschitz and Biger, 1990). However, the viewpoint of the Zionists during the 1920s had a great impact on the afforestation enterprise.

Etinger was convinced, like others,[12] that during antiquity the country was covered by forests, as in Europe. Only man's interference with the environment resulted in its deforestation, and consequently brought about 'soil disaster' by erosion of the mountain terraces. Etinger offered various solutions for different soil types (sands, swamps and mountains):

> the sand area along the Mediterranean coast is about 500,000–600,000 dunams (50,000–60,000 hectares) ... afforestation of sands is important for several reasons. But it confronts huge technical difficulties, high costs for planting of

> each dunam, and demands patience, consistency and persistency in order to stop the dunes from moving into the country and cover agricultural areas. For a long period of time a wrong concept on swamp afforestation was accepted. The Jews in the Diaspora and in Palestine were persuaded that plantations will dry the swamps. This is wrong. Even the eucalypt is not capable of drying the swamp. Swamps should be dried by drainage via trenches and pipes ... the drying by drainage turns the swamp to a very fertile soil ... since now onwards nobody will use the swamp-soils for forests. This does not mean that there is no need to plant in the future forest trees on fertile soils ... but this will not be the afforestation routine in the country. Most of the area – millions of dunams – waiting for future afforestation is found in the mountains. One of our main targets, as a nation of immigrants who are coming back home, is to afforestate the mountains of Judea, Samaria and Galilee.

He also raised the question of cooperation with the British regime in the protection of residual forests and new plantations. Etinger considered the afforestation enterprise as an additional task of the JNF included within the main goal of the Zionist Movement, the purchase of lands.

> JNF must consider the afforestation as one of the national missions that has to develop progressively, and more intensively than till now. We must release ourselves from the fears, that by involvement with the afforestation of our country we risk the main task of the JNF – raising enough money for the purchase of lands ... this is one of the most important ways to develop and strengthen the relationships between the Diaspora Jews and the purpose to improve the local conditions and facilitate the settlement. Tree-donations till now are collected by chance ... the time has come to organise in the Diaspora, by the local JNF departments, groups or com[m]ittees of friends of the national forests in Palestine.

According to his view '30% of the soils in European countries are covered by forests. If we will attend to the situation so that at least a tenth of this percentage of our national land will be covered by forests, our sons will be delighted'. This viewpoint expresses the popular thoughts and moods that affected the people who were involved in the future afforestation activities. A prominent change is obvious from the romantic concept of large-scale afforestation projects, as in Europe, towards a realistic target – to afforest about 3% of the area. This aim was achieved only by the State of Israel, at the beginning of the 1970s, when about 600,000 dunams were already planted. The idea of drying the swamps by afforestation was neglected. The main afforestation activities were concentrated in the mountains. Small-scale afforestation projects were carried out in a limited area along the coastal plain, in order to stabilize the moving dunes, and stop them from covering the adjacent fertile regions. He also recommended cooperation with the British regime and with the Jewish settlers, who started to take part in the conservation of the forest and its expansion by wide-scale plantations.

Otto Warburg, whose nomination as the president of the Zionist Movement ended in 1920, supported the afforestation concept on general lines, as presented by Bodenheimer and Etinger. In his speech before the 12th Zionist Congress, in 1921, he claimed that afforestation of the country, and especially of the mountains, was most important for the preservation of ground-water. According to his viewpoint the afforestation was aimed 'first to dry the swamps by planting of eucalypts, secondly to plant trees on the mountain slopes and third – to afforestate the numerous sandy regions.' However, he considered this enterprise as too heavy for the JNF that was involved in numerous other projects. As a well-trained botanist he was able to demonstrate how to plant trees on sand dunes (Warburg, 1948, pp. 144–145).

The Palestine Joint Survey Committee, under the direction of Lord Melchet, that visited Palestine in 1928, paid only little attention in its printed report to the subject of afforestation. The committee came to the conclusion that mountain afforestations would contribute to the development of agriculture by (i) avoiding soil erosion, (ii) uptake of runoff water, and (iii) the strengthening of welling forth. The forests would beautify the country and conceal the destruction caused by man to the landscape.[13]

A survey of the existing plantations in the JNF forests at that period shows that during the first years of planting, until 1925, only non-fruiting trees were planted. These included numerous species, mainly eucalypts, acacias, casuarinas, pines, cypresses, etc. Until 1920 pine trees constituted only 4.4% of the plantations (altogether 645 trees), while during the next 6 years (1920–1925) pine trees constituted 21% of the planted specimens, constituting altogether 62,882 trees. At the same time, i.e. until 1920, eucalypt trees constituted 77.9% of the forest trees, and in the next years, i.e. 1920–1925, its share attained only 53% of the plantations.[14]

These afforestation experiments led to the decision to plant mainly a single species – *Pinus halapensis* (Aleppo pine) – due to its successful establishment and rapid growth, especially in the mountains and on calcareous soils. From 1926 onwards, with the increase in the mountain lands purchased by the JNF, a prominent change in the composition of the plantations was prominent. The eucalypt was replaced by Aleppo pine, which become the dominant species, constituting above 85% of the planted trees each year. The failure of the sand-dune afforestation in Rishon le-Zion reduced to a great extent the number of eucalypts planted by the JNF (Weitz, 1970, pp. 162–164; Liphschitz and Biger, 1995). The planting of eucalypts in swampy areas stopped. From that period onwards the JNF forests became single-species forests, dominated by Aleppo pine (Liphschitz and Biger, 1994b).

Afforestation Policy – Spaces and Tree Species During the Period 1928–1948

In the second and third decades of organized afforestations in Palestine the planting of forests was accelerated and experiments to acclimatize numerous exotic plants were continued. The experiments with *Pinus brutia* were successful, and the tree adapted itself to various soil types. Moreover, planting of *Pinus halapensis* and *Pinus brutia* of the same age under the same conditions showed that *Pinus brutia* attained greater height and developed a straight trunk, whereas *Pinus halapensis* trees were shorter with a bent bole. Nevertheless, the percentages of *Pinus brutia* as a forest trees were very small until the 1940s (Biger and Liphschitz, 1991). Experiments conducted with *Pinus canariensis* were also successful. The tree had a rapid growth rate, developed a straight trunk and a nice canopy. Moreover, the tree developed as coppice both after cutting and after burning of the bark and branches. In spite of these values it was claimed that this species 'is unable to inherit *Pinus halapensis* because of its soft wood, which is of little value for the industry, because of its narrow canopy ... and its resistance to calcareous soils'.[15] Another pine species that was investigated was *Pinus pinea*, which was planted on Mt Carmel, at Qiriat Anavim and Ben Shemen in 1919–1920. The expenses involved in obtaining seed and its slow growth rate, together with its failure on calcareous soils, minimized its planting. Other conifers that were examined included *Cupressus sempervirens* – var. *pyramidalis* and var. *horizontalis*, *Cupressus arizonica* and *Cupressus macrocarpa*. *Cupressus sempervirens* was familiar, and was already successfully raised. It was claimed that var. *pyramidalis* produced small amount of xylem and therefore was not suitable as a forest tree, but the share of var. *horizontalis* in plantations was also very small. Casuarina proved very successful due to the rapid growth rates, wide canopy and easy growth in nurseries and adaptation to the area. In spite of all these advantages Casuarina was taken off the list of forest trees, because: 'the tree is confined to stoney mountain soils', and is not too successful on gravel-beds.

Experiments were also conducted with broadleaves. *Ailanthus altissima* and *Melia azedarach* seemed successful at the beginning, but developed quite slowly, and did not prevent the development of an herbaceous underlayer. The same held true for *Robinia pseudoacacia*. Celtis trees, which grow naturally in Palestine, were not successful in competition with *Pinus halapensis*. *Parkinsonia aculeata* seedlings had some difficulties in adaptation, but the mature specimens succeeded as fence-trees and were planted to mark the borders of forests. Experiments were carried out with native forest species – the oaks. However, their slow growth rates and the need to weed the underlayer made them difficult for plantations. Tests with *Acer negundo* and *Sophora japonica* were a complete failure, and the experiments with *Dalbergia sissoo* were inconclusive. However, the carob (*Ceratonia siliqua*) was introduced as a fruit tree to the plantations. In his conclusions Weitz wrote as follows:

> summing up this paragraph, i.e. choosing the tree species, I have to admit that until now the list of tree species suitable for afforestation is very limited. Among conifers – mainly *Pinus halapensis* and also *Pinus brutia* among broadleaves – mainly carob but also nuts, and *Parkinsonia* as a fence tree.

Neither *Pinus brutia* nor *Cupressus sempervirens* var. *horizontalis*, which proved successful for plantations, were considered important for afforestation. The same was true of other species like *Pinus canariensis* and *Pinus pinea*. Those who determined the afforestation policy were deeply impressed by the merits of *Pinus halapensis*, and consequently the forests became monocultures.

After 10 years of afforestation experiments, including species diversity and planning of plantations, afforestation policy was determined during the 1930s and 1940s, and turned out to be an important Zionist enterprise in itself. In a lecture '25 Years of Forest Formation'[16] given by Weitz in 1945 he presented the main afforestation guidelines. In addition to his declaration that *Pinus halapensis* was the preferred species and the mountains were the preferred region, he declared that:

> the forest is a creation for itself. A cultural creation and a cultural need. The forest is a harmony of colours, shadows and notes ... in the forest one can find tranquility and an environment which brings one closer to God ... one may sometimes find there something more important than food ... the JNF was determined to plant this forest in our country, a forest with three main trends: soil — agricultural, settlement — national and creation — social.

At that period the forest had the additional political–legal mission to obtain ownership of the land against the claims of the Arabs who settled on the lands purchased by the JNF. This mission was fulfilled by afforestation on Mount Carmel immediately after the First World War, and afterwards in the eastern Ezdraelon Valley, where tens of thousands of trees were planted along the plots which were purchased during the period 1923–1924. After the settlements were built and the demands for land ownership were fulfilled, the trees were cut down and the plantations were replaced by agricultural fields.[17] The use of afforestation in order to assure ownership on land was used very often in various districts of the country until the end of the British regime. This policy determined the planting of the forest near Mishmar Haemeq and the large areas in Zvulun Valley during the 1930s.

The creation of sources of supply for the local timber industry was achieved by planting eucalypts and pines. Many settlements were built at that time, and afforestation was used as an important form of employment for the early settlers. Forest nurseries, which were established in many settlements, offered work to the settlers for a long period of time. Such nurseries flourished at Tel Yoseph, Mishmar Haemeq, Hanita, Rosh Pinah, Eylon, Qiriat Anavim, Kfar Hahoresh and Haifa Bay. Mesheq Hapoalot (women worker farms) in Jerusalem, as well as in Tel Aviv, functioned as central nurseries at the beginning of the 1920s. The settlers of Kfar Hahoresh started as foresters at Shimron Hills. Employment at Qiriat Anavim was first based on afforestation activities, which terminated after the financial improvement of the settlement. The settlers at Nveh Ilan, which was founded in the autumn of 1946, were mainly employed in afforestation activities, as were the settlers of Maale Hahamisha which was established during the mid 1930s (Weitz, 1970, pp. 162–164).

In many cases forests were planted after it became clear that the soil was not suitable for agriculture, and that was the only way to maintain the Jewish ownership on the land. The same happened in areas of the western Ezdraelon Valley, near Nahallal, which were found to be unsuitable for agriculture. The Mishmar Haemeq forest was planted after it became obvious that about 8000 out of 18,000 dunams were unsuitable for agriculture. Another forest was planted near Genigar for similar reasons, after it was found that more than half of the land was unsuitable for agriculture.

Forests were planted from the early 1920s as memorials with money donated by Jewish communities in the Diaspora. These forests created continuous connections between the Jews in Palestine and abroad. The first planted memorial forest was Herzel Forest, but its failure did not encourage further experiments. Memorial forests became popular after the first successful projects of the JNF. In 1924 the Calvin family from Brazil donated money in order to plant a memorial forest. The impressive eucalypt forest was planted at Kfar Mallal on 500 dunams. Planting ended in 1927, and about 50,000 eucalypts, including *Eucalyptus resinifera*, were planted. However, this forest was cut down several years later, due to the development of settlements in this region (Liphschitz and Biger, 1994a). Nevertheless, the Calvin forest started a tradition, which continues, of memorial plantations, among which the most popular were the Balfour and King George V forests in Nazereth mountains.

The multi-purpose concept of the forest was defined during the British Mandate period. Afforestation of devastated regions, which are unsuitable for agriculture, was, and still is, the main target. Forests were sometimes also created for other reasons – economic, national and political. This concept is still the guideline for afforestation policy today.

Criticisms of the Afforestation Policy

Several protests were raised against afforestation targets, policy and methods. It was claimed, for example, that the JNF did not cooperate with the British regime and the Arabs.[18] Criticism against the principles of the JNF afforestation was expressed in a letter, published by S. Lavee, in the newspaper 'Davar'.[19] The writer criticized the afforestation of the mountain lands, which he thought suitable for settlement, and the creation of memorial forests that would avoid in the future their change into settlements. The director of the JNF forest department argued in response that much land was suitable only for afforestation. Moreover, he argued, forests produce immediate income and provide benefits to local agriculture; this was true for the whole world – and the same held true for Palestine. Any suggestion to plant a memorial forest is first considered by JNF experts.[20]

Many experts argued against the creation of monocultures of *Pinus halapensis*. In 1938 Gindel (who was for a long period the director of the station for forest experiments of the Zionist Organization) argued that mixed forests were superior.[21] A similar criticism against pine monocultures was raised by Professor Bodenheimer in 1939. Bodenheimer, a famous entomologist, came to his conclusions as a result of his numerous experiments on *Matsucoccus josephi* bast scale, which caused heavy damages to *Pinus halepensis* trees.[22] He summed up the subject as follows:

> It is now obvious that the disease raises numerous questions concerning the present afforestation policy. After the disease has spread to such extents, I doubt

whether we shall succeed to raise monocultural, densely planted Aleppo pine
forests ... We have to examine very seriously the possibility to replace *Pinus
halepensis* by other pines like *Pinus pinea* or *Pinus brutia* and to consider the
possibility to plant mixed forests ... We have to try mixed plantations of pines
with deep-rooted trees as oaks and carobs ... we have to end two phenomena:
fires and the disease. Both caused heavy damages to our forests during the last
years and both testify that we have to change our present afforestation policy.

The afforestation experiments carried out in Palestine over two decades led
Weitz to continue planting monocultures of Aleppo pine despite increasing
concern voiced by critics.[23][24] In his report *Afforestation Policy in Palestine*[25]
Weitz wrote that many examinations were carried out concerning the
introduction of numerous tree species, various afforestation methods,
improvement of natural forests, regional studies, and treatment of new
plantations. 'All these examinations were focused on one purpose: to find the
proper way or the safest and best methods to obtain the main goal – the
afforestation of the country.' These experiments determined the afforestation
policy in most of the country. Weitz added that:

> This conclusion does not necessarily means that there is no room for other
> experiments for the improvement of the methods ... but when we are coming
> now to lay the foundations to the afforestation policy in Palestine we have to
> learn from the available data ... If we wish to continue with a wide scale
> afforestation enterprise we must continue according to out experience, but to a
> greater extent. I cannot claim that we will never change anything in this policy
> in the future, but today the afforestation policy is based on the available
> information and not on the unknown.

Gindel's (1952) critique of the afforestation policy was based on his long
experience as a forester and a forest-researcher. A special chapter of this book
is devoted to the subject of 'Forest character and composition'. Gindel presents
the two afforestation concepts which were accepted among the foresters
during the last 200 years: the German concept which took into consideration
the economic, but not the ecological demands, and the French concept which
favoured natural regeneration. Conifers were preferred over broadleaves, and
therefore the Germans tended to plant monocultural coniferous forests.

> After several generations they succeeded to obscure the character of the forests,
> and instead of a normal composition constituted of 71% of broadleaves and 29%
> of conifers, the majority of specimens were conifers. Consequently, forest disease
> spread rapidly, which destroyed vast forest areas. The French foresters succeeded
> to preserve the natural forests, and avoid the tragic shocks that the German
> forests have passed. The river Rhine on the border between France and Germany
> functions as a border for the spreading of the pests.

Therefore, according to Gindel,

> the lesson of Europe and the United States can guide our attitude towards the
> afforestation of our country ... the forest composed of broadleaves should be the
> dominant one, due to the permanent danger of fires which threatens us for 6–8
> months each year. Conifers are burning much easier than broadleaves.

After a fire in a conifer forest, 20 years are required until this forest will attain its original dimensions. If conifers are burnt before ripening they should be sown or planted again. On the other hand, the renewal of a broadleaved forest is much quicker both after fire and after cutting. Young seedlings regrow easily by root-coppicing. He concludes:

> Planting of numerous species in great numbers, when mixed among each other, or on small patches, is important both for the protection against fungi or pests, which usually attack one or two species, and their progress is stopped on confronting a new species. These facts bring us to the conclusion that it is essential to plant a mixed forest composed of two thirds of broadleaves, of various species, planted or sown in a mixture. It is obvious that the adaptation of the various species to soil and microclimatic conditions should be taken into consideration, according to ecological principles.

In spite of these conclusions afforestation with monocultures of Aleppo pine continued for many years.

Expenses and Results of the Zionist Afforestation

The afforestation activities of the JNF between 1920 and 1948 resulted in the planting of about 3.7 million trees over an area of 2000 ha. These afforestation schemes of the JNF, although heavily publicized, were not by any means among the main operations of the Zionist Movement or JNF. Only 5% of the total budget of the JNF was spent on afforestation between 1921 and 1926.[26] This approach did not change in later periods. In a report published by the Zionist Executive and the Jewish Agency, presented to the 20th Congress of the Zionist Movement in Basel in 1937, less than a single page was devoted to afforestation compared with 600 pages of the total report, and 36 pages devoted to the JNF. Only 5% of the annual budget of the JNF was spent on afforestation in both 1934 and 1935.[27] During this period the JNF spent around 90% of its budget on land purchase.

At the end of the period under discussion the afforestation enterprise was presented as an important source for timber and fuel supply, and as a source of income for the settlements which were involved within the afforestation and the maintenance of the existing forests. Recreational developments within the pine forests and new hotels on Mount Carmel and the Judean Mountains were also based on JNF and private forests (Gurevitch and Gertz, 1947). The afforestation of the country started as a national enterprise with an economic aspect. The need to change the afforestation targets and the tree species planted was accepted long before the outbreak of the First World War. The change from planting fruit trees to planting forest trees started in 1918.

This policy, determined mainly during the British Mandate period, influenced the aims and methods of afforestation even after the establishment of the State of Israel, although the new spaces, financial sources and the targets of the free country influenced the process more and more. Today, about 80,000 ha of land, comprising 5% of the area of Israel, are covered by plantations established mainly by the JNF.

Acknowledgement

This research was carried out with the financial support and encouragement of the JNF.

References

Aaronson, A. (1913) *On the Deforestation of Eretz Israel and a Suggestion for its Renovation*. Submitted according to invitation by His Majesty Backri Bay, the General Governor of Beirut, on occasion of his visit to the agricultural research station, in October 1913. Beit Aaronson Pub., Zichron, Yaakov. (In Hebrew.)

Ben Artzi, Y. (1988) *Jewish Moshava Settlements in Eretz Israel (1882–1914)*. Yad Izhak Ben Zvi, Jerusalem, pp. 186–187. (In Hebrew.)

Biger, G. and Liphschitz, N. (1991) Early distribution of *Pinus brutia* – a reassessment. *Holocene* 1, 157–161.

Biger, G. and Liphschitz, N. (1994) Australian trees in Eretz Israel. *Research in Geography* 14, 108–121. (Hebrew).

Cohen, A. and Biger, G. (1987) British mandate actions to preserve the country's landscape and nature. In: Schiller, E. (ed.) *Zev Vilnay Jubilee Book, Vol. 2*, Ariel Pub., Jerusalem, 295–300. (Hebrew).

Gindel, I. (1952) *Forest and Afforestation in Israel*. The Laboratory for Afforestation Research, Rehovot. (Hebrew).

Gurevitch, D. and Gertz, A. (1947) *Statistical Handbook of Jewish Palestine, Jerusalem*. The Jewish Agency of Palestine, Jerusalem, p. 187.

Herzel, T. (1896) *The Diary. Vol. 1*, a note from 23 August 1896, 334. (Hebrew).

Kliot, N. (1993) Ideology and afforestation in Israel – man-made forests of the J.N.F. *Research in Geography* 13, 87–106. (Hebrew).

Levontin, Z. (1925) *'Le'eretz Avotainu' (To the Land of Our Forefathers)*. Eitan and Shoshani, Tel Aviv, p. 111. (Hebrew).

Liphschitz, N. (1986) Overview of the dendrochronology and dendroarchaeology in Israel. *Dendrochronologia* 4, 37–58.

Liphschitz, N. (1988) Dendrochronological and dendroarchaeological investigations in Israel as a means for the reconstruction of past vegetation and climate. *PACT* 22, 133–146.

Liphschitz, N. and Biger, G. (1990) The dominance of *Quercus calliprinos* (Kermes oak) – *Pistacia palaestina* (Terebinth) association in the Mediterranean territory of Eretz Israel during antiquity. *Journal of Vegetation Science* 1, 67–70.

Liphschitz, N. and Biger, G. (1994a) Afforestation policy of the British Regime in Palestine. *Horizons in Geography* 40–41, 5–16. (Hebrew).

Liphschitz, N. and Biger, G. (1994b) The rise and fall of *Pinus halepensis* (Aleppo pine) as a main afforestation tree in Eretz Israel. Division of Publications of the JNF, no. 24. (Hebrew).

Liphschitz, N. and Biger, G. (1995) Sand dune reclamation in British Palestine. In: Simmons, I.G. and Mannion, A.M. (eds) *The Changing Nature of the People–Environment Relationships: Evidence from a Variety of Archives*. Prague, pp. 47–55.

Livne, A. (1969) *Aharon Aaronson*. Bialik Institute, Jerusalem, p. 71. (Hebrew).

Myoraq, Y. (1992) Conference of the New Settlers in Jaffa 1894. *Cathedra* 66, 144–167. A note by Nyago on 162. (Hebrew).

Rupin, A. (1937) *Thirty Years of Building in Eretz Israel*. Jerusalem, p. 42. (Hebrew).

Shilony, Z. (1990) *Jewish National Fund and Settlement in Eretz Israel 1903–1914*. Yad Izhak Ben Zvi, Jerusalem, pp. 71–96. (Hebrew).

Tristram, H.B. (1863–1864) *The Land of Israel. A Journal of Travels in Palestine*. Copyright by Bialik Institute 1977, Israel.

Warburg, O. (1948) *Warburg Book*. (Hebrew).

Weitz, J. (1970) *Forest and Afforestation in Israel*. Massada, Israel. (Hebrew).

Notes

CZA — Central Zionist Archive, Jerusalem

1. 'The World', First year, copy no. 13, 27 March 1907, p. 138, and also copy no. 11, 13 March 1907, p. 131. (In Hebrew.)
2. Report from 4–5 January 1908, p. 17, File KKL/1/586, CZA. (In Hebrew).
3. Files KKL/1/578 and KKL/1/586 p. 100 in CZA. (In Hebrew).
4. A letter in File Z/2/646 in CZA. (In Hebrew.)
5. File Z/2/646 in CZA, no. 21, note no. 17. (In Hebrew.)
6. A document from 26 July 1921 in File KKL/5/1932, in CZA. (In Hebrew).
7. Files KKL/1/578 and KKL/1/586 in CZA. Reports on April 1911. (In Hebrew.)
8. A survey in File KKL/3/136 in CZA. (In Hebrew.)
9. Etinger, A., 'Herzel Forest' and Afforestation of the Country. A survey published in August 1920 in File KKL/3/1932, and also a document in File KKL/3/136. (In Hebrew.)
10. The National Library of the JNF, 'The Beginning of the National Afforestation', in File KKL/5/4285, CZA. (In Hebrew.)
11. A document published on 4 February 1921, in File KKL/3/136 CZA. (In Hebrew.)
12. A document in Etinger Archive A/111/21, CZA. (In Hebrew.)
13. Report on the Joint Palestine Survey Commission, London, 1928, p. 59.
14. Reports in File KKL/5/1929, CZA. (In Hebrew.)
15. Hassadeh, Vol. II (3), 1940. (In Hebrew.)
16. A lecture on 11 January 1945, in File KKL/5/13936, CZA. (In Hebrew.)
17. Acccording to planting reports of various years. Various files in KKL/5 in CZA. (In Hebrew.)
18. A letter from Weitz to Zinger, on 17 November 1929, in File KKL/5/4671, CZA. (In Hebrew.)
19. Lavee, S., a letter published on 15 May 1935 in the newspaper 'Davar': 'Don't let the afforestation be uncontrolled'. (In Hebrew.)
20. Letter in File KKL/5/7312, CZA. (In Hebrew.)
21. Letter of I. Gindel in File KKL/5/10524, CZA. (In Hebrew.)
22. Letter on 30 October 1939, in File KKL/5/10524, CZA. (In Hebrew.)
23. 'Hassadeh' Vol. II (3), 1940. (In Hebrew.)
24. A document from 24 November 1940 in File KKL/5/11825, CZA. (In Hebrew.)
25. A document issued by the JNF – the main office, Jerusalem 1940. (In Hebrew.)
26. Report of the Executive of the Zionist Organization to the XIV Zionist Congress, Vienna 1925.
27. Report of the Executive of the Zionist Organization to the XX Zionist Congress, Zurich 1937. Jerusalem. 1937.

CHAPTER 13
The expansion of the forest and the defence of nature: the work of forest engineers in Spain 1900–1936

Eduardo Rico Boquete
Universidade de Santiago de Compostela, Departamento de Historia Contemporánea, Praza da Universidade, I, 15703 Santiago de Compostela, Spain

Spanish forestry politics changed rapidly in the early part of the twentieth century. There were several administrative developments including the forest assessment carried out for the Catalogue of Forests in Public Use of 1901. This provided a great deal of information about the extent and type of forests and allowed plans to be made for the improvement of forests and the control of the environment. There was an increase in the number of magazines and publications, such as *The Forest Review, Forests and Industries, The Country, Forestry Renewal, Timber Industry, The Gamekeeper, Forests and Rivers*, which helped to promote and popularize the idea of forestry. Moreover in 1907 the first forest research station was established and this encouraged links with other European research stations and organizations.

Various new pieces of legislation, such as the Conservation and Encouragement of the Forest Law of 1908 and the Reafforestation Law of 1926, were introduced in this period to support forestry. In addition, and very importantly, the Defence of Private Property Law of 1918 limited the ability of private landowners to damage their woodland. It was introduced to control the huge tree fellings which were being carried out during the First World War. In 1924 this law was made more permanent and it was reaffirmed in 1933.

In practice it has to be pointed out that the main impulse given to reafforestation was the idea of creating large plantations of economically productive timber trees. These forests were promoted through an association between the Forest Administration and the Provincial County Councils. Special attention was paid to soil erosion and protection forestry and new Hydrological–Forest Divisions were created. Among their principal aims was the repopulation of dune areas (Guardamar, Rosas Gulf, Guadiana and Guadalquivir), the restoration of mountain streams (The Pyrenees, Central System) and the protection of several river basins (Ebro, Júcar, Segura, Guadalentín and Lozoya).

In this period serious concern was shown for the environment and one result of this was the establishment in 1917 of the first two Spanish National

Parks. Natural Areas of National Interest, with rather lower levels of protection than National Parks, were also introduced. In the same year it was decided to create a Catalogue of Notable Trees which would be protected by law. This process of awakening, the formation of a conservationist spirit, was brought about by forest engineers and other social groups convinced that there was a need to propagate a culture of admiration of trees and nature as a key to a healthier and freer society attached to the environment. The instigation and extension all over Spain of 'Arbor Day' (Fiesta del Arbol), from 1898, is a good example of the work developed by forest engineers such as Ricardo Codorníu and Rafael Puig Valls.

We can examine the general trend through the study of what happened in a specific geographical area, Galicia. Situated in the north-west of Spain, with Atlantic weather, it has an area of 29,432 km^2, the same as Belgium, with extensive experimental plantations of such species as *Pinus pinaster, Pinus insignis* and *Eucalyptus globulus*.

Popularization, Advertisement and Protection

The development of the new environmental forestry can be traced in the large number of newspaper articles and advertisements arguing in favour of woodlands. Forestry, industrial and scientific professionals proclaimed the necessity of restoring forests because they were a future source of progress and welfare, for public health and good weather, for the creation of jobs and for national wealth. The Agricultural Congress of 1906 had already reached the conclusion that it was necessary to start a replanting and afforestation process to improve the current situation. It also asked for the creation of a forestry school, similar to the ones in Germany, because the School for Forest Engineers at Madrid was the only forest school in the whole of Spain. Similar conclusions were also reached in all the various congresses and meetings held in those years. The enthusiasm for forestry can be demonstrated by three events: the restoration of the Arbor Day, the formation of the Catalogue of Notable Trees and the declaration of Protected Spaces.

Arbor Day

The first Arbor Day was held in Madrid in 1896. The celebration arrived in Galicia thanks to forest engineer Rafael Areses in 1910. From 1904 onwards it started having an official status and from 1914 it was declared compulsory: its aims were educational, instructive and aesthetically functional. It was held in order to try to educate youngsters to love and admire trees. At the same time leisure and picnic areas were created to be enjoyed by all the citizens. The organizers were forest engineers and technicians, teachers and other professionals; sometimes these activities were organized by public institutions, sometimes by private groups. Only limited public funding was available and many private individuals assisted. Arbor Day was always on a Sunday or a bank holiday so that more children and schools could take part. It took place between October and May, the wettest months, and in public places where there was a need for woodland. The species used, given by official nurseries free of charge, were very varied. Exotic trees such as Acacias, *Acer pseudoplatanus*, *Populus* and *Eucalyptus* were very popular because of their quick growth and their visual attractiveness.

The most important participants were schoolchildren who received presents and light meals, planted trees, and sometimes freed pigeons or balloons. The organizers tried to keep that day in the children's minds as an important event in order to instil concepts of the conservation of nature, trees, birds and so forth. The political authorities also attended, and it is interesting to note that the more important the Fiesta became the more they attended. Eventually the Fiesta ended up being a shop window in which politicians demonstrated their particular campaigns. This fact hit the original aims of the Fiesta hard and as a result it lost its spontaneous, instructive and amusing character. Eventually this development weakened the Fiesta and caused a negative effect on many people, especially during General Primo de Rivera's dictatorship (1923–1930).

To promote the Fiesta and spread the idea that woodlands were beneficial, Societies of Friends of the Tree were created; in addition other associations of land workers listed the annual celebration of the Fiesta in their statutes. By this means, the Fiesta spread rapidly: in 1910 there were only three Fiestas in Galicia; 3 years later there were 22, with more than 500 celebrations all over Spain; by the 1920s there were more than 100 Fiestas in Galicia. To sum up, the Fiesta had rather a late start in Galicia but it soon developed as an important means of instigating in children a love for trees, a respect for nature, and knowledge of the importance of the fight for the recovery of lost things. It was an example of a new naturalistic, conservation sense which emerged in Spanish and Galician society and which owed a lot to the work of forest technicians.

Catalogue of Notable Trees

In 1917 a catalogue of large, ancient trees was established (Decree 23-2-1917). This included trees characterized by their large dimensions and peculiar shapes; they were recognized as historical monuments which needed special protection to avoid being felled, burnt or attacked by fungi or disease. They were of different species and most were native (*Taxus, Quercus pedunculata, Quercus ilex, Castanea vesca* and so forth). The idea of the list was to protect the trees and to display them to the public in the interests of education and tourism. Although some of the trees were destroyed during General Franco's dictatorship (1939–1975), the majority of the catalogued trees still exist and are in a good condition. Some examples are illustrated in Figs 13.1–13.4.

Catalogue of Protected Areas

Under legislation introduced in 1916 relating to the protection of nature, unusual areas with an important landscape value were to be protected in a

Fig. 13.1. *Quercus suber*, A. Estrada, Pontevedra; circumference 9 m. E. Corbelle (amateur), 1996.

Fig. 13.2. *Quercus suber*, A. Estrada, Pontevedra; circumference 6 m, height 7 m, canopy diameter 30 m. E. Corbelle (amateur), 1996.

Fig. 13.3. *Pinus pinea*, Tomiño, Pontevedra, 1909. Circumference 5.25 m, height 25 m, canopy 425 m^2.

Fig. 13.4. *Castanea vulgaris*, Viana do Bolo, Ourense, 1909. Circumference 14.1 m, height 15 m.

special way. Forest engineers started developing a list of the most important places in each province which met the conditions necessary to be declared Natural Areas of National Interest. The Republican Constitution of 1931 confirmed that 'The Government will also protect the notable areas for their natural beauty or for their recognized artistic or historical value' (section 3, chapter 2 and article 45). This gave a big boost to the protection of the environment. The result was that the forest engineers made a list of more than 20 places which had the necessary conditions to be declared places of great national interest.

From 1931 to 1935 (Decrees 7-6-1931 and 5-7-1935) four places situated in the provinces of Pontevedra and La Coruña were declared Natural Areas of National Interest. They covered more than 300 ha and were on or near the coast. Their management was entrusted to the district forest engineers, to the local authority and to the administrative offices of the National Parks; in practice the management was by the forest engineer. During the dictatorship years, from 1939 to 1975, there was a lack of interest in environmental subjects and no new places of national interest were established. It was not until the new constitution of 1978 that the protection of nature again became a statutory obligation of the government.

Social Participation

In addition to the work of the forest engineers and professionals, various social organizations whose aim was the protection of nature were established in this period. Examples include the Society of Friends of the Trees in La Coruña and in Ortigueira, the Society of Friends of the Trees and Birds in Lugo, the Lovers of the Fields, the Lovers of the Galician Landscape, the Excursionist Society Lyceum and so on. In this respect the role of Galician emigrants in America is very important. Many Galicians had emigrated because of the lack of work and prospects at home but this did not necessarily result in a separation between the emigrants and their homes. Many emigrants tried to keep a close relationship with their families and home region and fought to improve the cultural level of their old country by, for example, building schools and teaching centres. They were also concerned about the protection of the countryside and its embellishment; perhaps the great distance made them value the landscape that they had abandoned.

The work carried out by emigrants' societies in La Habana and Santiago de Cuba (Cuba), in Buenos Aires (Argentina), and in Montevideo (Uruguay) was particularly remarkable. For example, the society *Pro Montibus* Santa Tecla in Tui (Pontevedra), which promoted archaeological digs and forest restoration, was mainly formed by emigrants living in La Habana, Puerto Rico and other places in the Caribbean. The work carried out by La Unión Hispano-Americana was also important because it built some schools and surrounded them with trees planted on Arbor Day: a perfect union between public education and nature conservation. In general all the societies had the intention of 'Celebrating Arbor Day every year according to the universally established manner'. In other cases, teachers boosted the creation of children's societies called Mutualities through which children developed an interest in natural history and drew up nature conservation plans.

Forest Protection

In 1918, as a consequence of the huge fellings made during the First World War, strong legal controls over the management of private forests were established to prevent further devastation. Provincial Commissions for the Conservation of Private Forests were created, presided over by the Provincial Governor with the forest district engineer as secretary and with representatives of wood industries and forest owners as members. The law did not contemplate a complete ban on the fellings but stipulated the way in which fellings should be carried out. It was forbidden to fell a whole wood at one time which resulted in a reduction in the number of tons cut yearly, and it also required the owner to replant the area felled.

The effectiveness of this measure was limited and its extension difficult. For one thing, insufficient funds were made available and there was a lack of suitable personnel in the forest administration to ensure the law was obeyed. Many owners opposed the law because they did not want their profits reduced or delayed; the measure was also disliked by sectors of the wood industry and wood exporters. Of course, in the 'public' forests which belonged to the villages and were managed by the government through districts, the level of state control, including the management of traditional woodland uses such as grazing and firewood collection, depended on the district forest engineer, who would try to stop people from felling more trees than they were allowed.

Other forest protection measures were developed in areas difficult to afforest. Thus, the reafforestation of certain areas where there had been important landslides (Las Ermitas, Ourense) was tried, as was the afforestation of coastal dunes (O Bao, Pontevedra). The aim was to restore the 'pre-existent' vegetation in order to stop the erosion of fertile lands and thus protect agriculture and land workers.

Reafforestation

The creation of large-scale plantations was always the forest engineers' main objective. The public forests were in a very poor state with few trees, and the forest professionals and other social groups made every effort to establish new plantations and stop what they perceived to be the slow deterioration of the forests.

The first step was to establish tree nurseries capable of supplying the necessary plants and seeds. Until 1906 there were no official nurseries in Galicia, but in that year one was created in Areas (Pontevedra), and from then until 1936 six permanent official nurseries were created in addition to some temporary ones (Fig. 13.5). The nurseries were places for experiments and the acclimatization of new species from America, Australia and other European countries. These plants were to be used later on for planting on Arbor Day and for making private and public plantations. The success of the nurseries was due to the hard work of the forest technicians; government support was minimal and the forest professionals did everything they could to assist. Engineer Rafael Areses even established a non-official forest museum in the Forest House of the tree nursery in Areas; this was in addition to his attempts to create a 'Memorial Forest' in the face of serious financial difficulties.

The result of all this work was the start of afforestation in 1910. This was a slow process, with many ups and downs and with considerable differences between the provinces, but every year more land was planted. Perhaps more importantly vital preconditions for the later large-scale afforestation of Galicia, such as the creation of provincial afforestation plans and the establishment of afforestation associations, were set in place. Between 1910 and 1925, 12,000 ha of public forests were planted in addition to considerable areas of afforestation on private land.

Conclusion

At the beginning of the twentieth century, forestry experiments in Spain and Galicia heralded very important changes in forest conservation. The practical and popularized work of forest engineers began to show results and these were

Fig. 13.5. Areas Nursery, the first nursery in Galicia, 1913.

demonstrated through a series of educational events, festivals and forest activities. This work had a big influence on society and made it more receptive to plans for the conservation of nature. Forest assistants and guards played an important role in boosting general knowledge about forestry and the particular tasks and practices required to develop and protect woodland. Forest conservation was greatly assisted by the establishment of a forest administration and the work of forest officers.

In the early part of the twentieth century many parts of Spanish and Galician society had an interest in and affection for woodlands. This was shown by the appearance of numerous organizations and different publications pledged to the protection of the forests. This sensibility was spread throughout all the social classes, in the city and in the country, inland and on the coast. Emigrant organizations played an especially important role because, despite being far away from their land, they fought for its conservation and for the improvement of children's education. These associations provided an ecological conscience for their time. The proliferation of books and articles, of meetings and congresses, of fiestas and professional magazines were other examples of the strength and vitality of new ideas about forestry.

Woodland was considered an inexhaustible source of richness, providing raw materials for industries, pasture for cattle and profits for land workers. Afforestation was also justified by the jobs which it generated. Thus forestry was seen as meeting ecological, economic and social criteria, a view not dissimilar to that espoused, for example, in Britain at the same time (James, 1990). This enthusiasm for forestry was to some extent discredited by the politicians who made used of festivals such as Arbor Day for their own ends.

Franco's dictatorship resulted in a clear step back for forest and nature conservation. The repressive system eliminated most social organizations including those which had been established to help conserve the forests. Only one professional forestry magazine continued to be published and this was completely controlled by the regime to the extent that there was no chance for differences of opinion or open debate. This journal helped to concentrate the work of forest engineers on productive forestry which benefited a particular industrial sector to the detriment of conservation. Further indications of the decline in nature conservation are that no new National Areas of Natural Interest were designated between 1940 and 1975 and that in the same period several of the Notable Trees owned by the government were felled.

References

Balboa, X.L. (1990) *O Monte en Galicia*. Edicións Xerais Universitaria, Vigo.
Bauer Manderscheid, E. (1980) *Los Montes de España en la Historia*. Ministerio de Agricultura, Madrid.
Corvol-Dessert, A. (1987) *L'homme aux Bois. Histoire des Relations de l'Homme et de la Forêt, 17–20 Siècles*. Eduard Fayard, Paris.
Fernández Leiceaga, X. (1990) *Economía (política) do Monte Galego*. Universidade de Santiago de Compostela, Santiago.

Fernández Prieto, L. (1992) *Labregos con Ciencia. Estado, Sociedade e Innovación Tecnolóxica na Agricultura Galega.* 1850-1939. Editorial Xerais Universitaria, Vigo.

Gómez Mendoza, J. (1992) *Ciencia y Política de los Montes Españoles (1848–1936).* Icona, Madrid.

Groome, H. (1990) *Historia de la Política Forestal en el Estado Español.* Agencia del Medio Ambiente, Comunidad de Madrid, Madrid.

Groome, H., Martín Montalvo, R. and Llorca, A. (1989) Historia forestal de España. *News of Forest History.* 9–10, Vienna.

James, N.D.G. (1981) *A History of English Forestry.* Blackwell, Oxford.

Villares Paz, R. (1982) *La Propiedad de la Tierra en Galicia, 1500–1936.* Editorial Siglo 21, Madrid.

CHAPTER 14
The promotion of participation in planning for soil and water conservation through reforestation: a case study of Guadalajara (Spain)

José D. García Pérez
Department of Environmental Management, University of Central Lancashire, Preston PR1 2HE, UK

Introduction

This chapter assesses the capacity of the existing governmental forestry agency in Guadalajara Province, and rural community organization to promote the planning and implementation of reforestation projects. The experience of Rapid Rural Appraisal and Participatory Rural Appraisal methods in planning reforestation, mainly in developing countries, are used to argue for their suitability to increase the level of success of such plans in a European context. The case of Guadalajara is examined in a historical perspective, to be able to detect to what degree past planning trends still impinge on the contemporary ability of the existing governmental forestry agency, to protect abandoned agricultural land through reforestation under the 'set-aside' programme.

Since the 1800s the case made by foresters for the protection of soil in Spain has been gaining greater relevance and influence in policy making. Forestry sciences, and the arguments made by foresters for the control of erosion were supported by the political and ideological structure of the State sanctioning technocratic strategies.

Technical approaches are generally seen as solutions to land degradation with little attention being paid to its socio-economic processes or to the problematic of planning and implementation through participation. This chapter examines how past trends in reforestation planning still impinge on the ability of the present forestry agency; it proposes an alternative approach in planning to promote the participation of land users as a means of reducing the rate of project failure. The methodologies of research are those of physical geography, written history and the collection of oral history data using anthropological methods of fieldwork.

Historically, land users in Spain have been perceived by foresters as the problem, rather than the solution, to land degradation. Influential forestry engineers, such as Codorniu (1915), who were greatly concerned about the deteriorating condition of land, argued in defence of forests, and put the blame on land users and their alleged blind irrational hatred of trees (p. 50). Although Codorniu was inaccurate and unjustified in blaming land users, he envisaged the need to gain the assistance of rural people for reforestation. Codorniu proposed to do this through education in schools, distributing seeds and plants among children, the propagation of birds, the encouragement of the creation of forestry societies, tax reductions for slow growing natural forests, and using State forests as examples of good management. These can be regarded as early Spanish examples of innovative proposals to promote the participation of rural people in conservation. These explanations and proposals, however, were not widespread among foresters during the first part of the century, and became less so after the take-over by the military regime in 1939.

The First Influence – the German School of Forestry

The influence of technical forestry measures can be divided into two, the first one from the nineteenth century until the 1950s and the second thereafter. The first technical influence reinforcing the view of reforestation as an engineering project, rather than as a project of rural development encompassing socio-economic dimensions in planning, derived from the German School of forestry. The *Escuela Especial de Ingenieros de Montes* (Spanish Forestry School) was created in 1848 with the purpose of finding a scientific or technical solution to the problem of deforestation and erosion. This School was formed after two natural sciences lecturers were sent to Germany to study forestry sciences.

The technocratic approach to reforestation which emerged as a result of the use of 'imported' forestry techniques, disregarded the value of ecologically well-adapted natural vegetation and the local knowledge of land users, in favour of fast growing tree species and 'scientific' fixes. Natural Mediterranean vegetation and land use techniques are well adapted to withstand the high summer temperatures and unpredictable rainfall. These are very different to those experienced in northern European countries such as Germany. Yet forestry techniques based on northern European circumstances were taught and unsuitably applied to Mediterranean conditions in Spain.

The classification of some forest areas as *montes de utilidad publica* (mountains of public utility, or 'protectors') in 1931 with the intention of stimulating protective management and reforestation, was an achievement of foresters in changing legislation and protecting forests. After 1931 legislation compelled private owners of land classified as 'of public utility' to sustain its permanent production and fertility. This, as recommended by foresters, often required reforestation to be carried out by private land owners. Failure to do so gave the Government the automatic right to expropriation. The 1931 law was, however, rarely put into effect before 1940. The high cost of expropriation, lack of political will, and the economic, social and political repercussions of a reduction in the amount of land available to marginal land users, would have made expropriation politically unattainable.

The main objective of the State forestry agency after the Civil War was to provide the politically influential and burgeoning wood manufacturing and pulp industries with cheap raw material of mainly coniferous wood. According to Groome (1990), profit was put before sound ecological considerations. The State forestry agency, Patrimonio Forestal del Estado (PFE) or at a later stage the Instituto para la Conservación de la Naturaleza (ICONA), were in charge of the technical and administrative aspects of reforestation.

Earlier recommendations from the 'German School', by foresters such as Burgers (1949), writing in the main Spanish forestry journal *Montes*, were ignored. Burgers' warnings were based on findings of the acidification effects of the soil by fast growing conifer species experienced in Holland and Germany. The advice he gave to Spanish foresters, to use mixed species for quality wood production and soil protection (p. 505), was not taken up.

Although the objective of the planting of conifers has been portrayed as mainly for protection, the argument about how much it is for that, or for wood production, is tenuous. As an example of the dual character of such claims Sagasta Azpeitia, a prominent Spanish forester, expected reforested areas (classified as predominantly for protection when managed by ICONA) to provide 35% of the national wood supply (Sagasta Azpeitia, 1979, p. 179). Claims for the need to protect water catchment areas sanctioned dubious reforestation projects for wood production, and this is a continuing strategy. In the case of extremely degraded areas which have been recolonized and protected by *matorral* for some years, it can be assumed that the need for speedy protection no longer arises and that wood production is the real 'hidden' objective of reforestation.

Successive governments after 1939 implemented a reforestation policy which disregarded the proposals of some foresters for restoration of existing vegetation, such as those made by Ceballos and Ximenez in 1939 (Groome, 1985). Representative of the Government's ideology after 1939 were Martinez Hermosilla who, according to Groome (1990), promoted the consortium arrangement for reforestation, and Foyo Panaleon who, adopting the fascist concept of the 'social function' of property, had a less tolerant attitude towards land users' duty to reforest their land for the benefit of the Nation (p. 62). Even well before that, another forester, Garcia Maceira in 1915 argued that pursuing individual interests converted forests into a means of production for the market, which neglected their protective functions (Groome, 1985, p. 83).

The implementation of a policy for the extensive planting of conifers, which previous governments had not attempted, was possible because of the disregard for the social and economic effects that it would have on local populations. González Bernáldez (1990, p. 442) explains that excessive abuses and authoritarian rule during the Franco period (1939–1976) created great resentment in rural areas which still persists today.

The majority of foresters, as a reflection of the political climate after 1939, favoured the planting of fast growing species for wood production. From 1939, the government of the Franco dictatorship was isolated from international trading, and was also bankrupt. This precluded it from purchasing wood in international markets, and in response a policy of economic autarky emerged, as described by Tamames (1976, Vol. II, p. 22). Groome (1981), explains that after 1940 foresters trying to achieve 'spectacular' reforestation results for job promotion within the administration, opted for massive reforestation (p. 31) with single species of conifers. Prieto Rodríguez (1995), explains that the choice of single species of fast growing conifers for reforestation was the result of lack of technical knowledge in the management of mixed plantations, and the pressure exercised by powerful groups interested in the production of wood (pp. 17, 20). Achieving the maximum possible rent from reforestation was openly defended by many foresters, such as Abreu y Pidal (1963), as a principal objective. This required 'efficient' forms of planting techniques and the maximum return in the shortest period using machinery and conifers. The next section evaluates the validity of technocratic methods as solutions to land degradation problems.

The Second Influence – the American School of Forestry

After the transfer of modern methods from the German School, a second wave of 'modernization' saw Spanish foresters being sent to the USA to learn what was perceived to be 'highly' scientific soil management and reforestation techniques. This second technical influence in the 1950s reinforced the view of reforestation as engineering work, rather than as a project of rural development.

The use of methods based on quantifiable measurement, such as the USLE, gave foresters greater scientific standing and credibility – and disregarded land users' knowledge even further. Montoya Oliver (1983) refers to the short-sighted vision of Spanish foresters who, blinded by modernization tenets, snubbed integrated agropastoral methods and the knowledge of land users (p. 34). Such scientific methods gave respectability to political decisions for reforestation. Hijacking scientific findings 'to disguise the political nature of planning decisions by making them appear scientifically or technically inevitable' (Richards, 1981, p. 8), is a common occurrence which enhances the apparent validity of reforestation policy in Spain.

Using heavy machinery for the construction of terraces was another technical method adopted by foresters in the pursuit of modern more 'cost-effective' forms of planting. This type of technology and approach was 'exported' to Spain via visits by influential people such as Bennett, the Head of the US Soil Conservation Service (SCS). Terracing techniques had been tried in the US before. To rectify damaging practices the SCS demonstrated Soil and Water Conservation (SWC) measures to farmers in the US. The application of terracing, tried in Navajo Indian reservations, and technicians' disregard for 'locally adapted and appropriate technologies ... provoked an intense negative reaction, not only to SWC but also to all government programmes' (Pretty and Shah, 1994, p. 6). The technocratic approach, Baker (1984) warns, sees the problem as 'a physical one ... amenable to a technical solution.' Rather than as 'symptomatic of a social and political crisis' (p. 53). It can be argued that in the

Spanish the belief in the 'superiority' of technical approaches, which still persists among many foresters, has been underpinned by the 'demonstration' of peasant's irrationality and ignorance. The social and political crisis was one the Government of the time triggered in forcing reforestation against the will of land users.

The reasoning and type of Spanish–USA cooperation strongly disregarded the reasons for land users' practices. During a 3-month research visit to Spain in 1956 on behalf of the International Cooperation Administration of the USA, Bennett studied the problems of erosion and admired the erosion control achieved by Spanish foresters. Commenting on landowners' reticence to give up their land to reforestation, he concludes that, although understandable, in the long term they would be better off by paying more attention to SWC in gentler sloping land than wasting their 'time and energy on land obviously unfit for farming' (Bennett, 1960, p. 72). In response to this opinion, the land users' view may have been that land is unfit for farming only if it cannot provide them with a sustained livelihood. Baker (1984) argues that the persistence of the technocratic approach 'even in the face of its self evident ineffectiveness' is because of its ability 'to shift the blame for its own failure onto its victims ... [and] tighten the screws on the marginalised poor' (p. 53).

The type of advice provided by Bennett to land users does not explore the reasons why SWC measures are not implemented. It implies land users' ignorance, and proposes as a solution greater conservation work by land users assuming this would yield a profitable return to their investment (financial and labour time) in SWC.

The Drive to Plant Conifers

The grandiose ideologies typical of fascist Spain largely ignored socio-economic and ecological circumstances. This was reflected in the PFE plans of 1952 to plant between 100,000 and 150,000 ha annually. This mentality has, to a certain extent, endured changing economic strategies and ecological concerns. Lingering technocratic measures continue to be devised for the plantation of fast growing species such as conifers. According to Groome (1990), in 1976 Martinez Hermosilla (an influential forester with financial interests in the wood pulp industry; Rico Boquete, 1995, p. 84), proposed the reforestation of 2 million ha in 10 years, and in 1986 there were proposals for the urgent reforestation of 5.5 million ha (p. 115).

Plans for reforestation continue today with the aid of the EU. The EU contemplates the reforestation of 1 million ha of agricultural land in Spain under the 'set-aside' programme, from 1993 to 2012 (Prieto Rodríguez, 1995, p. 19). The region of Castilla-la-Mancha (in central Spain) expects to reforest 132,000 ha between 1993 and 1998 and double its reforested area (1.9 million ha) in 60 years (López Carrasco, 1994, pp. 51, 57). Proposals for the form of reforestation vary from the continued use of mainly conifers, to those of mixed plantations, accounting for some change in perspectives. Araujo (1994), calls for a carefully planned reforestation, using techniques which are least disturbing to the existing vegetation, at a rate of 50,000 ha per year for 25 years (p. 60). Plans to reforest with mixed species of trees indicate changes in policy. However, the ambitious pace of reforestation, akin to past grandiose expectations, is not easily attainable.

One of the main arguments for Spanish foresters choosing fast growing and frugal species, such as *Pinus nigra* and *Pinus pinaster*, has been that they are able to succeed in highly degraded soils. Although this is a valid reason for these type of conditions, plantations with these species have been employed indiscriminately even in areas where soils are not too degraded, or in areas which have been stable for many years under *matorral*.

The Socio-economic and Physical Outcomes of the Planting Policy

Because the processes of land degradation are physical *and* socio-economic, solutions require a broad perspective of alternatives, including those advocated by land users. Planning for reforestation, to control land degradation or for wood production is not purely a technical matter but one of rural development. The planning of rural development require socio-economic assessments. In the short term, plantations imply the exclusion of pastoral and agricultural activities, so reducing the amount of land available to land users. In the long term, conifer species can have inhibiting effects on the growth of grasses or other vegetation which may be valuable to local economies. In the light of this, this section discusses how the control of erosion using mechanical methods of reforestation affected the condition of land and the ability of land users and foresters to work together.

Ignoring land users' circumstances, knowledge and perceptions, amounts to an exclusively technocratic reasoning, provoking distrust with detrimental repercussions for the promotion of participation and the success of project implementation. This Spanish reforestation policy based on technocratic solutions had three main outcomes. These are illustrated in the rest of the chapter by making use of data gathered in fieldwork in the Guadalajara Province (Fig. 14.1) between June 1993 and April 1994. The area of fieldwork is shown in Fig. 14.2.

The first outcome – the limited protection of conifers

The first outcome of the reforestation policy is that it did not provide better protection against erosion than if land was abandoned and recolonized by *matorral*. Although the policy of reforestation was defended on the grounds of the 'more advanced technical knowledge' of forestry science, in reality the use of scientific assessment in planning was minimal. Fieldwork tests and measurements which can be associated with reforestation for SWC were absent in all the reforestation projects shown in Fig. 14.2. Examination of the

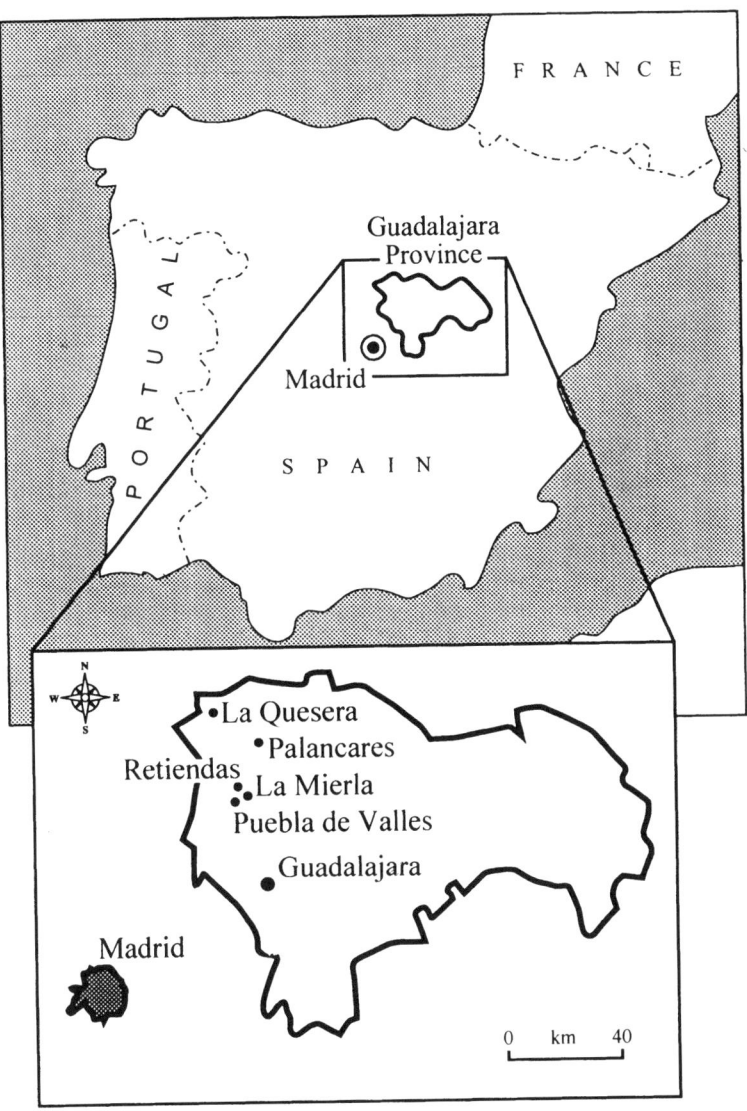

Fig. 14.1. The Guadalajara Province.

documentation of these reforestation projects shows the great dearth of physical data and socio-economic assessment of the conditions for reforestation. Examples of the projects under greater scrutiny in fieldwork are shown in Table 14.1.

The utilization of bulldozers for the construction of terraces in the area of study is an example of the growing popularity of mechanical/technical methods of planting, starting in the late 1960s and early 1970s in Spain as a result of the American influence. The construction of terraces, explained by Coelho *et al.* (1994, pp. 3–4), involves the creation of a new slope following

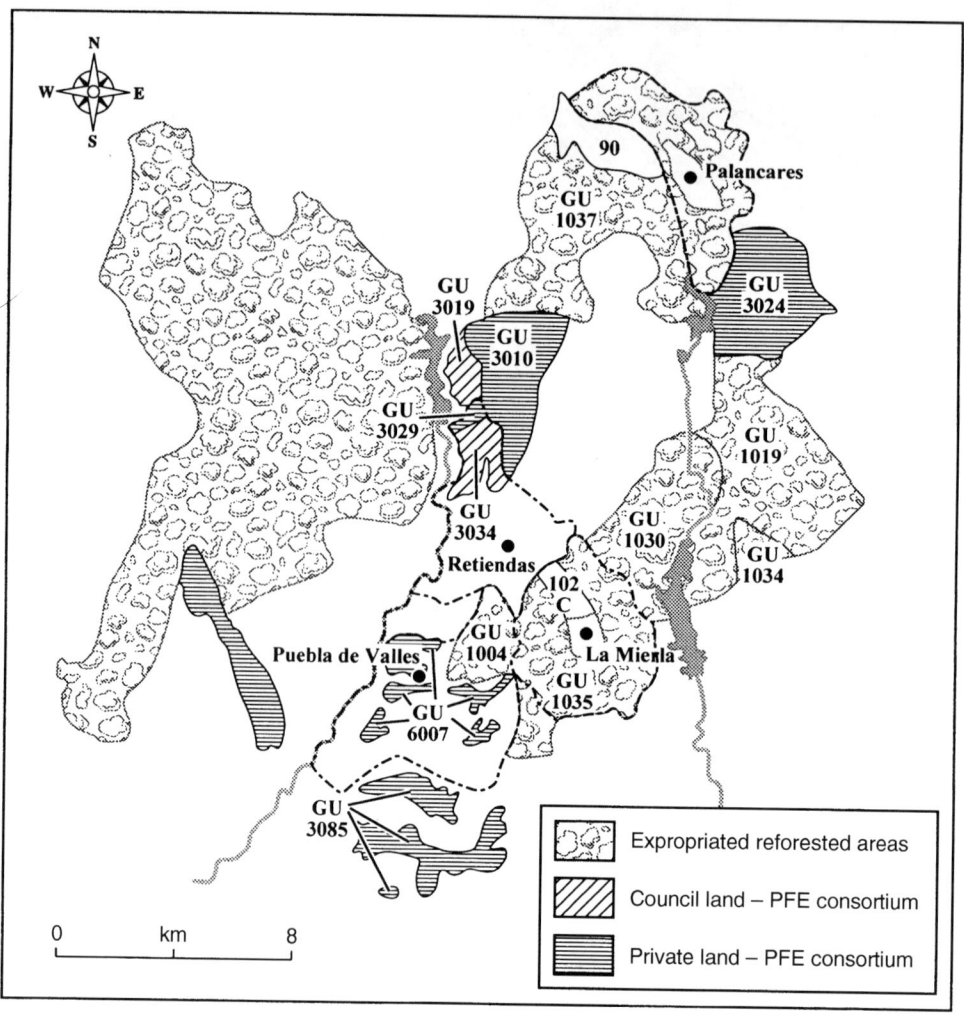

Fig. 14.2. Fieldwork areas of the Guadalajara Province.

contour lines with machines of more than 100 CV of power, with power shovels, angledozer or tilldozer, and rippers for subsoiling. The angledozer pushes the soil down the slope which will be the next bench terrace constructed by the machine. In many cases the construction of terraces, involving the use of heavy machinery destroys or inverts horizons with devastating effects for the recovery of soil fertility.

The experiments carried out in the area by IBERLIM, recorded in Coelho et al. (1994), under *matorral,* mature *Pinus,* bench terrace and gully conditions, point to these devastating effects. Runoff and soil loss measurements have been studied on several scales by IBERLIM's scientists – the gully scale, runoff plot scale, and the point scale for the period of June 1992–September 1994. Additional measurements were made for a range of soil properties influencing their erodibility.

Table 14.1. Survey and reforestation projects.

Area, year of project	Project no. and forest no.	Year of reforestation	Objectives	Method*	Tree species**
Puebla de Valles, 1946	Exprop. Docs. GU 1004	1948	Protection	LI, Sowing, LM	PL, PP
La Mierla, 1950	03/29/0005	***	Protection and regeneration	LI, Sowing, LM	M, PP, QI, P, A, W
Retiendas' 1952	Exprop. Docs. GU 3029	1953	Protection	LI, Sowing, LM	PP, PL, PS
Retiendas' 1952	Exprop. Docs. GU 3034	1958–1962	Protection	LI, Sowing, LM	PP, PL, PS
Retiendas and Tamajon, 1956	03/28/0013	1958	Protection and production	LI, LM, dynamite	PP, PL, PS, P, E
Retiendas and Tamajon, 1957	03/28/0014 GU 3010 GU 3019	1958?	Protection and production	LI, LM, dynamite	PP, PL, PS, P, E
Puebla de Valles, 1960	03/28/0018 GU 6007	?	Protection and production	LI, LM, dynamite	PP, PL, P
Puebla de Valles, 1964	03/28/0026 GU 3085	1960	Protection and production	LI, LM, dynamite	PP, PN, P
Palancares, 1968	03/29/0032 GU 1037	1973?	Protection and production	Mechanical, terraces	–
La Mierla, 1968	03/29/0032 GU 1030 GU 1035	1974–1979	Protection and production	Mechanical terraces	PP, PN

LI, labour intensive; LM, local materials.
*Methods: **Tree species: A, acacia; E, elm; M, *matorral*; P, poplars; PL, *Pinus laricio*; PN, *Pinus nigra*; PP, *Pinus pinaster*; PS, *Pinus sylvestris*; QI, *Quercus ilex*; W, willow. ***Project planned but not carried out.
Compiled from various sources.

At the gully scale the bench terrace areas, reforested using angledozers in the 1970s in La Mierla (GU 1035 in Fig. 14.2), show lower runoff than *matorral* gully areas. They did, however, record the highest sediment yields, and can therefore be classified as a 'highly active erosion system'. Sediment yield, measured in the weirs from the two *Cistus* areas, are inconclusive. The readings of one of the stations was affected by road construction and the other by visible animal disturbance in the bare gully walls (pp. 210–211).

At the plot scale, a total of nine bounded runoff plots were set on three different types of land use; three on bench terraced forest (in GU 1035), two in mature *Pinus* forest (reforested using traditional methods of sowing and planting young trees in hand dug holes) in Puebla de Valles (GU 1004 in Fig. 14.2), and four in *Cistus matorral*. The area around bench terrace *Pinus* runoff plots installed in La Mierla has been terraced for conifer planting in the 1970s. Sediment yield results, measured in these runoff plots, are higher than under *Cistus matorral* and *Pinus* mature forest conditions. The highest sediment yield predictably occurred on the most degraded bench terrace plot, where vegetation cover was more sparse. The greater vegetation cover of the other two plots managed to reduce sediment yield but not runoff flow because of the presence of expanding clays in the surface as a result of the use of heavy machinery to construct the terrace. This results in cumulative runoff flow along the terraces to lower, more vulnerable points such as bare gully walls or to bridging of terrace risers (p. 198). This area shows worse runoff and sediment yield conditions than that of mature pine forest of Puebla de Valles and those under *matorral*. The conditions of these last two areas of *matorral* and *Pinus* mature forest are very similar.

Analysis of soil properties from surface and subsurface horizons were carried out for diverse land use areas. 'Laboratory measurements of ... soils included bulk density, hydraulic conductivity, soil moisture retention characteristics, organic matter and organic carbon, aggregate stability, as well as an analysis of clay mineralogy' (p. 7). 'Sampling and measurements predominantly relate to the surface soil horizon as this has the major influence on infiltration, overland flow generation and erosion' (p. 161).

Measurements under *matorral* show that soil moisture falls from an average of 12% in June to as low as 4% in October. The rainfall absorption capacity of soils in the same areas at the beginning of October is considerable, '(on average 7 cm of rain for 50 cm depth of soil)'. Moisture levels indicate that 'overland flow under *matorral* is likely to be very infrequent'. Soil aggregates are highly stable due to their organic carbon content. However, at present, these areas of *matorral* are 'potentially vulnerable to adverse management practices [and] in a delicate state of equilibrium' (p. 171). Ternan et al. (1996) explain that 'soils with a higher clay content have a lower aggregate stability' and that 'The most stable aggregates occur under *matorral* and may represent a lag of more resistant aggregates surviving past land-use-related erosional processes' (p. 181). Soils under these *matorral* areas and the area of the mature *Pinus* runoff plots are similar, although organic carbon and clay contents are lower under *Pinus*.

Soil moisture content under mature *Pinus*, where reforestation started in 1948, averaged 30% in winter and 10–20% in summer, indicating interception and transpiration losses by vegetation. The aggregates are moderately resistant but less stable than those under *matorral*. Fungal hyphae present under *Pinus*, acting as a temporary binding medium holding particles together, impede the slaking of particles when rapidly wetted. Lower organic carbon content and poorer soil aggregate stability in these areas, indicate that 'afforestation may have increased soil erodibility'. The protection of tree canopy, low soil moisture and pine litter, however, can compensate for this increased erodibility (Coelho et al., 1994, p. 172). Ternan et al. (1996) explain that the least stable of aggregates are in the cultivated areas and claim that this can be caused by existing cultivation practices, when expandable clays are brought to the surface by the actions of deep ploughing (p. 181).

Soils in the terraced areas 'are compact [especially in the inner treads of the terraces], of low hydraulic conductivity, low organic carbon content, and highly dispersable'. The poor aggregate stability of these soils results in slaking when rapidly wetted. This can be a consequence of terracing, mixing soil horizons, resulting in low organic carbon and expandable clay contents also making vertical drainage slow. Soil moisture early in the summer was double that recorded under *Cistus matorral* and *Pinus* mature forest, while in September 1994, at the end of a dry summer, moisture levels were remarkably low. Soil moisture was greater at the base of the terrace riser and at the inner tread, with riser crests being considerably drier (pp. 172–173). Plot runoff coefficients and volumes are greater in these areas than in *matorral* or *Pinus* mature forest.

Point scale rainfall simulation experiments consisted of 1 m² bounded plots, with diverse vegetation and conditions of ground surface, to monitor sediment losses. The findings emphasize 'the importance of vegetation in protecting the soil surface against sealing and encouraging infiltration' (p. 190). The rainfall simulation experiments demonstrate the 'relatively slow overland flow response to ... simulated rain' under *matorral*, revealing that low soil moisture increases are due to the moisture retention capacity of the '*Cistus* canopy, and the litter and lichen cover on the ground surface' (p. 185).

Under the same experiment, the mature *Pinus* area produce substantially more runoff than *matorral*. Pine litter seems to be less capable of retaining rainwater than *matorral*. 'Cumulative infiltration under *Pinus* litter ... was less than under *matorral*' (p. 185). Rainfall simulation results show a high runoff coefficient and considerable increase in soil moisture under bench terrace conditions. Cumulative infiltration was comparable to *matorral*, but in comparison with the *matorral* area, lack of canopy and lichen crust resulted in low moisture retention (p. 185).

The second outcome – the planning team's constraints

The second main outcome of the Spanish reforestation policy was that it created a forestry agency staffed predominantly by forestry engineers. These, although well qualified to deal with the physical aspects of reforestation, climate, soil type, species to be planted, and infrastructure to be constructed (such as bridges, check dams and even housing for the forestry guards), lacked the knowledge and experience of dealing with the socio-economic aspects of planning. The professional composition of the reforestation office in Guadalajara is largely of foresters (engineers, technicians and forestry guards), and supporting administrative personnel. This is inadequate for the needs of forestry planning – a situation acknowledged by some officials in the office, who also point out the need for the involvement of sociologists in assessing the socio-economic constraints encountered in project planning (García Pérez, 1996).

Most agencies are aware of the dangers of single-discipline approaches to rural development planning, such as reforestation, especially when these disciplines are only partially suited for the task. Cernea, writing for the World Bank, stresses that:

> present day planners and technicians quite often do poor social engineering, unassisted by the professional competence derived from sociological and

anthropological knowledge. ... Planning agencies or policy bodies should be wary of relying on mechanical engineers to do social engineering

(Cernea, 1991, p. 30)

In Spain forestry planning has been done by forestry engineers with little training in gathering or analysing social science data. The reforestation plans for Guadalajara largely ignore the social conditions crucial to successful planning of a reforestation project, not an uncommon occurrence in the forestry sector.

The third outcome – resentment and distrust

The third main outcome of the Spanish reforestation policy is that past technocratic and autocratic approaches generated resentment and distrust among affected people. Although expropriation was less used as an implementation tool after the Franco regime ended in the mid 1970s, mistrust in the reforestation agencies of the State, ICONA and those of the Autonomous Governments, still persists. The threat of expropriation for reforestation projects is less used by the forestry agency today; however, resentment of past practices and distrust still disrupts the outcome of planning. Reforestation left to a non-compulsory arrangement using the financial incentives of the EU in arrangements such as the one for the reforestation of 'set-aside' land, requires participatory approaches to planning and implementation.

Although compulsion under the threat of expropriation is less common today, the existing distrust between forestry planners and land users is aggravated by the lack of multidisciplinary teams, and a considerable lack of organizational experience on the part of land users. The latter is a result of past political repression (until 1976) against organizations outside the control of the regime. These were excluded from negotiation, if not harassed, or persecuted where perceived as a threat. The seriousness of political persecution prevented the formation of independent organizations and the development of expertise among land users.

Multidisciplinary teams or specialists in the social sciences can defuse existing tensions between foresters and land users at the crucial initial planning stage. Attempts to reforest a large part of the village of Puebla de Valles under the 'set-aside' programme ended in failure because of shortcomings at the initial stages of planning.

The rural development project for Puebla de Valles proposed by the service of agrarian marketing (*Servicio de Comercializacion Agraria*) in 1994–1995, if not exactly autocratic, can still be considered a top-down one. This project provides an example of the constraints for negotiating compromises between the administration and rural people. The limitations of the administration are largely because of rigid procedures and officials' professional limitations regarding the promotion of their own projects and that of participation.

The transformation of the existing scattered, small plots owned by each household, into larger and less dispersed plots has been portrayed by the Government until the 1970s and land users alike as a measure to improve agrarian potential. The Regional Government of Castilla-la-Mancha approved a project for land consolidation in 1990 (DOCM, 1990, p. 3105), after a request by land owners in Puebla de Valles. Although the project for land consolidation was approved it was not put into effect. In November 1993, the Council of Puebla de Valles (CLCAPV, 1993 (documents)), and in October 1994 the newly formed Commission for Land Consolidation and Afforestation of Puebla de Valles (CLCAPV) (CLCAPV, Oct. 1994), were pressing the *Delegación* in Guadalajara for the consolidation of their land.

One of the main aims of the 'set-aside' reforestation project (Laso Rhodes, 1994) is to increase the amount of communally owned land for reforestation and hunting. The proposal for land consolidation (Laso Rhodes, 1994, p. 10) is as follows:

Dryland cereal production	600 ha
Irrigated land	150 ha
Pastoralism	300 ha
Reforestation	1000 ha

Government officials thought that Puebla de Valles could be used as a testing ground for the promotion of the reforestation plan. The 'innovative' governmental proposal consisted of changing the long-standing request for land consolidation into a project of reforestation. Planners proposed that individual landowners would own an indivisible share of the newly reforested area proportional to the amount of land they provided. That share could be sold at any time, or at the end of the minimum period of 20 years after reforestation took place, the wood could be sold. However, owning a share was not acceptable to landowners, making the project unaccomplishable.

The professional composition of the four-person team in charge of running the EU Reforestation Plan for the 'set-aside' of abandoned land in the Province cannot be considered the most ideal to promote or deal with the whole range of social, economic and political complexities of its implementation. Although specialists in their own field, the one veterinary expert, one forestry engineer and two agronomists lacked training and experience in dealing with the socio-economic complexities of planning. As a result they did not address conflicts of interest between landowners and land users, or between different land users.

The formation of CLCAPV emerged from a series of meetings in the village *municipio* to try to find ways to make use of the 'set-aside' funds. The committee was formed through an informal village meeting in the *municipio's* office where consensus, consent and nomination of candidates play a part as important as secret ballot. CLCAPV wrote to all landowners and 95% of them agreed to the land consolidation project. A major new obstacle for this project is the opposition of the 5% still against it and some ecologist pressure groups.

The ecologist pressure group 'De Raiz' published a small article in the journal *Quercus* (November 1994), asking its readers to send letters of complaint to the President of the Government of Castilla-la-Mancha and to the Commissioner in charge of agriculture in Guadalajara. As a result other ecologist pressure groups such as 'Grupo Ecologista Turon' and 'Casa de la Mata', and some individuals wrote letters of complaint in September–December 1994, against the proposed project of land consolidation and reforestation of the area. They argued that the reforestation of this land constituted an attack against the environment and its natural recovery through the natural evolution of the vegetation. Some of these letters, opposing reforestation with conifers are, according to a member of CLCAPV, inaccurate, and grossly distort the expectations of the majority of the villagers (CLCAPV, 1995).

Three main factors can be considered responsible for the failure of this project: (i) the pressure of ecologist groups (as discussed above), (ii) the semi legality of the use of funds, and (iii) the inappropriate way in which the project was initially proposed to land users. Land qualifying for EU funding under the 'set-aside' programme must have been used for agrarian production at least once in the last 10 years. Although this was a clear stipulation of EU conditions, most of the land proposed for reforestation has been abandoned for 20 or more years: in the case of Puebla de Valles the *Delegación* in Guadalajara would have turned a blind eye to this requirement. The meeting to promote the reforestation project (Laso Rhodes, 1994) at the *municipio's* hall, arranged by Government officials, took place in an atmosphere of suspicion between both parties, a viewpoint expressed by both officials and land users (García Pérez, 1996).

Villagers criticize the way the reforestation plan was presented to them, without previous consultation, or written proposals upon which they could reflect in view of a possible second meeting. They also resented officials not providing details of what reforestation methods and species would be used. Insistence by villagers that reforestation could take place only if land consolidation was implemented first, was rejected outright by officials. Other options were not suggested or discussed. Officials explain the rejection of their proposal as an example of the grave problems they have to face because of the irrational and backward character of land users and landowners – regarding them as the problem rather than as potential partners.

The outcome was a stalemate, and the opportunity to work out an answer in conjunction with the participation of the villagers was missed. Some members of CLCAPV (García Pérez, 1996) believe that the articles in the provincial, and national press and the letters of protest of those ecologist groups against reforestation, made the *Delegación* in Guadalajara revoke the reforestation plan – including that initially proposed by Laso Rhodes (1994). Some individual landowners, however, have plans to reforest their land with mixed species of trees without waiting for land consolidation.

The Plan provided the Government with a good opportunity for the encouragement of greater social cohesion between villagers and officials. In the case of Puebla de Valles the participation of members of ecologist pressure groups in planning and participation can be explained not only by the 'rights' they 'acquired' (some of them own land in the village) in the decision making process, but also as members of ecologist pressure groups, claiming concern with environmental management. Ignoring or challenging the 'rights' of these groups could be settled in two main forms: firstly in the arena of conflict and the struggle for power, on the basis of strength (members of these groups threatened to draw public attention by chaining themselves to the doors of the building of the forestry office in Guadalajara), or, secondly by integrating them in the decision making process, settling the extent of their 'rights' in the arena of debate and consensus finding through participation.

Whether the participation of such groups is acceptable to Government planners or not, is at this stage a matter of speculation which the outcome of Puebla de Valles (and although outside the scope of this chapter, other similar cases, such as that of La Quesera and Palancares in Guadalajara Province, in García Pérez, 1996), will demonstrate. Whether the *Delegación* and the Autonomous Government choose to incorporate the observations and advice of ecologist pressure groups or abandon the project, rather than be involved in negotiations with such groups, is a matter for further *a posteriori* research. The perseverance of CLCAPV and the results of participatory fieldwork in the area by the author, however, point to the advantages of this approach to planning to ensure the successful outcome of these type of reforestation projects.

Fieldwork Considerations in the Promotion of Participation

The fieldwork methods used, based on Rapid Rural Appraisal (RRA) and Participatory Rural Appraisal (PRA), are proposed as an initial stage in planning. The initial stages are especially important in cases of antagonism and distrust between planners and affected people because if they fail it can be extremely difficult to reverse the greater perception of distrust failure may exacerbate.

Although the need for multidisciplinary teams in planning and implementation for SWC through reforestation may have always been present, the recognition that there is more to reforestation than just planting trees according to soil type, climate, etc., is growing. The changing social, economic and political composition of rural society and expectations for greater regard for the environment require adjustments at government level and the adoption of a 'bottom-up' strategy in planning. Gil (1986) claims that cooperation by affected people in setting up rural development objectives and the elaboration of plans must be achieved through their active and free-will involvement. If cooperation is not accomplished plans are condemned to failure from the very start (p. 137).

A fundamental way of gaining the 'free-will' cooperation of Spanish rural people for the elaboration of a SWC project, is by finding out how they perceive the problems which affect them. According to Millington *et al.* (1989), the main criterion for the introduction of planning for soil conservation should be the perception of the problem by peasant communities and individuals rather than by experts (p. 288), if the objectives of such SWC programmes are to serve the needs of those communities.

RRA methods are quick and cost-effective fieldwork forms of data gathering used mainly in developing countries. While RRA methods mainly extract information from respondents, PRA strives to facilitate their *involvement* in the collection and analysis of data, often with the purpose of facilitating their empowerment. PRA emerged from the 'deficiencies' of RRA, the needs for refinement and the involvement of respondents, not only in parting with reliable information, but also as 'research partners' in the process (Chambers, 1994a). In this chapter the methods of RRA and PRA provide an example of data gathering which, if adopted in Spanish forestry planning, could improve the results of reforestation projects.

The improvement could arise firstly, because a greater amount of more reliable social data is made available to decision makers. The amount of data of the socio-economic conditions of the affected areas made available to decision makers is primarily descriptive and still very limited. Secondly, in terms of the participatory effects of gathering that data, extending the decision making process to affected people, could improve not only the planning, but also, in the medium and longer terms, the operations of implementation and maintenance of forestry projects.

In planning for SWC it is crucial that the reasons for land degradation are well understood, especially from the point of view of the land user. The RRA and PRA methods used in fieldwork in Guadalajara Province examine the factors involved in land users' decision making with respect to SWC. Blaikie and Brookfield (1987) identified land users' decision making processes as warranting further research. The awareness and perceptions of land users about degradation, erosion and conservation is part of the rationale affecting their decisions on land use practices, and must be considered relevant to policy formulation.

Land users' perceptions are addressed through the results of a field enquiry based on several interviews with each land user. The initial stages of these interviews, which were structured around respondents' perceptions of a series of 'factors' and 'effects' relating to land use and land degradation discussed in García Pérez et al. (1995), are those explored below.

The main aims of the initial interviews for the collection of these data were to test respondents' attitude and disposition to part with information, and to build up mutual trust for the elaboration of a list of causes and effects of land degradation. This list of causes and effects was used in a more exhaustive second stage of the enquiry about land users' perceptions of landscape changes resulting in respondents being interviewed several times. Questions in the initial and successive stages of the enquiry aimed at problematizing land users' predicaments in two main interrelated ways: one, by probing into their working agricultural practices using concrete examples of the effects these had on nature, and two, by debating the socio-economic, environmental and political effects of reforestation.

A study of the factors involved in decision making for conservation by land users is important for three main reasons. Firstly, a dialectical approach to discussion helps to break down barriers of distrust between land users and researchers. This contains the seeds from which a participatory strategy could grow. Secondly, the methodology and field methods of data gathering constitute the basis upon which to construct improved planning strategies. Thirdly, based on the two points above, it provides guidelines for effecting a participatory strategy for policy formulation, implementation and operations.

The answers expected in the interviews were, in many cases, specific or concrete in the sense that explanations often raised further questions in the minds of respondents or the researcher. The method of data collection can be divided into three phases: firstly, the probing or initial interviews, secondly, the construction of the list of 'factors' of land degradation and their 'effects', and thirdly, data gathering, using a finalized list of questions and interviews based on 'factors' accounting for land degradation and their 'effects' on the land. These three phases were not discrete but there was some overlapping between them. This is the reason for more than three activities being specified in Table 14.2a. The first two activities mainly fall within the first phase of probing during initial interviews. These overlap with some aspects of the third and fourth, while the third to fifth, as a group, mainly fall within the construction of the list of questions. Table 14.2a and b helps us to visualize the stages and main details of the initial attempts to collect data. These are subsequently explained and discussed in greater detail through the text below.

Taking into account that land degradation (LD) is a long-term process, the basic aim of the initial enquiry was to search for information which put these processes into an historical perspective. Research methods for qualitative data gathering were used in the enquiry. Quantitative data gathering was not considered feasible or necessary, especially, as argued by Devereux and Hoddinott, when the focus is on changes through time (1992, p. 36).

Showing genuine interest in people's problems, or as Francis puts it 'establishing a role', is difficult yet important because it affects the nature and quality of data collected (in Devereux and Hoddinott, 1992, p. 87). A common pitfall for officials is, according to Gil (1986, p. 19), their general tendency to give instructions instead of solving problems through debate and personal guidance. Giving instructions rather than listening does not contribute to a climate of mutual trust to generate reliable data, nor participation.

The initial stages of the enquiry used RRA methods. Taking part in the construction of the 'list' (of 'factors' of land degradation and their 'effects') encouraged the participation of respondents in the collection and analysis of data, a process which can be considered of progression from RRA to PRA.

Letting the people approached ask the questions first, usually about the reasons for the researcher's interest in their particular place or in them as individuals, helped to build up trust. This led to a relaxed climate of exchange, avoiding mechanical question and answer rigidity. The first difficulty was to convince respondents that their opinions, knowledge of working practices and perceptions of landscape changes, were relevant and important to the explanation of land degradation processes.

During initial interviews the great reverence respondents held for 'scientific knowledge' using complex measuring equipment was evident. This, it may be speculated, can be the result of the dissemination of technocratic ideological values by the better educated dominant social classes in Spain. At

Table 14.2. Sequential explanation of data collection.

Activities	Objectives	Results	Constraints
(a) Initial interviews in Retiendas			
Gradual introduction to other villagers by a respected member of their community	Establish credentials	Corroboration of genuine intentions of researcher and research	Little time available by 'respected' member for introductions
Explanation of IBERLIM project	Establish credentials	Building rapport	Initial distrust, precautions
Trial of first list of questions	Obtain basic information of socio-economic structure of households	Realization of little differentiation within the villages	Probing into economic circumstances generated suspicion
Trial of second list of questions	To obtain data on socio-economic and climatic events which were perceived as erosive	Realization of the graduality of LD processes, even in such rapidly eroding landscapes	Probing into personal details to detect dates of special weather events generated suspicion
(b) Further interviews in Retiendas and the other three villages			
Construction of a list of 'causal factors' of LD and their 'effects' in relation to land users' working practices	To test new list of 'factors' and 'effects' and encourage respondents' participation in the research	Build further rapport. Collect data smoothly. Perceived change from distrust to genuine interest	Establish credentials in Puebla de Valles. Build rapport. First trial of list of factors and effects. Annotation difficulties
First set of interviews using list of 'causal factors' and their 'effects'	To obtain data of gradual/progressive processes of LD and to encourage participation in the research	Data collection and interviewing. Involvement of a 'research partner' in Puebla de Valles	Difficulty in annotation with one interviewer only. Anecdotal and factual information discussed further to obtain better reasoning
Second and third sets of interviews in the three villages of study	To obtain data. Analysis of data from Puebla de Valles	Preliminary results from Puebla de Valles indicating suitability of method. Writing up article with research partner	Difficulty in annotation with one interviewer only in Retiendas and La Mierla

this initial stage respondents were more interested in an explanation of the physical measuring techniques used by the IBERLIM researchers than the socio-economic aspects of the project, or in giving their own explanations. It proved less helpful to start the discussions with a direct reference to the core objectives of the enquiry (i.e. their perceptions of landscape changes and socio-economic processes relating to land degradation), than to explain the physical measurements being carried out by the IBERLIM team. The reference to IBERLIM work, however, was only used in the introductory stages of interviewing.

Gradually these explanations, discussion of physical measurements and some of the initial results, were changed to a focus on the importance of respondents' detailed vernacular knowledge of the area and processes. The building up of mutual trust helped to break the danger of distorted data through 'ventriloquism', appearing in agreement with what the researcher may want to hear. If the researcher is perceived as an official, respondents may tailor their information to obtain more advantageous conditions rather than expressing their real opinions.

The 'entry' of the researcher into the community, and the way s/he is received is extremely important for building up essential rapport with potential respondents (Devereux and Hoddinott, 1992, p. 88). The credentials of the person who may be introducing the researcher into the community is also important. The author was introduced to the four villages in which data was collected in different ways, which had implications for the conduct and efficacy of field research. In Retiendas and Puebla de Valles the author was introduced by well known and respected members of the communities.

The first village where data was collected for the enquiry was Retiendas. At the end of the interview the first respondent was asked to continue to answer questions at a future date, to introduce the researcher into his community, and to comment on the contents of the list of questions asked. Respondents were told that interviews were informal and would consist of two or three sessions on different days. Arranging dates for following interviews proved complex and unworkable and it was more realistic to make respondents understand that they would be requested to discuss issues further at a later date and make loose arrangements for the second and third interviews. 'The same day next week at roughly the same time' was seldom remembered, but what respondents remembered was that they committed themselves to answer questions for two more sessions.

The result of not being able to arrange definite dates, was an *ad hoc* approach of visits to the village which was time consuming. Waiting for some one to turn up, however, was not a total waste of time. Conversation with people who did not accede to be interviewed or with those who already finished theirs, was about village life and their working practices, or about the work of the researcher. This type of 'disjointed' but, in the words of Chambers (1992) 'relaxed' interviewing often provided valuable data and most important, served to build rapport.

However, it was in Retiendas that the greatest difficulties were experienced. The reasons for this were twofold: firstly the list of questions was at that stage still in the process of being constructed, and sometimes generated suspicion and misunderstandings. Secondly, the author was at that stage unfamiliar with the idiosyncrasies of the local culture, such as probing into personal social or economic details, which although not important *per se*, were initially thought to be important to ascertain the contributing explanations to land degradation.

The first list of questions was constructed to gain basic information on the social and economic structure of the household of respondents. It was assumed that a mixture of questions of working practices, coupled to the circumstances of the household during specific periods, would produce more accurate information pointing to differentiation, intensification of production and land degradation. Attempts to obtain information linking socio-economic, climatic, rainfall, location and time-based events, initially expected to provide some answers clarifying the processes of land degradation, were fruitless and provoked suspicion because the questions probed too closely into their economic livelihoods. Respondents also found it difficult to remember singular erosive events or even the year of events which caused particular damage to their crops. Some of these questions were modified and eventually discarded.

Government officials for whom this type of data is crucial, should bear in mind that the information concerned is, according to Devereux and Hoddinott (1992), 'sensitive [meaning that] ... it can be used in a manner contrary to the interests or wishes of the informant' (p. 124). This type of data should only be gathered after achieving good rapport with respondents, and only if strictly necessary. Attempts to ascertain patterns of differentiation by examining the socio-economic evolution of households, such as those postulated by Chayanov's theories (1966), or for example in Harriss (1982), or those proposed in Ellis (1988), were abandoned also when it was realized that the social and economic structure of the village is fairly homogeneous.

A productive part of this initial pilot enquiry in Retiendas was a detailed explanation of respondents' working practices in combination with a series of 'factors' and 'effects' on the condition of land. When questions about their working practices started to generate repetitive answers, these questions were discarded and the 'factors' and 'effects' were used for the construction of a final list. The final version of the list of factors and effects was based on some of the questions which had been tried initially. The sequential explanation of data collection for this part of the interviewing process is synthesized in Table 12.2b.

The objectives of the enquiry and the use that was going to be made of information was explained, assuring respondents that such information was confidential and names of individuals would not be mentioned. Explaining the reason for the type of questions being asked, as da Corta and Venkateshwarlu experienced during fieldwork in India (in Devereux and Hoddinott, 1992, p. 106), helps to dispel initial distrust and makes the enquiry and further work more acceptable to the respondents. During the enquiry it was emphasized that there were no right or wrong answers to questions: all the respondents' knowledge and opinions were important.

The first stages of the research achieved the involvement of respondents in the construction of the final list of 'factors' and their 'effects'. The cooperative spirit could easily be used for future participation in the planning and implementation of a SWC project, although this would be subject to constraints, not so much of time since PRA methods are faster than conventional ones, but subject to the ability of officials to put this method into operation.

A crucial factor securing a cooperative spirit was achieved at the end of each initial interview when their opinion was requested. They were asked to put themselves into the hypothetical role of the researcher and, keeping the objective of the enquiry in mind (their opinion of the socio-economic factors accounting for land degradation), to advise on what kind of questions they would have asked, and which ones in the list they would have not used.

The reasons for this request were always explained to them beforehand. These involved a short and concise explanation, that to have the correct answers, which they knew better than anybody else, the researcher had to know the questions first. It was obvious that people knowing the answers would also know the right questions. Their participation was essential for the results to be accurate. Their reaction was still one of surprise, puzzlement and finally cooperation. Surprise because interviews are often about asking questions rather than asking for advice on what kind of questions should be asked. Some respondents initially said that their experience in this respect was limited, and were therefore puzzled, but when prompted with specific examples they soon were able to provide an opinion. Lastly, although understating the importance of their advice, their interest in the progress of interviewing increased.

Three main things were achieved with this method. Firstly their advice gave rise to modifications in the content of the list of initial questions and in the final list of 'factors' and 'effects' used in the second part of the research. The second achievement was that their participation contributed to the creation of a climate of mutual trust. The third achievement is that their participation culminated in the writing of an article (García Pérez, et al., 1995), with the direct participation of a villager (P. Martín Ruíz) helping to design the list of questions and the collection and analysis of data, making the method change from RRA to PRA. Another possible achievement which needs further research to verify it, is the influence the researcher may have had, through the problematization of respondents' predicaments involved in interviewing, in the formation of CLCAPV in Puebla de Valles. If this was the case the role of the interviewer could be considered one of facilitating their participation and also empowerment, as envisaged by Chambers (1994c).

Conclusion

This chapter has shown that the constraints still facing the planning and implementation of reforestation projects in Spain are rooted in past technocratic and autocratic trends. The detrimental effects of these constraints could be initially limited, and eventually phased-out, by adopting a participatory approach to planning. The participation of affected people, as the results of fieldwork for this project have shown, can be achieved by understanding what the processes of land degradation are as perceived by land users, through RRA and PRA methods of fieldwork. To involve affected people in the decision making process of planning, multidisciplinary teams of planners are needed and participatory methods facilitated.

The State reforestation agency in Guadalajara may have few options in trying a multidisciplinary approach. The two most feasible ones, which would benefit from further scientific testing, but could be tried by the agency are: firstly to put into effect an in-house training programme for the existing staff, or secondly, to make use of specialist consultants in the social sciences who could form part of an integrated and multidisciplinary team of planners. The likely benefits from these attempts could be a catalyst for changes in the professional composition of planning teams, from near monodisciplinary to multidisciplinary ones, resulting in more successful planning.

References

Abreu y Pidal, J.M. (1963) Hacia la obtención de la máxima renta en el cultivo del bosque. *Montes* 19(112), 347–348.

Araujo, J. (1994) Cultivar la cultura. *El Boletín* 15, MAPA, Madrid, 58–62.

Baker, R. (1984) Protecting the environment against the poor. The historical roots of the soil erosion orthodoxy in the Third World. *The Ecologist* 14(2), 53–60.

Bennett, H.H. (1960) Soil erosion in Spain. *The Geographical Review* 50, 59–72.

Blaikie, P. and Brookfield, H. (1987) *Land Degradation and Society*. Methuen, London.

Burgers, T.F. (1949) La regeneración natural y la mezcla de especies como ideal en la silvicultura. *Montes*. 30, ETSIM, Madrid, 505–508.

Cernea, M.M. (ed.) (1991) *Putting People First. Sociological Variables in Rural Development*, 2nd Edn. World Bank, Washington, DC.

Chambers, R. (1992) *Rural Appraisal: Rapid, Relaxed and Participatory*. Discussion Paper 311, Institute of Development Studies, Sussex.

Chambers, R. (1993) *Challenging the Professions. Frontiers for Rural Development*. Intermediate Technology Publications, London.

Chambers, R. (1994a) The origins and practice of Participatory Rural Appraisal. *World Development* 22(7), 953–969.

Chambers, R. (1994b) Participatory Rural Appraisal (PRA): Analysis of experience. *World Development*, 22(9), 1253–1268.

Chambers, R. (1994c) *Paradigm Shifts and the Practice of Participatory Research and Development*. Institute of Development Studies, Working Paper 2, University of Sussex, Brighton.

CLCAPV (1993) *Sanz Iruela, M. to Delegado provincial, 19 November*. Letter, CLCAPV, Guadalajara.

CLCAPV (1994) *CLCAPV to Delegado provincial, 13 October*. Letter, CLCAPV, Guadalajara.

CLCAPV (1995) *Sanz Azconas, N. to Hurtado Pérez, F., 8 March*. Letter, CLCAPV, Guadalajara.

Codorniu, R. (1915) La repoblación forestal en España. Medios de fomentarla y de convencer de su necesidad a las clases rurales I. *Montes*, 912–915, Madrid.

Coelho, C. et al. (1994) *IBERLIM. Land Management and Erosion Limitation in the Iberian Peninsula*. EU Project EV5V–0041.

De Raiz (1994) Valle de Jarama. *Quercus*, Noviembre, p8.

Devereux, S. and Hoddinott, J. (eds) (1992) *Fieldwork in Developing Countries*. Harvester Wheatsheaf, Hertfordshire.

DOCM (1990) *Documento Oficial de Castilla-la-Mancha*, No. 81, Toledo.

Ellis, F. (1988) *Peasant Economics. Farm Households and Agrarian Development*.

Cambridge University Press, Cambridge.

García Pérez, J.D. (1996) Socioeconomic processes of land degradation in Guadalajara Province (Spain). Unpublished PhD thesis. Plymouth University.

García Pérez, J.D., Charlton, C. and Martín Ruiz, P. (1995) Landscape changes as visible indicators in the social, economic and political process of soil erosion: a case study of the municipality of Puebla de Valles (Guadalajara Province), Spain. *Land Degradation and Rehabilitation* 6, 149–161.

Gil, N. (1986) Desarrollo de cuencas hidrográficas y conservación de suelos, *Boletin de Suelos de la FAO* 44, Roma.

González Bernáldez, F. (1990) Consideraciones ecológico-políticas acerca de la conservación y regeneración de la cubierta vegetal en España. *Ecología* (fuera de serie), ICONA, Madrid, 439–445.

Groome, H. (1981) Las sugerencias nunca atendidas del Plan Forestal español. *Quercus*, ETSIM, Madrid, 31–34.

Groome, H. (1985) El desarrollo de la política forestal en el Estado español: Desde el siglo 19 hasta la Guerra Civil. *Arbor* 121(474), Madrid, 59–89.

Groome, H. (1990) *Historia de la Política Forestal en el Estado Español.* Agencia del Medio Ambiente, Comunidad de Madrid, Madrid.

Harriss, J. (ed.) (1982) *Rural Development. Theories of Peasant Economy and Agrarian Reform.* Hutchinson, London.

Laso Rhodes, A. (1994) *Infórme Tecnico Sobre la Concentración Parcelaria in Puebla de Valles.* Delegación Provincial de Aricultura, Guadalajara.

López Carrasco, F. (1994) Castilla-la-Mancha, Región rural. *El Boletín*, 18, MAPA, Madrid, 50–57.

Millington, A.C., Mutiso, S.K., Kirby, J. and O'Keefe, P. (1989) African soil erosion – Nature undone and the limitations of technology. *Land Degradation and Rehabilitation* 1, 279–290.

Montoya Oliver, J.M. (1983) *Pastoralismo Mediterráneo.* Monografias 25, MAPA-ICONA, Madrid.

Pretty, J.N. and Shah, P. (1994) *Soil and Water Conservation in the Twentieth Century: a History of Coercion and Control.* Research series No. I, Rural History Centre, University of Reading.

Prieto Rodríguez, A. (1995) Gestión forestal. *Revista Forestal Española*, No. 12, 11–23.

Proyecto 03/29/0003, Cuarta División Hidrológico Forestal (1943) *Documento de Expropiación de Puebla de Valles.* Finca – Barranco del Lugar, Mego etc. Monte GU1004. Delegación de Agricultura, Guadalajara.

Richards, P. (1981) Appraising appaisal – Towards improved dialogue in rural planning. *IDS Bulletin* 12(4), 8–11.

Rico Boquete, E. (1995) *Política Forestal e Repoblacións en Galicia (1941–1971).* Servicio de Publicacións e Intercambio Científico, Universidade de Santiago de Compostela, Santiago de Compostela.

Sagasta Azpeitia, J.M. (1979) Coste social de las inversiones forestales. *Montes*, 193, ETSIM, Madrid, 175–182.

Tamames, R. (1976) *Estructura económica de España. Vol. 1: Introduccion, sector agrario.* Biblioteca Universitaria Guadiana, Madrid.

Ternan, J.L. *et al.* (1996) Aggregate stability of soils in Central Spain and the role of land management. *Earth Surface Processes and Landforms* 21, 181–193.

CHAPTER 15
Making the invisible visible: ancient woodlands, British forest policy and the social construction of reality

Judith Tsouvalis-Gerber
Department of Geography, University of Nottingham, Nottingham NG7 2RD, UK

Introduction

At a woodland conference at Monks Wood in 1972, Roger Parker-Jervis, a woodland manager, stressed that

> he and his contemporaries would do their best to manage ancient woods sympathetically if only someone would tell him *which woods* were ancient
> (Peterken, 1993, pp. 322–323, emphasis added)

The purpose of this chapter is to account for the curious state of affairs, whereby foresters of the 1970s could not '*see* the wood for the trees' in terms of ancient woodlands, while by the late 1980s they had not only learnt how to recognize them, but were obliged to treat them in particular ways. Ancient woodlands by that time had become institutionalized, and institutionalization always implies historicity and social control. In the words of the social constructionists Berger and Luckmann (1971, p. 72), institutions:

> always have a history, of which they are the products. It is impossible to understand an institution adequately without an understanding of the historical process in which it was produced.

In this chapter, the historical and social processes that led to the institutionalization of 'ancient woodlands' will be investigated. Although 'ancient woodlands' have external referents in the 'real world' in the form of certain spatio-temporal entities made up of certain living and non-living beings, it is their social nature that is of interest here. The choice of the term 'certain' is indicative of this social nature, for not all spatio-temporal entities qualify as 'ancient woodlands'. And here it must be asked: what does it mean for some 'thing' to be called an 'ancient woodland', in what ways is this some 'thing' socially constructed and produced, and how is it socially enforced? In order to answer these questions, we have to explore the processes by which subjective meanings of reality become objectified realities.

The Social Construction of Reality

The process of the social construction of social reality can be summarized as follows. First, subjective meanings of the world are negotiated socially via the system of language, a process in which the definition, classification, categorization and signification of the world is determined. Second, some of these subjective meanings are in turn institutionalized, and when taken-for-granted, become objective facts. Institutionalization implies that agreed upon meanings are subsumed under social control, that is, that they are enforced through laws, codes of practice, moral codes and the like. People then often forget that social reality 'originates in their thoughts and actions, and is maintained real by these' (Berger and Luckmann, 1971, p. 33). In other words, they mistake socially constructed reality for reality *per se* (Bourdieu, 1991).

Third, through processes of socialization (upbringing and education) institutionalized social reality is passed on and internalized by successive generations unless challenged and re-negotiated. The resulting way of thinking, perceiving and acting is what Bourdieu (1989) tries to capture with his notion of *habitus*, which I shall elaborate below. Fourth, through the system of language, which is the prime instrument for the legitimization of institutional order, logic is imposed on social reality.

Before turning to the case study below, it is important to consider what enables human beings to enact objectified subjective meanings. One notion that accounts for this ability is, as indicated above, Bourdieu's notion of *habitus*. *Habitus* not only embodies an individual's history, but is constitutive of an individual's particular social environment (its material conditions, cultural conventions, etc.); it is, according to Bourdieu (1985, p. 78), 'history turned into nature'. The reason why he describes *habitus* as history turned into nature, is that it is bodily enacted and expressed (ways of dressing, of moving, of employing gestures and expressing emotions all have their foundation in *habitus*). It is something a person *is*, rather than something a person *has*. Elsewhere, he defines *habitus* as:

> the system of structural and structuring dispositions which is constituted by practice and constantly aimed at practical – as opposed to cognitive – functions. It is a system of durable and transposable ... schemes of perception, appreciation and action that result from the *institution of the social in the body*.
> (Bourdieu, interviewed by Wacquant, 1989, pp. 42, 44)

Due to *habitus*, a human being is able to participate in the most complex social practices without thinking about what to do: action guided by the feel for the game, as Bourdieu calls it, is not based on reason (Bourdieu, 1992, p. 11). Rather, 'the feel for the game' is what we might also refer to as 'instinct'. Therefore, the logic of practice, rooted in *habitus*, cannot be explained by theoretical accounts of practice. This is due to the atemporal gaze of the analyst, who is unable to identify with the urgency and seriousness of a directly experienced situation. Practice, as Bourdieu (1992, p. 81) argues, unfolds in time and:

has all the correlative properties, such as irreversibility, that synchronization destroys. Its temporal structure, that is, its rhythm, its tempo, and above all its directionality, is constitutive of its meaning.

It is this immersedness in time that prevents a detached, reflective view. In order to overcome this dilemma and grasp the logic of practice, he argues, the analyst should focus not on practice itself, but on the principles of its production, rooted in the *habitus*; the cognitive and evaluative structures that organize the schemes of perception of the world in accordance with the objective structures of a given state of the social world. This view of practice and its logic is closely linked to Lefebvre's claim that 'all social practice ... is lived directly before it is conceptualized ... the unconscious level of lived experience *per se*' (1991, p. 34).

Although I share Bourdieu and Lefebvre's view that practice is most of the time based on a taken-for-granted view of the world, I would dispute the claim that it is necessarily pursued unconsciously. In order for change to occur, for the unconscious to be raised to the level of consciousness, and for the meaning of the taken-for-granted world to be questioned and re-negotiated, the potential for reflective activity must be re-introduced into the analysis. This potential has been well captured by Foucault in his notion of *problematization*. Problematization is that which transforms a given into a question, and in order for this transformation to occur, in order for a

> domain of action ... to *enter the field of thought*, it is necessary for a certain number of factors to have made it uncertain, to have made it lose its familiarity, or to have provoked a number of difficulties around it. These elements result from social, economic and political processes.
> (Foucault in Rabinow, 1984, p. 388, emphasis added)

I would add here that it also results from biophysical and material processes, which are where socio-economic and political processes find their manifestations. Problematization is what links what we might refer to, following Lefebvre (1991), as the lived and the perceived with the conceived; the former two of which according to him are unreflective in nature. The notions lived, perceived and conceived are used by Lefebvre to overcome the dualism between mind and body associated with Descartes and Kant. As he puts it:

> their dualism is entirely mental, and strips everything which makes for living activity from life, thought and society (i.e. from the physical, mental and social, as from the lived, perceived and conceived).
> (Lefebvre, 1991, p. 39)

The triad physical, mental, social, or lived, perceived, conceived, are a part of human *being*; they merge and mingle in complex ways. The distinction between them is conceptual and linguistic in nature, rather than ound *in* nature, and serves as a heuristic device to disentangle an otherwise incomprehensible mass of phenomena. Fig. 15.1 illustrates this point.

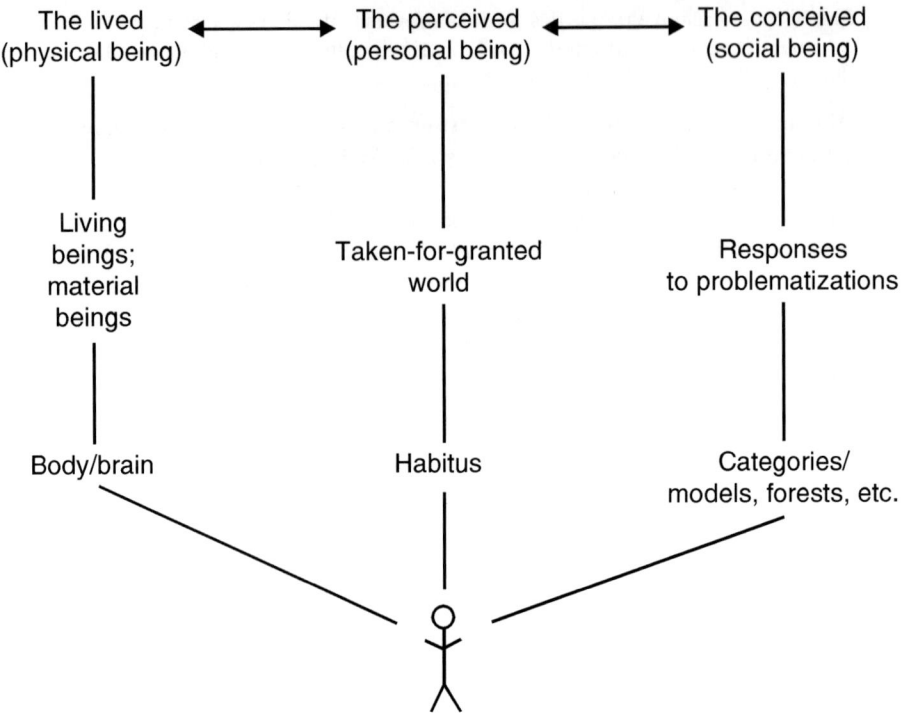

Fig. 15.1. The interrelationship between the lived, the perceived and the conceived in the constitution of *habitus*. Based on Harré (1979, 1984, 1991), Foucault (1984) and Lefebvre (1991).

When thought does intervene in practice through problematization:

> it doesn't assume a unique form that is the direct result or the necessary expression of ... difficulties; it is an original or specific response – often taking many forms.
>
> (Foucault in Rabinow, 1984, p. 389).

Examples of problematizations in forestry include the realization of the precarious state of woodland cover in Britain due to the severe wood shortages experienced as a result of the First World War, that led to the solution of a state forest service and the establishment of plantation forests being proposed. New obstacles and difficulties were encountered in the process of implementing these solutions (e.g. the unavailability of suitable land, inadequate technologies, 'pests' such as rabbits, deer, beetles, fungi and bacteria, etc.) (Forestry Commission Annual Reports), and once the forests came into view, resistance to them began to be expressed for various reasons but particularly in terms of their visual impact (Forestry Commission Annual Reports; Revill and Watkins, 1996). Furthermore, the forests themselves began to interact with the soil and the groundwater, while all sorts of biophysical factors interacted with the trees (Usher and Thompson, 1988; Hart, 1991). With other socio-cultural and economic changes taking place, the vision of the world as a 'global village' became prominent, and the interconnectedness of socio-economic and biophysical processes was increasingly recognized. Concerns over the disappearance of 'Nature' grew, and, in response, the concepts of 'sustainable development' and 'biodiversity conservation' were born (IUCN, 1980). In forestry, these problematizations were reflected in forest policy and forestry practice (Mather, 1991, 1994), and the conservation and indeed restoration of 'ancient woodlands' was but one response to these problematizations.

To summarize the processes of the social construction of reality outlined above, every human being carries within himself/herself his/her own history and that of his/her social environment. When perceiving 'the' world, physical/individual factors, and social factors come into play. The conceived world is the socially negotiated world – the official world, so to speak – but it is also deeply embedded in the lived and the perceived world (the latter of which is the taken-for-granted view of the world an individual or group holds). This is illustrated in Fig. 15.2. To investigate the complex interrelationships at work in the social construction of reality in more depth is, unfortunately, beyond the scope of this chapter (for a more thorough account see Gerber, 1997b).

The focus here will be primarily on the definition, categorization and institutionalization of a particular aspect of reality, namely, 'ancient woodlands', and the power struggles that ensued in that process. Before looking at how the woods that had faded from consciousness were made

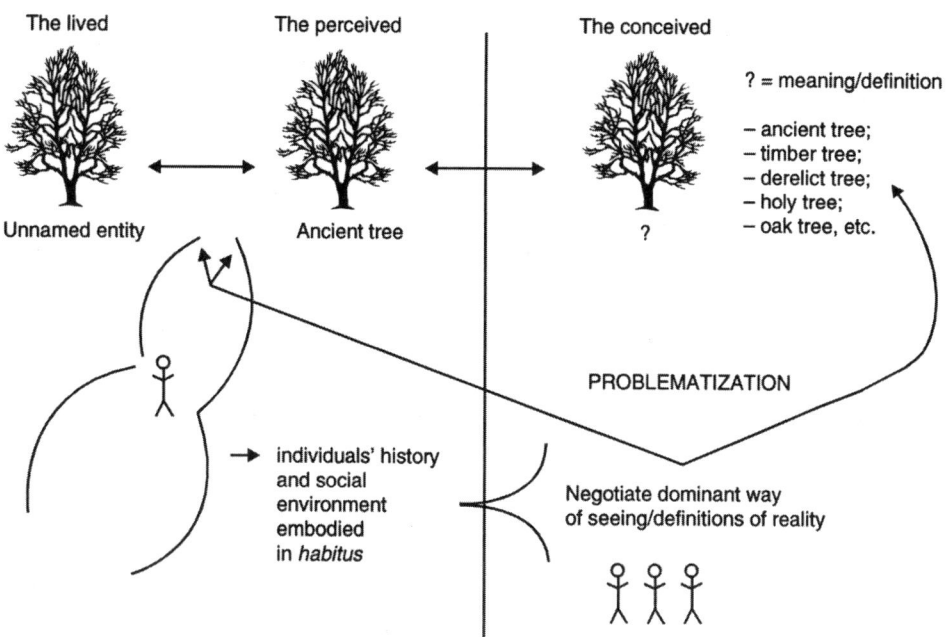

Fig. 15.2. Constructing reality: a dialectical activity rooted in a material world.

visible once more, a brief consideration of what had made their disappearance – in a cognitive and real sense – possible in the first place will be given.

The Rise of the Modern Forest – Forest Policy and Practice 1919–1985

One of the main reasons that led to the establishment of the Forestry Commission of Great Britain was the severe timber shortages experienced during the First World War. Britain, whose woodland cover by that time was less than 5%, relied on timber imports for most of its domestic demand (Ryle, 1969). Until the setting up of the Forestry Commission in 1919, Britain did not have a national forest policy. Neither, as the first Annual Report of the Commission of 1920 pointed out, did it have an organization of higher executive staff with British experience, forest officer personnel, foresters and foremen with state forestry experience or customs, a forestry code, or reliable statistics of Britain's woodlands apart from those provided by the Board of Agriculture and Fisheries (Forestry Commission, 1920; Peterken, 1993, p. 85).

Although many forest plantations of exotic softwood trees had been established during the seventeenth, eighteenth and nineteenth century (Rackham, 1976), modern forestry methods and practices on a large scale were introduced with the Forestry Commission. The practice of high-forest plantation management had originated in Germany, and it was through the Indian Forest Service that German ideas on forestry were disseminated in Great Britain. In 1862, the German botanist Dietrich Brandis had become Forestry Advisor to the Government of India, and it was a former student of Brandis, Sir William Schlich, who opened the first forestry school in Britain at Cooper's Hill, Surrey, in 1885. He was later to lecture at the Forestry Institute in Oxford. Schlich's first *Manual of Forestry* was published in 1889, based on German texts and experiences. One of his pupils, R.L. Robinson, was later to become a Commissioner of the Forestry Commission.

While the adverse effects of large-scale coniferous plantations had already been noted in Germany in the late nineteenth century, Britain was primarily interested in the quick establishment of just such plantations. This outlook was incorporated into policy. As Pryor (1992, p. 6) has pointed out:

> From 1919 until the mid-1980s forestry policy was remarkably consistent, being dominated by the expansion of the forest area and increasing timber production. However, the underlying justifications for this policy of expansion have changed over the years.

The key stages in the development of British forest policy are shown in Table 15.1. Although the justifications for practice changed over time, modern forestry practices themselves were firmly institutionalized and internalized by successive generations of foresters. These practices were characterized by:

- Plantations being established on bare ground (land without trees or where previous woodland had been clear-felled).

Table 15.1. Evolution of British Forest Policy 1919–1987.

Year	Basis	Implication
1919	Forestry Act	Forestry Commission (FC) set up. Aim: To build a strategic reserve of timber
1927	Forestry Act	FC enabled to make bylaws (e.g., concerning access on land, etc.)
1945	Forestry Act	FC to rebuild timber reserve, applying systematic techniques
1946	Forestry Act	FC to administer a Dedication Scheme to boost private forestry
1951	Forestry Act	No trees to be felled without a licence from the FC – continues system introduced in the Second World War
1957	Zuckerman Review	FC obliged to *balance employment* in remote rural regions
1967	Forestry Act	Consolidates previous Acts
1968	Countryside Acts	FC to have regard for *conservation, natural beauty and amenity; recreation*
1972	Unfavourable cost–benefit study of the Treasury	FC to *maintain employment and enhance the environment*. Grant Aid Schemes closed
1980	Policy Statement	FC to continue and *expand afforestation*, especially private afforestation
1981	FC disposal of land begins	
1981	Wildlife and Countryside Act	FC to *balance forestry and environment* interests
1985	Broadleaves Policy	FC to encourage *broadleaved* planting; special provisions made for *ancient woodlands*
1987	Countryside Commission Forest Policy	FC to create *Community Forests/Multi-Purpose Forests*

- Seedling trees raised in nurseries being planted on bare ground when 2–3 years old.
- Species planted being selected according to commercial judgement and site conditions (mostly coniferous species).
- Most planting being arranged in regularly spaced rows (efficient use of plants; easy care, cheaper).
- Ground vegetation ('weeds') being removed ('cleaned') by hand, machines or selective herbicides so as not to destroy seedlings (woody crop from previous woodland or self-sown trees referred to as 'scrub').
- Plants that die and leave 'blanks' in the plantation being 'beaten up', that is, removed and replaced with new plants.
- 'Thinning' being undertaken at intervals to clear out surplus stock (those trees exceeding the stock of mature trees ('final crop') required.
- Mature plantations being allowed to stand until growth rates decline, cash is needed by the owner, a good market for timber prevails, or other reasons.

- The final crop being clear-felled, leaving the ground bare once more, available for the cycle to be repeated.

(after Peterken, 1993, pp. 66–67).

In terms of institutionalized practices, it is worth considering Ryle's (1969, p. 81) observation made in relation to the monetary shortages faced by the Forestry Commission during the Second World War:

> Old practices and old ambitions die hard, if they ever die at all, and despite unavoidable restrictions and shortages they [FC] were able to maintain a very substantial rate of new planting.

Most Forestry Commission plantations were established on what was seen as 'unproductive' land such as upland moors and lowland heaths. Indeed, one of the main ideas behind afforestation was to make 'good use' of this land (Revill and Watkins, 1996). Partly because softwood trees were best suited for 'poor' land and partly because they matured much quicker than hardwood trees, coniferous species, particularly Sitka spruce, became the dominant trees planted by the Forestry Commission. With advances in science and technology, the time–space limitations posed by nature could increasingly be overcome and areas could be planted that were previously unsuitable for forestry. In 1995, the Forestry Commission's productive woodlands consisted of 763,000 ha of coniferous forests and 51,000 ha of broadleaved trees (Forestry Commission, 1995). It had long before become the biggest landowner in the country. The 'nature' so produced was one consisting primarily of monocultural, coniferous plantations with straight boundaries and it was when they came into sight – that is, when they started to become visible components of the landscape – that opposition to them began to grow (Revill and Watkins, 1996).

What I would like to stress here, however, is that what prevented the foresters of Great Britain from recognizing ancient woodlands is that they were brought up in the dominant tradition of modern forestry, in other words, high-forest plantation management. Their *habitus*, or what they had been taught to *see* through language and practice (Bourdieu, 1992), was such that to convert what is now called an ancient woodland into a conifer plantation, a common practice from the 1950s onwards, seemed nothing other than the good practice of turning a derelict wasteland to productive use (Watkins, 1990; Peterken, 1993). Still in 1992, Wright, writing for *The Guardian*, could observe in a leading article:

> Foresters have been *trained to look* at their plantations and *see* little more than cellulose or potential paper-pulp: Oliver Lucas [author of *The Design of Forest Landscapes*] urges them to seek out paintings and works of literature that can *teach* them about 'the spirit of place', and to consider what their plantations look *from afar*, and to open differentiated *views* within the block itself.
>
> (Wright, 1992, p. 8, emphasis added)

Similarly, before 1985, 'ancient woodlands' simply did not exist for the majority of foresters. However, that was soon to change.

The Idea of 'Ancient Woodland' – the Definition and Categorization of Reality

The idea of ancient woodland, as Watkins (1988) has shown, is not a new one. Some woodlands were recognized as distinct from plantations very early on and, in 1877, the New Forest Act gave protection to 'ancient and ornamental' woods. During much of the twentieth century, for the reasons outlined above, ancient woodlands as an idea as well as a reality faded from consciousness. It was during the 1960s, when historical approaches to woodland ecology were developed (see Steven and Carlisle, 1959 and Tubbs, 1964) that they became popular once more. At that time, the distinction made was between *primary* and *secondary* woodlands. Primary woodlands had never been cleared of trees and were, therefore, modified remnants of the original woodland cover of Britain. Secondary woodlands had grown up on sites that had been completely cleared of trees at some time. The major problem with that division was that it was extremely difficult to prove whether a wood was primary or not, and that was one of the reasons why the terms 'ancient' and 'recent' were introduced by Rackham in 1971 and developed by Rackham and Peterken subsequently (Peterken, 1994, pers. comm. 1995).

The term *ancient* woodland was to refer to woods that had been occupying a site since at least 1600. The term does not refer to the age of the trees; they might have been cut down several times over the centuries. Neither does it imply that these woods have remained untouched by human interference. What it does mean, is that woodland has been continuously present on the site, that is, the land was never used for other purposes (Peterken, 1994). The reason for the choice of AD 1600 as a threshold date is simply due to the fact that it marks a time when cartographic evidence of the existence of woodlands began to be more widely available, as well as a time when the planting of trees began to be widespread (Spencer and Kirby, 1992, p. 78). When ancient woodland was thus defined, two parameters were of importance: first, the choice of the wording – the term 'ancient' would 'mean more to people' (Peterken, pers. comm. 1995) – and second, the way it was going to be defined – the existence of these woods would be easier to prove.

The term *recent* woodland refers to woods that have grown up on unwooded land after 1600. While all recent woods are secondary and all primary woods ancient, some ancient woods might be secondary, which means they have grown up on unwooded ground before 1600. Another term frequently used and not to be confused with ancient woodland is that of *semi-natural* woodland, a notion introduced by Tansley in 1939, which refers to stands that have not been planted and consist primarily of native trees and shrubs.

Ancient woodland, apart from referring to a specific space containing certain lifeforms (e.g. bluebells, lichens, trees, etc.) and particular human artefacts (e.g. ditches, walls, etc.) existing in the real world, is a category. As Keith Thomas (1984, p. 52) in his study of perceptions of nature in early modern England has pointed out:

all observation of the natural world involves the use of mental categories with which we, the observers, classify and order the otherwise incomprehensible mass of phenomena around us; and it is notorious that, once these categories have been learned, it is very difficult for us to see the world in any other way. The prevailing system of classification takes possession of us, shaping our perception and thereby our behaviour.

It is primarily with the naming of an entity that that which is named comes into existence for us. It is not possible to *think* about *that* for which there is no word, and it is in that sense that words create worlds. For ancient woodlands to become part of social reality, they had first to be defined and categorized. According to Bourdieu,

> The categories of perception, the systems of classification, that is, essentially the names which construct social reality as much as they express it, are the crucial stakes of political struggle, which is a struggle to impose the legitimate principle of vision and division.
>
> (Bourdieu, 1990, p. 134)

Objectifying Subjective Reality – the Institutionalization of 'Ancient Woodland'

In 1978, Peterken drew up the Nature Conservancy Council forestry policy, to which the then Director General of the Forestry Commission, George Holmes, responded in uncompromising terms, reflecting the attitude of senior Forestry Commission personnel at the time: the notion of 'so-called ancient woodlands' was to be rejected (Peterken, pers. comm. 1996), and with it, of course, the existence of the ancient woodlands themselves was denied. In 1979, a House of Lords Select Committee under Lord Sherfield began to investigate *Scientific Aspects of Forestry*. In spite of the Forestry Commission's negative response, the Nature Conservancy Council proposed four treatment classes for the still existing ancient semi-natural and broadleaved woods to the Committee:

1. For 22,000 ha of 660,000 ha a policy of minimum intervention.
2. For 167,000 ha a policy of restoration to coppice.
3. For 197,000 ha a policy of restricted high forest of native species.
4. For 274,000 ha unrestricted high forest.

The following recommendations were made by the Committee in consequence:

> Section 125:
> ... the Committee accept the view put forward by the NCC ... that a proportion of [woodlands], of very special interest to nature conservation or of particular historical importance, should be managed as nature reserves; and presumably surveys ... will soon provide a sound basis for *choosing which of these are most valuable*. Some old broadleaf plantations may warrant *inclusion in this category*
> (House of Lords Select Committee on Scientific Aspects of Forestry, 1979–1980, pp. 42, 43, emphasis added)

And section 126:

> The proper objective for those woodlands and old broadleaved plantations *which are not specially selected* for nature conservation is to manage them *productively and profitably* in a way that is compatible with maintaining a value for wildlife and amenity.
> (House of Lords Select Committee on Scientific Aspects of Forestry, 1979–1980, pp. 42, 43, emphasis added)

Shortly after, the seminal conference on *Broadleaves in Britain*, was held at Loughborough in 1982. It had been organized by the Forestry Commission and the Institute of Chartered Foresters. Interestingly, it was at this conference that the tide turned against the dominant forestry practices in favour of 'ancient woodlands'. The conference attracted over 250 participants, and its purpose was to investigate the future of broadleaves in Britain and the steps necessary to improve their management. The Nature Conservancy Council, drawing on Forestry Commission statistics, was able to present in 'numbers' the dwindling number of broadleaved woodlands, and it proposed the four management classifications referred to above. Using numbers, it was able to render more real what hitherto might have been rejected as a mere fancy, a pure object of thought. To get a flavour of the attitudes of the Forestry Commission at the time, two quotes are given below. The first one is from George Holmes's introductory address to the conference:

> Everyone has their point of view of what is of value and what needs to be done: the *forester* wants to *improve productivity by more intensive management, perhaps involving fast growing broadleaved species where appropriate.*
> (Forestry Commission, 1982, p. 5, emphasis added)

The second one is from his closing remarks:

> We spent a lot of time talking about the *classification* of woodlands ... The NCC is concerned primarily with identifying the areas that need to be managed in the *national* interest. We shall certainly endeavour to help with the classification.
> (Forestry Commission, 1982, p. 35)

The Loughborough conference was evidently a key moment in policy development; a moment when the Forestry Commission recognized that they were out of step with the rest of forestry thinking, and their attitudes changed considerably thereafter.

In accordance with the recommendations of the House of Lords Committee, the Nature Conservancy Council began its Ancient Woodland Inventory in 1981. Its aim was 'to provide a strong factual basis for woodland nature conservation' (Spencer and Kirby, 1992, p. 78). In order to *see* ancient woodlands, various characteristics and sources for their identification were used. For example, 'ancient woodland indicators', in other words, plant species indicative of 'ancient woodlands', such as bluebells, were used, as well as historical sources and indicators, like Anglo-Saxon charters, old maps, estate papers, wood boundaries such as wood banks, walls, ditches, and so forth.

Meanwhile, the 1979–1980 report of the House of Lords Committee had been published on 12 November 1980 and debated in the House of Lords on 23 February 1981. On 8 February, the Government made its Response which was subsequently published. For the 4th Report, published on 21 July 1982, the Director General of the Forestry Commission amongst others gave evidence, and again, the classification and categorization of broadleaved woodlands was at the centre of the debate. Some of the comments made are worth considering. First, the Earl of Cranbrook asked:

> Mr Holmes (Director General, FC) has been talking mostly about the planting of broadleaved woodland and plantations of broadly broadleaved character, but at the Loughborough conference the Nature Conservancy Council put forward a management policy recognising four categories ... Effectively, if the Nature Conservancy Council's management categories were to be agreed within the terms of the policy, there would be only 274,000 hectares, not 660,000, of broadleaved woodlands on which the sort of plantation policy to which Mr. Holmes was referring just now could be carried out and therefore I wish to ask Mr. Holmes whether within the framework of his policy *he recognises* the four treatment classes proposed by the NCC?
> (Forestry Commission, 1982, p. 32, emphasis added)

The response to this question made by Holmes was as follows:

> I think the answer very broadly speaking is yes, that is to say we would support the idea of *some such classification system*. I must say we would also have I would think fairly *considerable reservations on some aspects of the classification* and in particular on the *quantitative* side of the classification ... Our intention ... when we have our updated Census figures ... is to get together with the NCC, and *discuss their classification against the data we have*
> (Forestry Commission, 1982, p. 32, emphasis added)

Then Lord Dulverton enquired:

> On this question of what the Nature Conservancy Council *have come to refer to as semi-ancient woodland – it is category one of the division of four categories* we were talking about which they really want not to be managed for timber production at all – I just wonder whether, first of all, *they ever could be classified and agreed between all the authorities*, and secondly if they were whether they should come outside the remit of the Forestry Commission
> (Forestry Commission, 1982, p. 33, emphasis added)

Eventually, the Broadleaves Policy was announced in 1985, and in the Annual Report of the Forestry Commission of 1986 it is stated that the Forestry Commission must ensure that, apart from caring for broadleaved woodlands for wildlife conservation, recreation, amenity and other purposes, that:

> the special interests of the ancient semi-natural woodlands is recognised and maintained.

With this policy, the category of 'ancient woodland' became firmly institutionalized, although no commitment was made to the four management classes identified by the Nature Conservancy Council before. In a discussion paper of the Forestry Commission, the voluntary approach to nature conservation in Britain was even stressed (Forestry Commission, 1985, p. 8).

Nevertheless, 'ancient woodlands' were to be identified through the Nature Conservancy Council's Ancient Woodland Inventory and their management governed by guidelines. In 1992, management grants were introduced for woods of high environmental quality, and Codes of Practice had been developed for their management by the Forestry Commission in collaboration with Peterken as part of the Broadleaves Policy. 'Ancient woodlands' became part not only of the vocabulary, but of the national heritage *per se*; they became, in other words, national institutions.

Concluding Remarks

What I have been trying to show in this chapter, is how closely perceiving and acting upon 'the world' are related to definitions and categorizations of it. Value judgements are an inherent characteristic of the latter two. Inclusion and exclusion of living and non-living beings and what can and cannot be done to that which is included and to that which is excluded were crucial stakes in the struggle for the 'ancient woodland' category. This has also been recognized by Peterken, who observed in 1993 that:

> Another change [in recent years] has been towards a *broader definition* of which woods *are important* and a *practical* breakdown of the distinction between woodland nature reserves and other woods. Any list of important woods *carries the implication that other woods are not important.*
> (Peterken, 1993, p. 322, emphasis added)

This implication was confirmed by the NCC, pointing out that:

> Throughout the country, where it is uncertain whether a site is ancient or recent it has been classed as ancient.
> (Spencer and Kirby, 1992, p. 83)

Whether for better or the worse, once the category 'ancient woodland' had been introduced into the common vocabulary and delineated in terms of what it includes and what it excludes; once it had come out on top in the symbolic power struggle over the legitimate vision of the world, it could no longer be denied as a reality. 'Ancient woodlands' are now part of a taken-for-granted world that only two decades earlier had lacked their presence; although the various entities constituting them had been there for longer or shorter periods of time, they had been undefined and thus for human beings non-existent. The meaning 'ancient woodlands' have for people does not stem primarily from the trees on the site, which after all, do not have to be very old; nor does it stem from the species present at the site (which might be rare, but far from ancient). It does not stem from the ditch surrounding it, nor from the bluebells that have just sprung up. The meanings that ancient woodlands have for people stem from the subjective meanings people assign to any of these entities or a combination of them all. They might also be rooted in invisible factors, such as the longing for a time gone by and the wish to preserve the traces left behind by our own species in a landscape so created. If, as I indeed believe, we do play a part in bringing about the world we live in, then the work of Oliver Rackham and George Peterken has no doubt been instrumental in bringing about a richer one.

References

Berger, P. and Luckman, T. (1971) *The Social Construction of Reality.* Penguin, Harmondsworth.
Bourdieu, P. (1985) *Outline of a Theory of Practice.* Cambridge University Press, Cambridge.
Bourdieu, P. (1989) Social space and symbolic power. *Sociological Theory* 7, 14–25.
Bourdieu, P. (1990) *In Other Words – Essays Towards a Reflexive Sociology.* Polity Press, Cambridge.
Bourdieu, P. (1991) *Language and Symbolic Power.* Polity Press, Cambridge.
Bourdieu, P. (1992) *The Logic of Practice.* Polity Press, Cambridge.
Forestry Commission (1920) *First Annual Report.* Her Majesty's Stationery Office, London.
Forestry Commission (1982) *Broadleaves in Britain: Addresses, Supplementary Papers and Discussions.* Grayson, A.J. (ed.) *Occasional Paper No. 13.* Her Majesty's Stationery Office, London.
Forestry Commission (1985) *Broadleaves in Britain – Review of Policy. A Discussion Paper.* Her Majesty's Stationery Office, London.
Forestry Commission (1995) *Forestry Commission Facts and Figures 1994–1995.*
Foucault, M. (1984) Polemics, politics and problematizations. In: Rabinow, P. (ed.) *The Foucault Reader.* Panthenon Books, New York.
Gerber, J. (1997a) The social construction of nature: the case of forestry in Great Britain since the turn of the twentieth century. Unpublished DPhil thesis, University of Oxford.
Gerber, J. (1997b) Beyond dualism – the social construction of nature and the natural and social construction of human beings. *Progress in Human Geography* 21(1), 1–17
Harré, R. (1979) *Social Being – A Theory for Social Psychology.* Basil Blackwell, Oxford.
Harré, R. (1984) *Personal Being – A Theory for Individual Psychology.* Harvard University Press, Cambridge, Massachusetts.
Harré, R. (1991) *Physical Being – A Theory for Corporeal Psychology.* Basil Blackwell, Oxford.
Hart, C. (1991) *Practical Forestry – For the Agent and Surveyor,* 3rd Edn. Alan Sutton, Stroud.
House of Lords Select Committee on Science and Technology, Session 1979–1980, 2nd Report, *Scientific Aspects of Forestry.* Confidential – Final Revise [to be published as House of Lords Paper 381–1]. Her Majesty's Stationery Office, London.
House of Lords Select Committee on Science and Technology, Session 1981–1982, 4th Report. *Scientific Aspects of Forestry – Supplementary Report.* Her Majesty's Stationery Office, London.
IUCN (International Union for the Conservation of Nature) (1980) *World Conservation Strategy.* IUCN.
Lefebvre, H. (1991) *The Production of Space* (Trans. Nicholson-Smith, D.). Blackwell, Oxford.
Mather, A.S. (1991) Pressures on British forest policy: prologue to the post-industrial forest? *Area* 23(3), pp. 245–253.

Mather, A.S. (1994) Policy reform and institutional restructuring: the case of the Forestry Commission. In: Gilg, A.W. (ed.) *Progress in Rural Policy and Planning.* John Wiley & Sons, London.

Peterken, G.F. (1981) *Woodland Conservation and Management.* Chapman & Hall, London.

Peterken, G. (1993) *Woodland Conservation and Management,* 2nd edn. Chapman & Hall, London.

Peterken, G.F. (1994) The definition, evaluation and management of ancient woods in Great Britain. *NNA-Berichte* 3(94), pp. 102–114.

Popper, K. and Eccles, J.C. (1977) *The Self and its Brain.* Springer-Verlag, Berlin.

Pryor, S. (1992) *Future Forestry – A New Direction for Forest Policy.* Report prepared for Wildlife Link.

Rackham, O. (1976) *Trees and Woodland in the British Landscape.* Dent, London.

Rackham, O. (1980) *Ancient Woodland: Its History, Vegetation and Uses in England.* Edward Arnold, London.

Revill, G. and Watkins, C. (1996) Educated access: interpreting Forestry Commission Forest Park Guides. In: Watkins, C. (ed.) *Rights of Way: Policy, Culture, and Management.* Pinter, London, pp. 100–128.

Ryle, G.B. (1969) *Forest Service – The First Forty-five Years of the Forestry Commission of Great Britain.* David & Charles, Newton Abbott.

Spencer, J.W. and Kirby, K.J. (1992) An inventory of ancient woodland for England and Wales. *Biological Conservation* 62, pp. 77–93.

Steven, H.M. and Carlisle, A. (1959) *The Native Pinewoods of Scotland.* Oliver & Boyd, Edinburgh.

Thomas, K. (1984) *Man and the Natural World. Changing Attitudes in England 1500–1800.* Allen Lane, London.

Tompkins, S. (1989) *Forestry in Crisis – The Battle for the Hills.* Christopher Helm, London.

Tubbs, C.R. (1964) Early encoppicements in the New Forest. *Forestry* 37, pp. 95–105.

Usher, M.B. and Thompson, D.B.A. (eds) (1988) *Ecological Change in the Uplands.* Special Publication Number 7 of the British Ecological Society. Blackwell Scientific Publications, Oxford.

Wacquant, J.D. (1989) Towards a reflexive sociology – a workshop with Pierre Bourdieu. *Sociological Theory* 7, pp. 26–63.

Watkins, C. (1988) The idea of ancient woodland in Britain from 1800. In: Fabio Salbitano (ed.) *Human Influence on Forest Ecosystems Development in Europe.* ESF FERN-CNR. Pitagora Editrice, Bologna, pp. 237–246.

Watkins, C. (1990) *Woodland Management and Conservation.* David & Charles, London.

Wright, P. (1992) The disenchanted forest. *The Guardian,* Weekend Edition, 7 November 1992, p. 8.

REFERENCES

Aaronson, A. (1913) *On the Deforestation of Eretz Israel and a Suggestion for its Renovation*. Submitted according to invitation by His Majesty Backri Bay, the General Governor of Beirut, on occasion of his visit to the agricultural research station, in October 1913. Beit Aaronson Pub., Zichron, Yaakov. (In Hebrew.)

Abel, W. (1974) *Massenarmut und Hungerkrisen im Vorindustriellen Europa. Versuch einer Synopsis*. Hamburg, Berlin.

Abelshauser, W. (ed.) (1994) *Umweltgeschichte. Umweltverträgliches Wirtschaften in Historischer Perspektive*. Acht Beiträge, Göttingen.

Abreu y Pidal, J.M. (1963) Hacia la obtención de la máxima renta en el cultivo del bosque. *Montes* 19(112), pp. 347–348.

Albion, R.G. (1926) *Forests and Sea Power: The Timber Problem of the Royal Navy, 1652–1862*. Harvard University Press, Cambridge, Massachussetts.

Allmann, J. (1989) *Der Wald in der Frühen Neuzeit: eine Mentalitäts – und Sozialgeschichtliche Untersuchung am Beispiel des Pfälzer Raumes 1500–1800*. Berlin.

Araujo, J. (1994) Cultivar la cultura. *El Boletín* 15, MAPA, Madrid, pp. 58–62.

Aslet, C. (1979) Thoresby Hall, Part I. *Country Life*, p. 165.

Balboa, X.L. (1990) *O Monte en Galicia*. Ediciòns Xerais Universitaria, Vigo.

Baker, R. (1984) Protecting the environment against the poor. The historical roots of the soil erosion orthodoxy in the Third World. *The Ecologist* 14(2), pp. 53–60.

Barberis, G., Peccenini, S. and Paola G. (1992) Notes on *Quercus ilex* L. in Liguria (NW Italy). *Vegetatio* 99–100, pp. 35–50.

Baring, F.H. (1909) *Domesday Tables for the Counties of Surrey, Berkshire, Middlesex, Hertford, Buckingham and Bedford, and for the New Forest*. London.

Bauer, E. (1962) *Der Soonwald im Hunsrück. Forstgeschichte eines Deutschen Waldgebietes*. Freiburg.

Bauer, E. (1981) *Unsere Wälder im Historischen Kartenbild. Beitrag zur Geschichte des Forstkartenwesens in Rheinland-Pfalz*. Grünstadt.

Bauer Manderscheid, E. (1980) *Los Montes de España en la Historia*. Ministerio de Agricultura, Madrid.

Bazeley, M.L. (1921) The extent of the English forests in the thirteenth century. *Transactions of the Royal Historical Society*, 4th series 4, pp. 140–172/146–148.

Beastall, T.W. (1974) *A North Country Estate*. Phillimore, Chichester.

Beckett, J.V. (1986) *The Aristocracy in England 1660–1914*. Basil Blackwell, Oxford.

Ben Artzi, Y. (1988) *Jewish Moshava Settlements in Eretz Israel (1882–1914)*. Yad Izhak Ben Zvi, Jerusalem, pp. 186–187. (In Hebrew.)

Bennett, H.H. (1960) Soil erosion in Spain. *The Geographical Review* 50, pp. 59–72.

Berger, P. and Luckman, T. (1971) *The Social Construction of Reality*. Penguin, Harmondsworth.

Biger, G. and Liphschitz, N. (1991) Early distribution of *Pinus brutia* – a reassessment. *Holocene* 1, pp. 157–161.

Biger, G. and Liphschitz, N. (1994) Australian trees in Eretz Israel. *Research in Geography* 14, pp. 108–121. (Hebrew).

Birch, W. de Gray (1885–1899) *Cartularium Saxonicum*. Whiting and Co., London; reprint 1964, Johnson Reprint Co., New York and London.

Bird, E.A.R. (1987) The social construction of nature: theoretical approaches to the history of environmental problems. *Environmental Review* 11, pp. 255–264.

Birrell, J.R. (1980) The English medieval forest. *Journal of Forest History* 24, pp. 78–85.

Birrell, J.R. (1996) Hunting and the royal forest. *L'Uomo e la Foresta Secc. 12–18*, Atti delle Settimane di Studi, Prato 1995, 27 (Prato), pp. 437–457.

Blaikie, P. and Brookfield, H. (1987) *Land Degradation and Society*. Methuen, London.

Blore, W.P. and Harvey, J.H. (1945) Recent discoveries in the archives of Canterbury Cathedral. *Archiv. Cantiana* 58, pp. 28–39.

Bode, W. and von Hohnhorst, M. (1994) *Waldwende. Vom Försterwald zum Naturwald*. München.

Boulton, H.E. (1959) The forest books of the Royal Forest of Sherwood. Unpublished MA thesis, University of Nottingham.

Bourdieu, P. (1985) *Outline of a Theory of Practice.* Cambridge University Press, Cambridge.
Bourdieu, P. (1989) Social space and symbolic power. *Sociological Theory* 7, 14–25.
Bourdieu, P. (1990) *In Other Words – Essays Towards a Reflexive Sociology.* Polity Press, Cambridge.
Bourdieu, P. (1991) *Language and Symbolic Power.* Polity Press, Cambridge.
Bourdieu, P. (1992) *The Logic of Practice.* Polity Press, Cambridge.
Brown, A.G. (1997) *Alluvial Geoarchaeology: Floodplain Archaeology and Environmental Change.* Cambridge University Press, Cambridge.
Burgers, T.F. (1949) La regeneración natural y la mezcla de especies como ideal en la silvicultura. *Montes.* 30, ETSIM, Madrid, pp. 505–508.
Butler, R.M. (1954) Archaeology in Nottinghamshire – achievements and prospects. *Transactions of the Thoroton Society of Nottinghamshire* 58, pp. 1–20.
Byng, J. (1789) *The Torrington Diaries. Tour of the Midlands.* Andrews, C.B. (ed.) (1954) London.
Cameron, A. (1975) Some social consequences of the dissolution of the monasteries in Nottinghamshire. *Transactions of the Thoroton Society of Nottinghamshire,* 79.
Cantor, L. (1982) Forests, chases, parks and warrens. In: Cantor, L. (ed.) *The English Medieval Landscape.* Croom Helm, London, pp. 56–85.
Cantor, L.M. and Wilson, J.D. (1964) The medieval deer-parks of Dorset: 3. *Proceedings of the Dorset Natural History and Archaeological Society for 1963* p. 86, pp. 141–152.
Carlowitz, H.C. von (1713) *Silvicultura oeconomica, oder hausswirthschaftliche Nachricht und naturmässige Anweisung zur wilden Baum-Zucht...,* Leipzig.
Carter, J. (1850) *A Visit to Sherwood Forest including the Abbeys of Newstead, Rufford, and Welbeck ... With a Critical Essay on the Life and Times of Robin Hood.* Longman, London.
Cave, T. and Wilson, R.A. (eds) (1924) *The Parliamentary Survey of the Lands and Possessions of the Dean and Chapter of Worcester Made in or About the Year 1649 in Pursuance of an Ordinance of Parliament for the Abolishing of Deans and Chapters.* Worcester Historical Society, London.
Cernea, M.M. (ed.) (1991) *Putting People First. Sociological Variables in Rural Development,* 2nd Edn. World Bank, Washington, DC.
Chambers, J.D. (1932) *Nottinghamshire in the Eighteenth Century.* King, London.
Chambers, R. (1992) *Rural Appraisal: Rapid, Relaxed and Participatory.* Discussion Paper 311, Institute of Development Studies, Sussex.
Chambers, R. (1993) *Challenging the Professions. Frontiers for Rural Development.* Intermediate Technology Publications, London.
Chambers, R. (1994a) The origins and practice of Participatory Rural Appraisal. *World Development* 22(7), pp. 953–969.
Chambers, R. (1994b) Participatory Rural Appraisal (PRA): Analysis of experience. *World Development,* 22 (9), pp. 1253–1268.
Chambers, R. (1994c) *Paradigm Shifts and the Practice of Participatory Research and Development.* Institute of Development Studies, Working Paper 2, University of Sussex, Brighton.
Chapman, F.R. (1907) *The Sacrist Rolls of Ely.* 2 vols. Cambridge University Press, Cambridge.
CLCAPV (1993) *Sanz Iruela, M. to Delegado provincial, 19 November.* Letter, CLCAPV, Guadalajara.
CLCAPV (1994) *CLCAPV to Delegado provincial, 13 October.* Letter, CLCAPV, Guadalajara.
CLCAPV (1995) *Sanz Azconas, N. to Hurtado Pérez, F., 8 March.* Letter, CLCAPV, Guadalajara.
Clifton, J.M. (1979) 'An enchanted palace', Clumber Park, the Newcastle family seat. Unpublished BArch dissertation, University of Nottingham, Nottingham.
Coates, R. (1989) *The Place-Names of Hampshire.* Batsford, London.
Cohen, A. and Biger, G. (1987) British mandate actions to preserve the country's landscape and nature. In: Schiller, E. (ed.) *Zev Vilnay Jubilee Book, Vol. 2,* Ariel Pub., Jerusalem, pp. 295–300. (Hebrew).
Cole, R.E.G. (ed.) (1917) Chapter Acts, Lincoln Cathedral, pp. 1547–59. *Lincoln Record Society* 15, pp. 51–52.
Collins, E.J.T. (1992) Woodlands and woodland industries in Great Britain during and after the charcoal iron era. In: Metaille, J.P. (ed.) *Protoindustries et histoire des forêts, (Les Cahiers de l'ISARD* 3). University of Toulouse, Toulouse, pp. 109–120.
Copley, S. and Garside, P. (eds.) (1994) *The Politics of the Picturesque.* Cambridge University Press, Cambridge.
Codorniu, R. (1915) La repoblación forestal en España. Medios de fomentarla y de convencer de su necesidad a las clases rurales I. *Montes,* pp. 912–915, Madrid.
Coelho, C. et al. (1994) *IBERLIM. Land Management and Erosion Limitation in the Iberian Peninsula.* EU Project EV5V–0041.
Corvol-Dessert, A. (1987) *L'homme aux Bois. Histoire des Relations de l'Homme et de la Forêt, 17–20 Siècles.* Ed. Fayard, Paris.
Cowell, B. (1998) Patrician Landscapes Plebian Culture: Parks and Society in Two English Counties, c. 1750–1850. Unpublished PhD thesis, University of Nottingham.

Cox, J.C. (1905) *The Royal Forests of England*. Methuen, London.
Cox, J.C. and Whitworth, R.H. (1910) Forestry. In: Page, W. (ed.) *The Victoria History of the County of Nottinghamshire*, vol. 1. London, pp. 365–381.
Crook, D. (1979) The struggle over forest boundaries in Nottinghamshire, 1218–1227. *Transactions of the Thoroton Society of Nottinghamshire* 83, pp. 35–45.
Crumlin-Pedersen, O. (1986) Aspects of wood technology in medieval shipbuilding. In: Crumlin-Pedersen, O. and Vinner, M. (eds.) *Sailing into the Past*. Proceedings of the International Seminar on Replicas of Ancient and Medieval Vessels, Roskilde, 1984. Roskilde.
Cummins, J. (1988) *The Hound and the Hawk, the Art of Medieval Hunting*. St Martin's Press, New York.
Daniels, S. (1988) The political iconography of woodland in later Georgian Britain. In: Cosgrove, D. and Daniels S. (eds) *The Iconography of Landscape*. Cambridge University Press, Cambridge, pp. 43–82.
Daniels, S. and Watkins, C. (1991) Picturesque landscaping and estate management: Uvedale Price at Foxley, 1770–1829. *Rural History* 2, pp. 141–169/170.
Daniels S. and Watkins, C. (eds) (1994) *The Picturesque Landscape. Visions of Georgian Herefordshire*. University of Nottingham. Nottingham.
Darby, H.C. (1977) *Domesday England*. Cambridge University Press, Cambridge.
Darby, H.C. (1979) Dorset. In: Darby, H.C. and Welldon Finn, R. (eds) *The Domesday Geography of South-West England*. Cambridge University Press, Cambridge, pp. 67–131.
Darby, H.C. and Campbell, E.M.J. (ed.) (1962) *The Domesday Geography of South-East England*. Cambridge University Press, Cambridge.
Davies, D. (1988) The evocative symbolism of trees. In: Cosgrove, D. and Daniels S. (eds) *The Iconography of Landscape*. Cambridge University Press, Cambridge, pp. 32–42.
Day, S.P. (1989) Reconstructing the environment of Shotover Forest, Oxfordshire. *Medieval Settlement Research Group Annual Report* 4, p. 6.
Demeritt, D. (1994) Ecology, objectivity and critique in writings on nature and human societies. *Journal of Historical Geography* 20, pp. 22–37.
De Raiz (1994) Valle de Jarama. *Quercus*, Noviembre, p. 8.
Devereux, S. and Hoddinott, J. (eds) (1992) *Fieldwork in Developing Countries*. Harvester Wheatsheaf, Hertfordshire.
DOCM (1990) *Documento Oficial de Castilla-la-Mancha*, No. 81, Toledo.
Doughty, R. (1996) Not a koala in sight: promotion and spread of eucalyptus. *Ecumene* 3, pp. 200–214.
Duckworth, A.M. (1971) 'Mansfield Park' and estate improvements: Jane Austen's grounds of being. *Nineteenth Century Fiction* 26, pp. 25–48.
Eckardt, H.W. (1976) *Herrschaftliche Jagd, bäuerliche Not und bürgerliche Kritik. Zur Geschichte der fürstlichen und adligen Jagdprivilegien vornehmlich im südwestdeutschen Raum*. Göttingen.
Eckstein, D., Bauch, J., Klein, P. and Wazny, T. (1986) New evidence for the dendrochronological dating of Netherlandish paintings. *Nature* 320, pp. 465–466.
Eddison, E. (1854) *History of Worksop; With Historical, Descriptive, and Discursive Sketches of Sherwood Forest*. Longman, London.
Edwards, J.F. and Hindle, B.P. (1991) The transportation system of medieval England and Wales. *Journal of Historical Geography* 17(2), p. 12.
Ellis, F. (1988) *Peasant Economics. Farm Households and Agrarian Development*. Cambridge University Press, Cambridge.
Epperlein, S. (1993) *Waldnutzung, Waldstreitigkeiten und Waldschutz in Deutschland im hohen Mittelalter: 2 Hälfte 11. Jahrhundert bis ausgehendes 14. Jahrhundert*, Stuttgart.
Ernst, C. (1995) Ein neuer Umgang mit Natur? Der Kondelwald im 18. Jahrhundert. In: Klaus Freckmann (ed.), *Sobernheimer Gespräche 3. Das Land an der Mosel – Kultur und Struktur*. Köln, pp. 21–32.
Erskine, A. (1981) The Accounts of the Fabric of Exeter Cathedral, part 1, 1279–1326. *Devon and Cornwall Record Society* n.s. 24.
Erskine, A. (1983) The Accounts of the Fabric of Exeter Cathedral, part 2, 1328–1353. Devon and Cornwall Record Society n.s. 26.
Esling, J., Howard, R.E., Laxton, R.R., Litton, C.D. and Simpson, W.G. (1989) List 29: Nottingham University Tree-Ring Dating Laboratory Results. *Vernacular Architecture* 20, p. 39.
Evans, S. (1973) *The Medieval Estate of the Cathedral Priory of Ely*. Ely.
Evelyn, J. (1670) *Sylva, or a Discourse of Forest-Trees, and the Propogation of Timber in His Majesties Dominions*. Royal Society, London [2nd Edn; 1st Edn 1667].
Evernden, N. (1992) *The Social Creation of Nature*. Baltimore and London.

Fairbrother, J.R. (1984) *Faccombe Netherton. Archaeological and Historical Research* 1, City of London Archaeological Society.

Farey, J. (1813) *General View of the Agriculture of Derbyshire, with Observations on the Means of its Improvement.* London.

Fearn, K. and Simpson, W.G. (1997) Worcester Cathedral as a source of forest and woodland history. In: Barker, P. and Guy, C. (eds) *Archaeology at Worcester Cathedral: Report of the Seventh Annual Symposium.* Worcester, pp. 6–16.

Fernández Leiceaga, X. (1990) *Economía (política) do Monte Galego.* Universidade de Santiago de Compostela, Santiago.

Fernández Prieto, L. (1992) *Labregos con Ciencia. Estado, Sociedade e Innovación Tecnolóxica na Agricultura Galega.* 1850–1939. Editorial Xerais Universitaria, Vigo.

Firth, C.H. and Rait, R.S. (eds.) (1911) *Acts and Ordinances of the Interregnum, 1642–1660.* v.2.

Foot, N.D.J., Litton, C.D. and Simpson, W.G. (1986) The high roofs of the east end of Lincoln Cathedral. In: Heslop, T.A. and Sekules, V.A. (eds) *Medieval Art and Architecture at Lincoln Cathedral* (BAA Conf. Trans., 8), pp. 47–74.

Forestry Commission (1920) *First Annual Report.* Her Majesty's Stationery Office, London.

Forestry Commission (1982) *Broadleaves in Britain: Addresses, Supplementary Papers and Discussions.* Grayson, A.J. (ed.) *Occasional Paper No. 13.* Her Majesty's Stationery Office, London.

Forestry Commission (1985) *Broadleaves in Britain – Review of Policy. A Discussion Paper.* Her Majesty's Stationery Office, London.

Forestry Commission (1995) *Forestry Commission Facts and Figures 1994–1995.*

Foucault, M. (1984) Polemics, politics and problematizations. In: Rabinow, P. (ed.) *The Foucault Reader.* Panthenon Books, New York.

Fowkes, D.V. (1967) Nottinghamshire parks in the eighteenth and nineteenth centuries. *Transactions of the Thoroton Society of Nottinghamshire* 71, pp. 72–89.

Fowkes, D.V. (1977) The breck system of Sherwood forest. *Transactions of the Thoroton Society of Nottinghamshire* 81, pp. 55–61.

García Pérez, J.D. (1996) Socioeconomic processes of land degradation in Guadalajara Province (Spain). Unpublished PhD thesis. Plymouth University.

García Pérez, J.D., Charlton, C. and Martín Ruiz, P. (1995) Landscape changes as visible indicators in the social, economic and political process of soil erosion: a case study of the municipality of Puebla de Valles (Guadalajara Province), Spain. *Land Degradation and Rehabilitation* 6, pp. 149–161.

Gentles, I. (1973) The Sales of Crown Lands during the English Revolution. *Economic History Review,* 2nd series 26, pp. 614–635.

Gerber, J. (1997a) The social construction of nature: the case of forestry in Great Britain since the turn of the twentieth century. Unpublished DPhil thesis, University of Oxford.

Gerber, J. (1997b) Beyond dualism – the social construction of nature and the natural *and* social construction of human beings. *Progress in Human Geography* 21(1), pp. 1–17.

Gerber, J. (1998) Beyond the primordial silence: the voice of reason and the voice of nature. *Environment and Planning A* (in press).

Gil, N. (1986) Desarrollo de cuencas hidrográficas y conservación de suelos, *Boletin de Suelos de la FAO* 44, Roma.

Gilpin, W. (1791) *Remarks on Forest Scenery, and Other Woodland Views (Relative Chiefly to Picturesque Beauty)...* R. Blamire, London.

Gindel, I. (1952) *Forest and Afforestation in Israel.* The Laboratory for Afforestation Research, Rehovot. (Hebrew).

Girouard, M. (1981) *The Return to Camelot. Chivalry and the English Gentleman.* Yale University Press, New Haven and London.

Glück, P. and Ottitsch, A. (1995) *Entwicklung eines Rauminformationssystems Wienerwald. Evaluierung der Wienerwalddeklaration.* Institut für Forstliche Betriebswirtschaft und Forstwirtschaftspolitik, Universität fur Bodenkultur, Wien, pp. 34–38.

Gómez Mendoza, J. (1992) *Ciencia y Política de los Montes Españoles (1848–1936).* Icona, Madrid.

González Bernáldez, F. (1990) Consideraciones ecológico-políticas acerca de la conservación y regeneración de la cubierta vegetal en España. *Ecología* (fuera de serie), ICONA, Madrid, pp. 439–445.

Goodchild, J. (1996) Lionell Copley: a seventeenth century capitalist. In: Jones, M. (ed.) *Aspects of Rotherham: Discovering Local History.* Wharncliffe Publishing, Barnsley, pp. 16–22.

Green, M.A.E. (ed.) (1870) *Calendar of State Papers Domestic, 1601–1603,* London.

Grees, H. (ed.) (1969) *Der Schönbuch. Beiträge zu seiner landeskundlichen Erforschung.* Konkordia, Bühl/Baden.

Grimm, J. and Grimm, W. (1877) *Deutsches Wörterbuch von Jacob Grimm und Wilhelm Grimm, 4, 2.* Leipzig.
Groome, H. (1981) Las sugerencias nunca atendidas del Plan Forestal español. *Quercus,* ETSIM, Madrid, pp. 31–34.
Groome, H. (1985) El desarrollo de la política forestal en el Estado español: Desde el siglo 19 hasta la Guerra Civil. *Arbor* 121(474), Madrid, pp. 59–89.
Groome, H. (1990) *Historia de la Política Forestal en el Estado Español.* Agencia del Medio Ambiente, Comunidad de Madrid, Madrid.
Groome, H., Martín Montalvo, R. and Llorca, A. (1989) Historia forestal de España. *News of Forest History* 9–10, Vienna.
Gurevitch, D. and Gertz, A. (1947) *Statistical Handbook of Jewish Palestine, Jerusalem.* The Jewish Agency of Palestine, Jerusalem, p. 187.
Halkin, J. and Roland, C.G. (eds) (1909) *Recueil des Chartes de l'abbaye de Stablo-Malmédy.* Kiessling, Bruxelles.
Hall, T.W. (1914) *Descriptive Catalogue Forming the Jackson Collection.* J.W. Northend, Sheffield.
Hammersley, G. (1973) The charcoal iron industry and its fuel. *Economic History Review. Second Series* 26, pp. 593–613.
Harré, R. (1979) *Social Being – A Theory for Social Psychology.* Basil Blackwell, Oxford.
Harré, R. (1984) *Personal Being – A Theory for Individual Psychology.* Harvard University Press, Cambridge, Massachusetts.
Harré, R. (1991) *Physical Being – A Theory for Corporeal Psychology.* Basil Blackwell, Oxford.
Harriss, J. (ed.) (1982) *Rural Development. Theories of Peasant Economy and Agrarian Reform.* Hutchinson, London.
Hart, C. (1991) *Practical Forestry – For the Agent and Surveyor,* 3rd Edn. Alan Sutton, Stroud.
Hasel, K. (1985) *Forstgeschichte. Ein Grundriß für Studium und Praxis.* Hamburg and Berlin.
Heal, F. (1973) The Tudors and church lands; economic problems of the bishops of Ely during the sixteenth century. *Economic History Review,* 2nd series 26, pp. 198–217.
Heal, F. (1980) *Of Prelates and Princes: A Study of the Economic and Social Position of the Tudor Episcopate.* Cambridge University Press, Cambridge.
Hendry, G.A.F, Bannister, N. and Toms, J. (1984) The earthworks of an ancient woodland. *Bristol and Avon Archaeology* 3, pp. 47–53.
Herzel, T. (1896) *The Diary. Vol. 1,* a note from 23 August 1896, 334. (Hebrew).
Hewett, C.A. (1985) *English Cathedral and Monastic Carpentry.* Phillimore, Chichester.
Hey, D. (1977) The ironworks at Chapeltown. *Transactions of the Hunter Archaeological Society* 10, pp. 252–259.
Hill, D. (1981) *An Atlas of Anglo-Saxon England.* Blackwell, Oxford.
Hilton, R.H. (1966) *A Medieval Society.* Weidenfeld & Nicholson, London.
Holdsworth, C.J. (ed.) (1974) Rufford Charters, 2. *Thoroton Society Record Series* 30.
Holland, J. (1828) *The History, Antiquities and Description of the Town and Parish of Worksop.* Sheffield.
Holt, J.C. (1982) *Robin Hood.* Thames & Hudson, London.
Hooke, D. (1981) *Anglo-Saxon Landscapes of the West Midlands: The Charter Evidence.* British Archaeological Reports, British Series 95. Oxford.
Hooke, D. (1982) The Anglo-Saxon landscape. In: Slater, T.R. and Jarvis, P.J. (eds) *Field and Forest: An Historical Geography of Warwickshire and Worcestershire.* Norwich, pp. 79–104.
Hooke, D. (1985) *The Anglo-Saxon Landscape, The Kingdom of the Hwicce.* Manchester University Press, Manchester.
Hooke, D. (1988) The Warwickshire Arden: the evolution and future of an historic landscape. *Landscape History* 10, pp. 51–59.
Hooke, D. (1989) Pre-Conquest woodland: its distribution and usage. *Agricultural History Review* 37, pp. 113–129.
Hooke, D. (1990) *Worcestershire Anglo-Saxon Charter-Bounds.* Boydell Press, Woodbridge.
Hooke, D. (1993) Woodland in the peasant economy of England. In: Brandl, H. (ed.) *Geschichte der Kleinprivatwaldwirtschaft, Geschichte des Bauernwaldes,* Mitteilungen der Forstlichen Versuchs – und Forschungsanstalt Baden-Württemberg, Heft 175. Freiburg, pp. 202–210.
Hooke, D. (1994) The woodlands of England in Domesday Book. In: Billen, C. and Vanrie, A. (eds) *Les Sources de l'Histoire Forestière de la Belgique.* Archives et Bibliothèques de Belgique, Numéro spécial 45. Brussels, pp. 35–51.
Hooke, D. (1998) *The Landscapes of Anglo-Saxon England.* Leicester University Press, London and Washington (in press).
Hopkinson, G.G. (1963) The charcoal iron industry in the Sheffield region 1550–1775. *Transactions of the Hunter Archaeological Society* 8, pp. 122–151.

House of Commons (1793) The Fourteenth Report of the Commissioners appointed to enquire into alienate Fee Farm and other Unimproveable Rents. *House of Commons Journal* 48, pp. 467–511.

House of Lords Select Committee on Science and Technology, Session 1979–1980, 2nd Report, *Scientific Aspects of Forestry*. Confidential – Final Revise [to be published as House of Lords Paper 381–1]. Her Majesty's Stationery Office, London.

House of Lords Select Committee on Science and Technology, Session 1981–1982, 4th Report. *Scientific Aspects of Forestry – Supplementary Report*. Her Majesty's Stationery Office, London.

Hunter, J. (1994) *Everyone Who Ever Mattered is Dead and Gone*. Institute of Environmental History, University of St Andrews (Unpublished paper from conference on Cultural Environments 2), St Andrews.

Irving, W. (1835) *Abbotsford and Newstead Abbey*. John Murray, London.

IUCN (International Union for the Conservation of Nature) (1980) *World Conservation Strategy*. IUCN.

Jacques, D. (1983) *Georgian Gardens: The Reign of Nature*. Batsford, London.

James, N.D.G. (1981) *A History of English Forestry*. Basil Blackwell, Oxford.

James, N.D.G (1996) A history of forestry and monographic forestry literature in Germany, France, and the United Kingdom. In: McDonald, P. and Lassoie, J. (eds) *The Literature of Forestry and Agroforestry*. Cornell University Press, Ithaca, pp. 15–44.

Jones, M. (1986a) Coppice wood management in the eighteenth century: an example from County Wicklow. *Irish Forestry* 43(1), pp. 15–31.

Jones, M. (1986b) *Sheffield's Ancient Woods: Some Notes on Their History and Past Management with Special Reference to Woods Owned by the City Council*. Unpublished report for Sheffield City Council.

Jones, M. (1989) *Sheffield's Woodland Heritage*. Sheffield City Libraries, Sheffield.

Jones, M. (1989–1992) *Inventory Survey of Ancient Woods in Rotherham Metropolitan Borough*. Progress reports 1 (1989), 2 (1990) and 3 (1992). Unpublished reports for Rotherham Metropolitan Borough Council Planning Department.

Jones, M. (1993) *South Yorkshire Forest: Historic Landscapes Study*. Unpublished report for South Yorkshire Forest Project.

Jones, M. (1995) *Rotherham's Woodland Heritage*. Rotherwood Press, Rotherham.

Jones, M. (1996) Deer in South Yorkshire: an historical perspective. In: Jones, M., Rotherham, I.D. and McCarthy, A.J. (eds) *Deer or the New Woodlands?* (*Journal of Practical Ecology and Conservation*, Special Publication No. 1).

Jones, M. and Talbot, E. (1995) Coppicing in urban woodlands: a progress report on a multi-purpose feasibility study in the City of Sheffield. *Journal of Practical Ecology and Conservation* 1, pp. 46–52.

Kaspers, H. (1957) *Comitatus Nemoris. Die Waldgrafschaft zwischen Maas und Rhein*. Dürener Geschichtsverein, Düren and Aachen.

Kiernan, D. (1989) *The Derbyshire Lead Industry in the Sixteenth Century*. Derbyshire Record Society, Chesterfield.

Kiess, R. (1958) *Die Rolle der Forsten im Aufbau des württembergischen Territoriums bis ins 16. Jahrhundert* (Veröffentlichungen der Kommission für geschichtliche Landeskunde in Baden-Württemberg Reihe B Forschungen 2. Band). Kohlhammer, Stuttgart.

Kiess, R. (1992) Forst-Namen als Spuren frühmittelalterlicher Geschichte in Württemberg. *Zeitschrift für Württembergische Landesgeschichte* 51, pp. 11–116.

Kiess, R. (1995) Der Bietigheimer Forst im Rahmen der allgemeinen Geschichte der Forsten. *Blätter zur Stadtgeschichte* 12, pp. 7–28. Hg. vom Archiv der Stadt Bietigheim-Bissingen.

Kiess, R. (1996) Forst-Namen als Spuren frühmittelalterlicher Geschichte II: Beispiele aus Baden und angrenzenden Territorien. Folgerungen für die Forstgeschichte. *Zeitschrift für die Geschichte des Oberrheins* 144, pp. 47–123.

Kirby, K.J. and Watkins, C. (eds) (1998) *The Ecological History of European Forests*. CAB International, Wallingford.

Kitching, C.J. (1986) Re-roofing old St Paul's Cathedral, 1561–66. *The London Journal* 12 (2), pp. 123–133.

Kliot, N. (1993) Ideology and afforestation in Israel – man-made forests of the J.N.F. *Research in Geography* 13, pp. 87–106. (Hebrew).

Laird, F.C. (1810) *A Topographical and Historical Description of Nottinghamshire*. Sherwood, Neely & Jones, London.

Laso Rhodes, A. (1994) *Infórme Tecnico Sobre la Concentración Parcelaria in Puebla de Valles*. Delegación Provincial de Aricultura, Guadalajara.

Lefebvre, H. (1991) *The Production of Space* (Trans. Nicholson-Smith, D.). Blackwell, Oxford.
Lemon, R. (ed.) (1856) *Calendar of State Papers Domestic, 1547–1580.* London.
Leser, H. (1991) *Landschaftsökologie: Ansatz, Modelle, Methodik, Anwendung.* Hartmut Leser; Ulmer, Stuttgart, p. 37.
Levontin, Z. (1925) *'Le'eretz Avotainu' (To the Land of Our Forefathers).* Eitan and Shoshani, Tel Aviv, p. 111. (Hebrew).
Lexer, M. (1992) *Mittelhochdeutschen Taschenwörtebuch* 38. Stuttgart.
Lindsay, J. (1974) The use of woodland in Argyllshire and Perthshire between 1650 and 1850. Unpublished PhD thesis, University of Edinburgh.
Linnard, W. (1982) *Welsh Woods and Forests: History and Utilization.* National Museum of Wales, Cardiff.
Liphschitz, N. (1986) Overview of the dendrochronology and dendroarchaeology in Israel. *Dendrochronologia* 4, pp. 37–58.
Liphschitz, N. (1988) Dendrochronological and dendroarchaeological investigations in Israel as a means for the reconstruction of past vegetation and climate. *PACT* 22, pp. 133–146.
Liphschitz, N. and Biger, G. (1990) The dominance of *Quercus calliprinos* (Kermes oak) – *Pistacia palaestina* (Terebinth) association in the Mediterranean territory of Eretz Israel during antiquity. *Journal of Vegetation Science* 1, pp. 67–70.
Liphschitz, N. and Biger, G. (1994a) Afforestation policy of the British Regime in Palestine. *Horizons in Geography* 40–41, pp. 5–16. (Hebrew).
Liphschitz, N. and Biger, G. (1994b) The rise and fall of *Pinus halepensis* (Aleppo pine) as a main afforestation tree in Eretz Israel. Division of Publications of the JNF, no. 24. (Hebrew).
Liphschitz, N. and Biger, G. (1995) Sand dune reclamation in British Palestine. In: Simmons, I.G. and Mannion, A.M. (eds) *The Changing Nature of the People–Environment Relationships: Evidence from a Variety of Archives.* Prague, pp. 47–55.
Litton, C.D. and Simpson, W.G. (1996) Dendrochronology in cathedrals. In: Tatton-Brown, T. and Munby, J. (eds) *The Archaeology of Cathedrals.* Oxford.
Livne, A. (1969) *Aharon Aaronson.* Bialik Institute, Jerusalem, p. 71. (Hebrew).
López Carrasco, F. (1994) Castilla-la-Mancha, Región rural. *El Boletín,* 18, MAPA, Madrid, pp. 50–57.
Lowe, R. (1798) *General View of the Agriculture of the County of Nottingham.* London.
Madge, S.J. (1938) *The Domesday of Crown Lands: a Study of the Legislation, Surveys and Sales of Royal Estates under the Commonwealth.* Routledge, London.
Malden, H.E. (ed.) (1905) *The Victoria History of the County of Surrey.* Constable, London.
Mantel, K. (1990) *Wald und Forst in der Geschichte.* M. and A. Schaper, Hannover.
Marshall, W. (1785) *Planting and Ornamental Gardening.*
Marshall, W. (1803) *Planting and Rural Ornament,* 3rd Edn.
Marson, T. (1798) On the method of planting, as practised at Clumber, by the Duke of Newcastle. In: Lowe, R. (1798) *General View of the Agriculture of the County of Nottingham.* London.
Mastoris, S.N. and Groves, S.M. (1997) Sherwood Forest in 1609. A Crown Survey by Richard Bankes. *Thoroton Society Record Series* 40, Thoroton Society, Nottingham.
Mather, A.S. (1991) Pressures on British forest policy: prologue to the post-industrial forest? *Area* 23(3), pp. 245–253.
Mather, A.S. (1994) Policy reform and institutional restructuring: the case of the Forestry Commission. In: Gilg, A.W. (ed.) *Progress in Rural Policy and Planning.* John Wiley & Sons, London.
Merchant, C. (1987) The theoretical structure of ecological revolutions. *Environmental Review* 11, pp. 265–274.
Metaille, J.P. (ed.) (1992) *Protoindustries et histoire des forets (Les Cahiers de l'ISARD* 3), University of Toulouse, Toulouse.
Metz, W. (1954) Das 'Gahagio regis' der Langobardan und die deutschen Hagen-Ortsnamen. In: *Beitrage zur Namenforschung in Verbindung mit Ernst Dickenmann, herausgegeben von Hans Krahe,* Band 5. Winter, Heidelburg, pp. 39–51.
Milisauskas, S. (1986) *Early Neolithic Settlement and Society at Olszanica.* University of Michigan Press.
Millington, A.C., Mutiso, S.K., Kirby, J. and O'Keefe, P. (1989) African soil erosion – Nature undone and the limitations of technology. *Land Degradation and Rehabilitation* 1, pp. 279–290.
Mills, C.M. (1988) Dendrochronology in Exeter and its applications. Unpublished thesis for the degree of PhD, University of Sheffield.
Montoya Oliver, J.M. (1983) *Pastoralismo Mediterráneo.* Monografias 25, MAPA-ICONA, Madrid.
Moore, J.S. (ed. and trans.) (1982) *Domesday Book, 15, Gloucestershire.* Phillimore, Chichester.
Mooser, J. (1986) Property and wood theft: agrarian capitalism and social conflict in rural society, 1800–50. A Westphalian case study. In Moeller, R. (ed.) *Peasants and Lords in Modern Germany.* Allen & Unwin, Boston, pp. 52–80.

Moreno, D. (1990) *Dal Documento al Terreno. Storia, e Archeologia dei Sistemi Agro-silvo-pastorali.* Il Mulino, Bologna.

Myoraq, Y. (1992) Conference of the New Settlers in Jaffa 1894. *Cathedra* 66, 144–167. A note by Nyago on 162. (Hebrew).

Nairne, D. (1890–1891) Notes on Highland woods, ancient and modern. *Transactions of the Gaelic Society of Inverness,* 17, Inverness.

Neeson, J.M. (1993) *Commoners: Common Right, Enclosure and Social Change in England, 1700–1820.* Cambridge University Press, Cambridge.

Neilson, N. (1940) The forests. In: Willard, J.F. and Morris, W.A (eds) *The English Government at Work, 1327–1336,* Vol 1. Medieval Academy of America, Cambridge, Massachusetts.

Olwig, K.R. (1996) Environmental history and the construction of nature and landscape: the case of the 'landscaping' of the Jutland heath. *Environment and History* 2, pp. 15–38.

Page, W. (ed.) (1910) *The Victoria History of the County of Yorkshire.* Constable, London.

Page, W. and Willis-Bund, J.W. (eds) (1913) Worcestershire, 3. *Victoria County History.*

Paola, G., Barberis, G. and Peccenini S. (1991) *Pinus halepensis* formations in Liguria (NW Italy). *Botanika Chronika* 10, pp. 609–615.

Paola, G. and Minuto, L. (1996) Indigenous and exotic species as markers of the climatic limits of the Mediterranean Region in Liguria (North Western Italy). *Proceedings of the International Colloquium. Mediterranean: Climatic Variability, Environment and Biodiversity,* Montpellier 6–7 April 1995, Maison de l'Environnement, Montpellier, pp. 180–184.

Pearson, C.H. (1887) *Historical Maps of England During the Thirteenth Century.* London, 10.

Peluso, N.L. (1995) Whose woods are these? Counter-mapping forest territories in Kalimantan, Indonesia. *Antipode* 27, pp. 383–406.

Peterken, G.F. (1981) *Woodland Conservation and Management.* Chapman & Hall, London.

Peterken, G.F. (1993) *Woodland Conservation and Management,* 2nd edn. Chapman & Hall, London.

Peterken, G.F. (1994) The definition, evaluation and management of ancient woods in Great Britain. *NNA-Berichte* 3(94), pp. 102–114.

Peterken, G.F. (1996) *Natural Woodland. Ecology and Conservation in Northern Temperate Regions.* Cambridge University Press, Cambridge.

Pfeiffer, J.F. von (1781) *Grundriss der Forstwissenschaft, zum Gebrauch dirigierender Forst- und Kameralbedienten, auch Privatguthsbesitzern.* Mannheim.

Pickersgill, A.C. (1979) The agricultural revolution in Bassetlaw, Nottinghamshire 1750–1873. Unpublished PhD thesis, University of Nottingham, Nottingham.

Popper, K. and Eccles, J.C. (1977) *The Self and its Brain.* Springer-Verlag, Berlin.

Porter, W.S. (ed.) (1910) *Handbook and Guide to Sheffield.* J.W. Northend, Sheffield (for the British Association for the Advancement of Science).

PPRS (1914) *Publications of the Pipe Roll Society.* London.

Pretty, J.N. and Shah, P. (1994) *Soil and Water Conservation in the Twentieth Century: a History of Coercion and Control.* Research series No. I, Rural History Centre, University of Reading.

Price, U. (1794) *Essay on the Picturesque, as Compared with the Sublime and the Beautiful; and on the Use of Studying Pictures for the Purpose of Improving Real Landscape.* London.

Priestley, C. (1958) The life and career of Henry Fiennes Pelham-Clinton, 1720–1794. Unpublished MA thesis, University of Nottingham.

Prieto Rodríguez, A. (1995) Gestión forestal. *Revista Forestal Española,* No. 12, pp. 11–23.

Proyecto 03/29/0003, Cuarta División Hidrológico Forestal (1943) *Documento de Expropiación de Puebla de Valles.* Finca – Barranco del Lugar, Mego etc. Monte GU1004. Delegación de Agricultura, Guadalajara.

Pryor, S. (1992) *Future Forestry – A New Direction for Forest Policy.* Report prepared for Wildlife Link.

Purdum, J. (1978) Profitability and timing of parliamentary land enclosures. *Explorations in Economic History* 15, pp. 313–326.

Rackham, O. (1975) *Hayley Wood; Its History and Ecology.* Cambridgeshire and Isle of Ely Naturalists' Trust, Cambridge.

Rackham, O. (1976) *Trees and Woodland in the British Landscape.* Dent, London.

Rackham, O. (1980) *Ancient Woodland: Its History, Vegetation and Uses in England.* Edward Arnold, London.

Rackham, O. (1986) *The History of the Countryside.* Dent, London.

Rackham, O. (1986b) *The Woodlands of South-East Essex.* Rochford District Council, Rochford.

Rackham, O. (1989) *The Last Forest. The Story of Hatfield Forest.* Dent, London.

Rackham, O. (1990) *Trees and Woodland in the British Landscape,* 2nd Edn. Dent, London.

Radkau, J. (1983) Holzverknappung und Krisenbewußtsein im 18. Jahrhundert. *Geschichte und Gesellschaft* 9, pp. 513–543.

Radkau, J. (1986) Zur angeblichen Energiekrise des 18. Jahrhunderts. Revisionistische Betachtungen über die 'Holznot'. *Vierteljahrschrift für Sozial- und Wirtschaftsgeschichte* 73, pp. 1–37.

Radkau, J. (1996) Wood and forestry in German history: in quest of an environmental approach. *Environment and History* 2, pp. 63–76.

Radkau, J. (1997) The wordy worship of nature and the tacit feeling for nature in the history of German forestry. In: Teich, M., Porter, R. and Gustafsson, B. (eds) *Nature and Society in Historical Context.* Cambridge University Press, Cambridge.

Radkau, J. and Schäfer, I. (1987) *Holz. Ein Naturstoff in der Technikgeschichte.* Reinbek.

Rahtz, P. (1979) *The Saxon and Medieval Palaces at Cheddar.* British Archaeological Reports, British Series 65. Oxford.

Ramsay, J.H. (1925) *History of the Revenues of the Kings of England.* Oxford.

Repton, H. (1790) *Plans, Hints and Views for the Improvement of Welbeck in Nottinghamshire, a Seat of His Grace the Duke of Portland.* Welbeck Record Office.

Repton, H. (1791) *Thoresby Park in Nottinghamshire, a Seat of Charles Pierrepont, Esq.* NUMD Ma 4P21.

Repton, H. (1793) *Welbeck in Nottinghamshire, a Seat of His Grace the Duke of Portland.* Welbeck Record Office.

Revill, G. and Watkins, C. (1996) Educated access: interpreting Forestry Commission Forest Park Guides. In: Watkins, C. (ed.) *Rights of Way: Policy, Culture, and Management.* Pinter, London, pp. 100–128.

Richards, P. (1981) Appraising appraisal – Towards improved dialogue in rural planning. *IDS Bulletin* 12(4), pp. 8–11.

Rico Boquete, E. (1995) *Política Forestal e Repoblacións en Galicia (1941–1971).* Servicio de Publicacións e Intercambio Científico, Universidade de Santiago de Compostela, Santiago de Compostela.

Rocheleau, D. and Ross, L. (1995) Trees as tools, trees as text: struggles over resources in Zambrana-Chaucey, Dominican Republic. *Antipode* 27, pp. 407–428.

Rodgers, J. (1908) *The Scenery of Sherwood Forest with an Account of Some Eminent People Once Resident There.* Fisher Unwin, London.

Rooke, H. (1790) *Descriptions and Sketches of some Remarkable Oaks, in the Park at Welbeck..., a Seat of His Grace the Duke of Portland. To Which are Added, Observations on the Age and Durability of that Tree. With Remarks on the Annual Growth of the Acorn.* Nichols, London.

Rooke, H. (1799) *A Sketch of the Ancient and Present State of Sherwood Forest in the County of Nottingham.* Tupman, Nottingham.

Rubner, H. (1965) *Untersuchungen zur forstverfassung des mittelalterlichen Frankreichs.* Wiesbaden.

Rubner, H. (1967) *Forstgeschichte im Zeitalter der industriellen Revolution.* Berlin.

Rupin, A. (1937) *Thirty Years of Building in Eretz Israel.* Jerusalem, p. 42. (Hebrew).

Russell, E.W.B. (1997) *People and Land Through Time. Linking Ecology and History.* Yale University Press, New Haven and London.

Ryle, G.B. (1969) *Forest Service – The First Forty-five Years of the Forestry Commission of Great Britain.* David & Charles, Newton Abbott.

Sagasta Azpeitia, J.M. (1979) Coste social de las inversiones forestales. *Montes,* 193, ETSIM, Madrid, pp. 175–182.

Sagl, W. (1993) Dimensionen der Nachhaltigkeit. In: *20 Tagung der Fachgruppe Wald – und Holzwissenschaft 27 und 28 Okt.* Universität für Bodenkultur, Wien, pp. 1–20.

Salzman, L.F. (1952) *Building in England Before 1540.* Oxford.

Savill, P.S and Spilsbury, M.J. (1991) Growing oaks at closer spacing. *Forestry* 64, pp. 373–384.

Sawyer, P.H. (1968) *Anglo-Saxon Charters: An Annotated List and Bibliography.* Royal Historical Society, London.

Schäfer, I. (1992) *'Ein Gespenst geht um': Politik mit der Holznot in Lippe 1750–1850. Eine Regionalstudie zur Wald- und Technikgeschichte.* Detmold.

Schama, S. (1995) *Landscape and Memory.* Harper Collins, London.

Schenk, W. (1996) *Waldnutzung, Waldzustand und regionale Entwicklung in vorindustrieller Zeit im Mittleren Deutschland. Historisch-Geographische Beiträge zur Erforschung von Kulturlandschaften in Mainfranken und Nordhessen.* Stuttgart.

Schmithüsen, J. (1934) *Der Niederwald des Linksrheinischen Schiefergebirges. Ein Beitrag zur Geographie der Rheinischen Kulturlandschaft.* Bonn.

Schwind, W. (1983) *Der Wald in der Vulkaneifel in Geschichte und Gegenwart. Übernutzung der Vulkaneifel bis 1840.* Göttingen.

Searle, J. (1850) Leaves from Sherwood Forest. In: White, R. (1875) *Worksop, 'The Dukery', and Sherwood Forest.* Simpkin Marshall, London, pp. 242–244.

Selter, B. (1995) *Waldnutzung und ländliche Gesellschaft. Landwirtschaftlicher 'Nährwald' und neue Holzökonomie im Sauerland des 18. und 19. Jahrhunderts.* Paderborn.

Seymour, S. (1988) Eighteenth century parkland 'improvement' on the Dukeries estates of north Nottinghamshire. Unpublished PhD thesis, University of Nottingham.

Seymour, S. (1989) The 'Spirit of Planting': eighteenth-century parkland improvement on the Duke of Newcastle's north Nottinghamshire estates. *East Midland Geographer* 12, pp. 5–13.

Seymour, S. (1993) The Dukeries estates: improving land and landscape in the later eighteenth century. *Transactions of the Thoroton Society of Nottinghamshire* 97, pp. 117–128.

Sherrat, A.G. (1965) Hayman Rooke, F.S.A. – An eighteenth century Nottinghamshire antiquary. *Transactions of the Thoroton Society of Nottinghamshire* 69, pp. 4–18.

Shilony, Z. (1990) *Jewish National Fund and Settlement in Eretz Israel 1903–1914.* Yad Izhak Ben Zvi, Jerusalem, pp. 71–96. (Hebrew).

Shore, T.W. (1988) Ancient Hampshire forests. *Hampshire Field Club and Archaeological Society* 1, pp. 40–60.

Simmons, I.G. (1993) *Interpreting Nature. Cultural Constructions of the Environment.* London.

Simon, C. (1993) *Environmental History Newsletter*, Special issue No. 1 (1993): Umweltgeschichte heute: Neue Themen und Ansätze der Geschichtswissenschaft – Beiträge zur Umwelt-Wissenschaft.

Simpson, W.G. (1996a) *Ely Cathedral: The Nave Roof Archive Report.* Unpublished report, Historic Buildings Research Unit and Tree-Ring Dating Laboratory, University of Nottingham for English Heritage.

Simpson, W.G. (1996b) Documentary and dendrochronological evidence for the building of Salisbury Cathedral. In: Keen, L. and Cocke, T. (eds) *Medieval Art and Architecture at Salisbury Cathedral* (BAA Conf. Trans., 17), pp. 10–20.

Simpson, W.G. (1996c) Master-builders: fresh research on cathedrals and other medieval buildings by the Historic Buildings Research Unit. In: Wilson, R.J.A. (ed.) *From River Trent to Raqqa: Nottingham University Archaeological Fieldwork, 1991–95.* University of Nottingham, Nottingham, pp. 87–92.

Simpson, W.G., Howard, R.E. and Guilding, E. (1994) *The Survey, Dating and Analysis of the Roof of the Nave, Worcester Cathedral.* Unpublished report, Historic Buildings Research Unit and Tree-Ring Dating Laboratory, University of Nottingham for English Heritage.

Sissons, F. (1888) *Beauties of Sherwood Forest. A Guide to the Dukeries and Worksop with Many Maps and Illustrations.* Sissons, Worksop.

Smith, A.H. (1970) *English Place-Name Elements, Part 1.* English Place-Name Society, Vol. 25. 2nd Edn. Cambridge University Press, Cambridge.

Smout, T.C. (1993) Woodland history before 1850. In: Smout, T.C. (ed.) *Scotland since Prehistory: Natural Change and Human Impact.* Scottish Cultural Press, Aberdeen, pp. 40–49.

South Yorkshire Forest (1994) *Forest Plan.* SYF Project, Sheffield.

Speechley, W. (1775) Account of the plantations upon the estate of his Grace the Duke of Portland, by Mr Speechly, gardener to his Grace. In: Lowe, R. (1798) *General View of the Agriculture of the County of Nottingham.* London, pp. 57–69.

Speechley, W. (1779) *A Treatise on the Culture of the Pineapple and the Management of the Hothouse.*

Speechley, W. (1790) *A Treatise on the Culture of the Vine.*

Speechley, W. (1820) *Practical Hints on Domestic Rural Economy, with an Appendix Containing Several Original Agricultural Essays.*

Spencer, J.W. and Kirby, K.J. (1992) An inventory of ancient woodland for England and Wales. *Biological Conservation* 62, pp. 77–93.

Squires, A. and Jeeves, M. (1994) *Leicestershire and Rutland Woodlands Past and Present.* Kairos Press, Newtown Linford.

Stenton, F.M. (1971) *Anglo-Saxon England,* 3rd Edn. Clarendon Press, Oxford.

Steven, H.M. and Carlisle, A. (1959) *The Native Pinewoods of Scotland.* Oliver & Boyd, Edinburgh.

Sutherland, L.S. (1958–1978) *The Correspondence of Edmund Burke.* Cambridge.

Swanton, M.J. (ed. and trans.) (1996) *The Anglo-Saxon Chronicle.* Dent, London.

Tamames, R. (1976) *Estructura económica de España. Vol. 1: Introduccion, sector agrario.* Biblioteca Universitaria Guadiana, Madrid.

Ternan, J.L. *et al.* (1996) Aggregate stability of soils in Central Spain and the role of land management. *Earth Surface Processes and Landforms* 21, pp. 181–193.

Thomas, K. (1984) *Man and the Natural World. Changing Attitudes in England 1500–1800.* Allen Lane, London.

Thompson, A.H. (1938) *The Premonstratensian Abbey of Welbeck.* Faber and Faber, London.

Thorn, C. and Thorn, F. (ed. and trans.) (1979) *Domesday Book, 6, Wiltshire.* Phillimore, Chichester.

Thorn, C. and Thorn, F. (ed. and trans.) (1982) *Domesday Book, 16, Worcestershire.* Phillimore, Chichester.
Thorn, C. and Thorn, F. (ed. and trans.) (1983) *Domesday Book, 17, Herefordshire.* Phillimore, Chichester.
Thorn, C. and Thorn, F. (ed. and trans.) (1986) *Domesday Book, 35, Shropshire.* Phillimore, Chichester.
Thorpe, B. (ed.) (1846) *Colloquium ad Pueros Lingua Latinae Locutiane Exercendis ab Ælfrico Compilatum, Analecta-Saxonica,* London.
Thorpe, B. (1865) *Diplomatarium Anglicum Ævi Saxonici.* London.
Tompkins, S. (1989) *Forestry in Crisis – The Battle for the Hills.* Christopher Helm, London.
Trinder, B.S. (1973) *The Industrial Revolution in Shropshire.* Phillimore, Chichester.
Tristram, H.B. (1863–1864) *The Land of Israel. A Journal of Travels in Palestine.* Copyright by Bialik Institute 1977, Israel.
Tubbs, C.R. (1964) Early encoppicements in the New Forest. *Forestry* 37, pp. 95–105.
Turberville, A.S. (1939) *A History of Welbeck Abbey and its Owners,* Vol. 2, 1755–1879. Faber & Faber, London.
Turner, M. (1986) Parliamentary enclosures: gains and costs. *ReFRESH* 3, 5–8.
Tyson, B. (1996) Some Cumbrian builders, 1670–1780. *Trans. C. & W. A. & A. Soc.* 96, pp. 161–186.
Usher, M.B. and Thompson, D.B.A. (eds) (1988) *Ecological Change in the Uplands.* Special Publication Number 7 of the British Ecological Society. Blackwell Scientific Publications, Oxford.
Venables, Rev. E. (1884) The Vicars' Court, Lincoln, with the Architectural History of the College and an Account of the existing buildings. *Associated Architectural Societies Reports* 17, pp. 235–250.
Vicario Di Tenda (1710) Lettera del Vicario di Tenda a Sua Altezza Reale concernente la vendita fatta dalla Communità di Pigna all'Abbate Sardi di San Remo di tre milla alberi per uso di barche et altri navigli da mare. Con uno stato delle qualità e prezzi di detti alberi esistenti nei boschi di Pigna. Archivio di Stato di Torino, materie economiche, caccie e boschi, 77, mazzo I.
Villares Paz, R. (1982) *La Propiedad de la Tierra en Galicia, 1500–1936.* Editorial Siglo 21, Madrid.
Wacquant, J.D. (1989) Towards a reflexive sociology – a workshop with Pierre Bourdieu. *Sociological Theory* 7, pp. 26–63.
Walpole, H. (1777) Letter to Lady Ossory, 24 August. In: Lewis, W.S. (ed.) (1965) *Correspondence,* Vol. 32 *The Countess of Upper Ossory,* pp. 374–375.
Warburg, O. (1948) *Warburg Book.* (Hebrew).
Watkins, C. (1981) An historical introduction to the woodlands of Nottinghamshire. In: Watkins, C. and Wheeler, P.T. (eds) *The Study and Use of British Woodlands.* Department of Geography, University of Nottingham, Nottingham, pp. 1–24.
Watkins, C. (1988) The idea of ancient woodland in Britain from 1800. In: Fabio Salbitano (ed.) *Human Influence on Forest Ecosystems Development in Europe.* ESF FERN-CNR. Pitagora Editrice, Bologna, pp. 237–246.
Watkins, C. (1990) *Woodland Management and Conservation.* David & Charles, London.
Weitz, J. (1970) *Forest and Afforestation in Israel.* Massada, Israel (Hebrew).
Welldon Finn, R. (1979) Wiltshire. In: Darby, H.C. and Welldon Finn, R. (eds) *The Domesday Geography of South-West England.* Cambridge University Press, Cambridge, pp. 1–66.
Wenzel, I. (1962) *Ödlandentstehung und Wiederaufforstung in der Zentraleifel.* Bonn.
West, J. (1964) The forest offenders of medieval Worcestershire. *Folk Life* 2, pp. 80–115.
Wickham, C. (1994) European forests in the early Middle Ages: landscape and land clearance. In: *Land and Power. Studies in Italian and European Social History, 400–1200.* British School at Rome, London, pp. 155–199.
Williams, M. (1989) *Americans and their Forests. A Historical Geography.* Cambridge University Press, Cambridge.
Williams, M. (1994) The relations of environmental history and historical geography. *Journal of Historical Geography* 20, pp. 3–21.
Williamson, D.M. (1956) *Lincoln Muniments.* Friends of the Cathedral, Lincoln.
Williamson, T. (1995) *Polite Landscapes: Gardens and Society in Eighteenth-Century England.* Alan Sutton, Stroud.
Willis-Bund, J.W. and Page, W (eds) (1906) *The Victoria History of the County of Worcester,* Vol. 2. Constable, London.
Witney, K.P. (1976) *The Jutish Forest: A Study of the Weald of Kent from 450 to 1380 AD.* Athlone Press, London.
Wren Society (1940) *Publications of the Wren Society* 17, pp. 76–77.
Wright, P. (1992) The disenchanted forest. *The Guardian,* Weekend Edition, 7 November 1992, p. 8.
Yeomans, D. (1992) *The Architect and the Carpenter.* RIBA, London.

Young, A. (1771) *The Farmer's Tour Through the East of England,* Vol. 1. London.
Young, A. (1799) *General View of the Agriculture of the County of Lincoln.* London.
Young, C.R. (1979) *The Royal Forests of Medieval England.* University of Pennsylvania Press, Pennsylvania/Leicester University Press, Leicester.
Zeuss, C. (ed.) (1842) *Traditiones possessionesque Wizenburgenses.* Kranzbühler, Speyer.

INDEX

Aaronson, Aharon 166, 168, 169
Abel, Wilhelm 85
acorn, as symbol 116, 121
afforestation
    participatory approach 191–196, 201–212
    as royal prerogative 11
    and Schlagwaldwirtschaft 84
    and soil and water conservation 8, 191–212
    in Spain 7, 182, 187–189, 191–212, *199*
    and Zionist Movement 3, 5, 165–178
    see also conifer plantations; pine; plantations
agriculture
    control 7
    and enclosure 5, 94, 117, 123, 125–126, 128–129, 143
    and forest clearance 1, 8, 11–12, 60
    and forests 4–5, 11–12, 15, 122–123
    improvements 119, 122, 125, 128, 143
    vs. timber production 64, 75, 80, 85, 86–88, 91, 194, 196
Amery, John 93
Appin of Dull, Highland estate 139–143, *139*, *140*, 144–151, *149*
Araujo, J. 195–196
Arbor Day, Spain 182–183, 186–187, 188, 189
Areses, Rafael 182, 188
aristocracy
    German medieval 11–12, 18
    and tree-planting 5, 116, 118
    see also Dukeries
Austria, sustainability and over-utilization 3, 4, 73–81
avalanches 74, 81
avenues, planting 102
Azpeitia, Sagasta 193

Baker, R. 194, 195
Baltic, and timber exports 43, 50
Bankes, Richard 1–2
bark
    peeling 62, 137, 142, 145, 147, 150–151, 152–153
    for tanning 63, 64, 144
barony courts, Scotland 3, 136, 137, 138–150, *149*, 151–152
beech (*Fagus sylvatica*) 64, 65, 88, 157, 161
Bennett, H.H. 194–195
Berger, P. and Luckmann, T. 215–216
Berman, M. 167–168
Bernáldez, González 193
Bietigheim Forest *16*, 17
Bilhagh Wood 93, 94–101, *99*, 102, 123

birch (*Betula pendula, B. pubescens*) 65, 94, 97, 121, 122, 123
Birkland Wood 93, 94–101, *98*, *99*, 102, 107, 110
Black Forest, place-name evidence 14, *14*
Blaikie, P. and Brookfield, H. 206
Board of Agriculture, report 118, 119, 122, 129
Bodenheimer, M. 168–169, 170, 171, 176–177
Bourdieu, P. 216–217, 224
Brandis, Dietrich 220
brecks 94, 118
Britain
    forest policy 4, 8, 218–228, *221*
    Georgian woodland management 115–132
    see also England; Scotland
Broome, John 45, *48*
Brown, Lancelot 'Capability' 126
building, demand for timber 5, 50, 63, 85, 117, 128, 145–146, 171
Burgers, T.F. 193
Burke, Edmund 108, 115–116

Campbell, Duncan 143
Carlowitz, H.C. von 84
Carter, James 110
Catalogue of Notable Trees, Spain 182, 183, 189
Catalogue of Protected Areas, Spain 183–186
cathedrals, English 3, 4, 39–51
Cave, T. and Wilson, R.A. 47–49
Cernea, M.M. 201–202
Chambers, R. 209, 211
charcoal
    and coppicing 6, 55, 60–63, 67, 70
    in Dukery estates 125, 128
    industry 5, 50, 60–63, 80
chase, right of 34, *35*, *36*, 60
Civil War, destruction of forests 49–50
Clarke, George 94–97, *99*, 101
Close Rolls, England 40
Clumber Park 101, 102, 118, 121, 126–131, *127*, *130*, *131*
Codorníu, Ricardo 182, 192
Coelho, C. *et al.* 197–298
Collins, E.J.T. 6, 55, 63
conifer plantations 2, 64, 68, 118, 123, 174–178, 193–204
    Britain 220, 221–222
    Germany 220
    see also masts; pine
conservation
    in Britain 219–228
    forests 75–80, 93, 162
    in Scottish Highlands 136–138
    soil 4, 191–212
    in Spain 181–189, 191–212

conservation *continued*
  timber supplies 44, 47, 60, 75, 79–80, 84–86, 90–91
  water 4, 191–212
Copley, S. 108
coppice 6, 123, 174, 178, 224
  decline 8, 55, 57, 63–67
  and 'Schlagwaldwirtschaft' 83
  in Scotland 144
  in south-west Yorkshire 55–57, 58, 60–67, 70
  with standards 8, 47–49, 60–62, 65, 67, 68–69
  see also spring wood management
Cowley, Abraham 116
Cowper, William 115
Cox, Richard, bishop of Ely 44
culture, and woodland 1, 7–8, 68, 91, 93, 175
Cummins, J. 20

Daniels, Stephen 7
deer-hunting 20–21, 25, 27, 45, 86, 102
deforestation
  effects 74, 79, 81
  extent 85–86, 165
  and forest income 35, *36*, 37
dendrochronology
  and English cathedral timbers 3, 39, 43
  and Sherwood oaks 106–107
dens 28, 32
Devereux, S. and Hodinott, J. 207, 210
Devon, and ecclesiastical estates 42
disease, and monoculture 176–177
Domesday Book 3, 20, 23, 25–28
  and south-west Yorkshire 57–60, *59*
Dorset, medieval forests 23–25, *24*
Druids, and oak trees 105, 107
Dukeries, Nottinghamshire
  Georgian woodland management 3, 115–132
  and Sherwood Forest 93, 101–107
  see also Clumber Park; Thoresby Park; Welbeck Park; Worksop Manor

ecology
  ecological revolution 84, 90–91
  and forest preservation 80, 88–90, 189
  historical 2, 33, 83–84, 223
  of Sherwood Forest 93
  see also environment; nature
Eddison, E. 110
Eifel region, Germany 83, 85–86, *87*, 88, 91
Ely cathedral
  timbers 41–42, 43
  and woodland management 39, 44
enclosure
  and agriculture 5, 94, 117, 123, 125–126, 128–129, 143
  and hunting 19, 20, 21–25
  and woodland protection 97, 100–101, 137, 151
engineers, forest 3, 181–189, 191–194, 201–202, 203

England
  cathedrals as historical sources 3, 39–51
  medieval forests and parks 3, 4, 19–29
  spring wood management 3, 55–70
  see also Britain; Dukeries; monarchy; royal forests; Sherwood Forest
environment
  and forest management 80, 181–182, 186–189, 203–205
  and society 37, 74–75, *75*, 83–84, 91, 153, 186–187, 201–206, 219
  see also ecology; nature
Essex, royal forests 34, 37, 136
Essex, James 50
estates
  Dukeries estates 118–132
  ecclesiastical, woodland management 44–45, *46*, 50
  improvement 115–118, 119, 131–132
  Scottish Highland 135–153
  and woodland management 2, 3, 5, 7, 101–107
Etinger, Aqiva 170, 171–172
eucalypts
  in afforestation of Palestine 166, 169, 170, 172, 173, 175–176
  in afforestation of Spain 182–183
Evelyn, John 101, 102, 106–107, 108, 116, 119, 121, 123
Exeter cathedral, timbers 41, 42–43
Exmoor, England 1, 19

Farey, J. 115
feld names 25
felling
  clear-felling 64, 220, 222
  control 44, 75–77
  in Dukery estates 121
  sectional 3, 6, 47, 83, 84, 88, 90
  selective 62, 161–162, 187
fencing
  and demand for timber 117, 123, 125, 129, 143–144
  quick 128–129
field-names 3, 11, 13, 18
fines
  and forest income 34, *35*, *36*, 37, 151, 152
  for timber cutting 137, 141–142, *141*, 144, 146–151, 152
fire damage 86, 137, 145, 147, 176–178
firewood 12, 63, 84, 85, 90, 96–97, 187
floods, damage to forests 74, 79, 81, 88
Forest of Dean 1–2, 34, 40, 45, 220–222
  and Crown Commissioners 95, 97
foresters
  census and firma 34, *35*, *36*
  on Highland estates 150–151, 152
  see also engineers, forest
forestry
  American School 7, 194–195
  French School 177
  German School 7, 177, 192–194, 220
  professionalization 2, 6–7, 64, 182, 220
  see also management; sustainability

Index

Forestry Commission 2, 70, 93, 218, 220, 222, 224–227
forests
    and administrative boundaries 2, 12–13, 18, 21
    communal 88, 90–91, 187, 203
    definitions 1–2, 4
    expansion 181–189
    high 64–67, 68, *68–69*, 79, 83, 220, 222, 224
    as indicator of fiscal property 11–18
    memorial 176, 188
    see also management; plantations; royal forests; woodland
'forst' 13–14, *16*, 17–18, 21
Foucault, Michel 217–218, *218*, *219*
Fowkes, D.V. 119
Frankish settlements
    and forest hunting reserves 21
    place-name evidence 11, 14, 17
fuel
    wood as 50–51, 60–64, 67, 85, 125, 162
    see also charcoal

game 4, 21, 45, 119, 121, 129
gardening, landscape 116–117, 119, 128
Garside, P. 108
geology, and forests 23, 27, 55, 58, 94
Germany
    place-name evidence for fiscal property 3, 4, 11–18
    'Schlagwaldwirtschaft' system 3, 4–5, 6, 83–91
Gil, N. 205, 207
Gilpin, William 108–109, 126
Gindel, I. 176, 177–178
goats, damage by 142, 144, 151
grazing
    effects 86, 97, 142–143, 151
    forest 2, 12, 34, 74, 158–159, 187
    heathland 25, 94
    woodland 5, 6, 7, 25, 47, 60, 62, 135–153
    see also pasturing
Greendale Oak, Welbeck 101–102, *103*, *104*, 105, 121
Groome, H. 193–194, 195
Guadalajara, conservation and reforestation 4, 196–212, *197*, *198*

habitus, and social construction of reality 216–217, *218*, *219*, 222
haga 21, *22*, 23–28, *24*, *26*, *29*
Hampshire, medieval forests 25–28, *26*, 34
Harley, Edward, 2nd Duke of Oxford 101–102, 113
hayes/hays 20–21, 25, 28
heathland 25, 93, 94
hedges
    and hunting 20, 21
    quick 5, 62, 128–129
Henry II, and forest income 33, 34
Henry III, and ecclesiastical building 40, 45
Henry VIII, and timber imports 43
Hermosilla, Martinez 193, 195
Herzel, T. 166, 167–168, 176

hill-forts, German 16–17
Hilton, R.H. 33
history
    approaches to 1, 2–4, 81, 135
    cathedrals as sources 3, 39–51
    names as sources 3, 11–18, 19–29
    oral 3–4, 205–212
Hohe Tauern National Park, Carinthia 73, 80, 81
Holmes, George 224, 225–226
hop industry, demand for timber 117, 125, 128
House of Lords, Select Committee on forestry 224–226
Hunsrueck forests 83, 85, 88, 90–91
hunting
    control over 4–5, 7, 25
    and forest income 7, 34
    and forest protection 25–27, 86
    royal forests 1, 11–12, 19, 20–28
    Sherwood Forest 102
Hwicce kingdom 20, 25, 27–28

improvement
    agricultural 119, 122, 125, 128, 143
    landscape 115–118, 119, 125, 126, 131–132
industrialization 5–6, 50–51
    and estate management 117
    and fuel supplies 60–63, 67, 85
    and sustainable forestry 77–78
Irving, Washington 109–110
Italy, Ligurian woodland management 3, 4, 6, 157–162

Jacques, D. 126
James, N.D.G. 7
John of England, and forest income 33, 34

Kent, Nathaniel 6
Kremnizki, J. 166

Laird, F.C. 108–109
land degradation, Spain 191–212
land use
    diversity 84–85
    intensity 2, 74, 76, 77
    see also afforestation; agriculture; hunting
landownership
    and access to woodland 136–153
    and environmental management 135–136, 187
    and management 27, 44, 94, 100–101, 115–117, 181, 192, 195, 203
    in Palestine 169, 175
    and Sherwood Forest 5, 100, 101–107, 110
    status and control 4–5, 115–116
landscape
    and afforestation 2, 80, 165, 169, 171, 173, 222
    belted 55
    cultural 91, 116
    of Dukeries 101–102, 116–117, 119–125, 228
    landscape gardening 116–117, 119

landscape *continued*
    and the picturesque 108–111, 119–122, 125
    potential 74
Laso Rhodes, A. 204
Lavee, S. 176
leah, in place-names *22*, 23–25, *24*, 27
Lefebvre, H. 217, *218*, *219*
Lincoln cathedral
    timber 39–40, *40*, *41*, 42, 43, 44, 50
    and woodlands 44–45, 47
Linnard, W. 61
Lowe, Robert 118, 119, 123, 128–129

Maceira, Garcia 193
management
    and commercial value 85, 137
    in Georgian woodland 3, 115–132
    high forest 64–67, 68, *68–69*
    and industrialization 5–6, 77
    landowner/tenant relationships 135–153
    and royal forests 95–97, 100–101
    scientific 1, 2, 6–7, 8, 90
    and timber preservation 44, 47, 75–80, 91
    and utilization conflicts 75–80, 85, 162
    see also coppice; Schlagwaldwirtschaft;
        spring wood management;
        sustainability
Marshall, William 119, 122
masts, timber supplies for 3, 6, 157–162
Menzies of Weem, Highland estate 3, 138–153,
    139
Merchant, Carolyn 84
Millington, A.C. *et al.* 205
monarchy
    and control of forest rights 4, 25
    German, and Crown property 11–12, 14–18,
        83
    income 3, 33–37, *35*, *36*, 44–45
    see also royal forests
monasteries, English
    dissolution 43–44, 118
    German 14–16
    as sources of forest and woodland
        history 39, 41–42, 43
    and spring wood management 61
    and wood pasture 60
monoculture 7, 88, 174, 176–178, 193–194, 222
Montoya Oliver, J.M. 194
moorland
    and agriculture 119, 128–129
    and tree-planting 19, 121, 123–125, 138, 222
Moreno, Diego 6

nature
    social construction 4, 8, 83–91, 182,
        215–228
    sustainability and over-utilization 73–81
    see also ecology; environment
Nature Conservancy Council 224–227
nemus/nemora 17–18
New Forest 2, 27, 34
    and Crown Commissioners 95, 97
Newcastle, 2nd Duke 101, 102, 109, 118, 126–131

Nottinghamshire
    and forest timbers 41, 44–45, 47
    see also Dukeries; Sherwood Forest
nurseries 7
    Britain 122, 129, 143, 221
    Palestine 167–170, 175
    Spain 187–188

oaks
    age 105–107
    bark 63–64, 150
    cultural meanings 93–111, 116
    and Druids 105, 107
    in ecclesiastical buildings 39–43, 49
    and the picturesque 108–111, 120, 122
    planting 65, 88, 120, 122, 123
    of Sherwood Forest 3, 8, 94, 95–105, *99*,
        103, *104*, 108–111, 119
    shortage 50
    value 117, 141
olive (*Olea europaea*) 157, 167–170
Ordnance Survey
    and Anglo-Saxon woodland 19
    and industrialization 50
    and Sherwood Forest 110
Ottisch, A. 74
over-utilization 73–81, 162

Palestine, afforestation 3, 4, 5, 8, 165–178
Panaleon, Foyo 193
Parker-Jervis, Roger 215
parks
    and ancient oaks 101–102, 107
    and landscape gardening 116
    in medieval England 19–29
    national 80, 181
    see also Dukeries
Parliament *see* regulation
pasturing
    and forest income 34, *35*, *36*
    in German forests 85, 90
    in Ligurian forests 159, 161–162
    rights 27, 45–47, 136
    in Scottish Highlands 4, 144
    in Sherwood Forest 97, 101
    in south-west Yorkshire 58–60, *59*
    see also grazing
patriotism, and trees 107, 116, 117–118, 123
Peterken, G.F. 215, 223, 224, 227–228
Pfeiffer, J.F. von 84
picturesque, the
    and estate improvement 119–122, 125
    and Sherwood oaks 108–111
Pierrepont, Charles 123–126
pine 50, 88, 159–62, 169, 173–175
    Pinus brutia 173–174, 177
    Pinus canariensis 174
    Pinus halapensis 7, 166, 173–175, 176–178
    Pinus pinea 170, 174, 177
    Pinus sylvestris 50, 64, 65
    see also conifer plantations
Pipe Rolls, and English royal forests 3, 33–34
pitwood 50, 63, 67, 117, 125

place-name evidence
  of coppicing 60
  of forests 3, 11–17, *13*, *14*, 23–27
  of hunting 19, 21, 25, *29*
plantations
  in Britain 2–3, 5, 7, 115–132, 218, 220–222
  in Palestine 5, 7, 166–178
  in Spain 181, 187–188, 194–196
  spread 8, 64, 68
  thinnings 67, 117, 122–123, 221
policy
  British 4, 215–228, *221*
  German *see* Schlagwaldwirtschaft
  Israeli 165–178
pollarding 6, 96, 101, 108
population increase, effects 1, 60, 73, 74, 85, 117
Portland, Duke of 100–101, 105, 110, 118, 119–122
power structures 1, 4–5, 8, 37, 102, 111, 115–116, 135, 205, 219–228
Pretty, J.N. and Shah, P. 194
Price, Uvedale 108, 109
productivity, and tree species 7
profit
  and royal forests 33–37
  and woodland management 137, 169, 171, 176, 178, 187, 193
property, fiscal, as indicated by 'forest' 11–18
protection *see* conservation
Pryor, S. 220

Rackham, Oliver 6, 25, 31, 39, 55, 57–58, 135, 223, 228
Radkau, J. 7
ramel/ramelia 62
Ramsay, J.H. 34
Rannoch, Highland estate 139–142, *139*, *140*, 144–151, *149*, 152
reafforestation *see* afforestation
recreation, forest and woodland use 2, 64, 68, 70, 80, 178
regulation
  in Austria 75, 77, 79–80, *79*
  in England 44, 45–47
  and Schlagwaldwirtshaft 83, 85–86, 90
  in Scottish Highlands 136–143, 145
  in Spain 181–182, 183–186, 187, 192–193
Repton, Humphrey 116, 119, *120*, 121, 125
Richards, P. 194
rides, forest 102, 109, 118, 129
rights
  communal 1, 4–5, 7, 90–91, 94–97, 100, 118
  pasturing 4, 27, 45–47, 136
Robin Hood, and Sherwood Forest 93, 109–111
Robinson, R.L. 220
Rodgers, J. 101–102
Rodríguez, Prieto 194
Rooke, Hayman 97, *98*, 102–107, *106*, 108, 109, 118–119
royal forests
  England 1–2, 4, 5, 118
    distribution 21–28, *22*, 34
    and ecclesiastical building 4, 40, 42, 53
    haga features 21, *22*, 23–28, *24*, *26*, *29*
    location 19–21
    and royal income 3, 4, 33–37, *35*, *36*

  see also Sherwood Forest
  Holy Roman Empire 11, 14–18
Rubner, H. 34
Rupin, Arthur 168–169
Ryle, G.B. 222

Sagl, W. 73
St Paul's cathedral, and forest timber 40–41
Salisbury cathedral, timbers 39, 40, 43
scaffolding, timbers 42–43, 63
Schama, Simon 7
Schenk, Winfried 85
Schlagwaldwirtschaft, Germany 3, 4–5, 6, 83–91
  and ecological revolution 90–91
  ecological and social consequences 88–90
  measures 86–88
  motives for implementation 85–86
Schlich, Sir William 220
Scotland
  central Highlands 3, 4, 5, 135–153
  and hunting 20
Scott, Walter 109, 110
Searle, January 110
Selter, B. 85
'set-aside' programme 191, 195, 202–204
Sherwood Forest
  ancient oaks 3, 8, 93–111
  and cathedral timbers 39–40, 53
  and ducal landowners 5, 7, 101–107
  and regeneration 97–100, 118–119
  and Robin Hood 93, 109–111
  and Rooke 102–107, *106*, 108, 109
  and royal forest 5, 7, 94–101
  surveys 1–2, 94–101, *99*, 105
shipping
  and demand for timber 44, 50, 96–97, 100, 116, 117–118
  in western Liguria 3, 4, 6, 157–162
shooting 129–131
silva minuta see coppice
silva modica 58
silva pastilis see pasturing
silvae (woods) 11, 15, 17–18, 23, 58
society, and environment 37, 74–75, *75*, 91, 186–187, 201–206, 219
soil
  conservation 4, 169, 171–172, 191–212
  erosion 74, 79, 81, 88, 171, 173, 181, 187
Soskin, Zelig 166
Spain
  conservation and reforestation 4, 7, 191–212
  and forest engineers 3, 181–189
species
  diversity 88, 121, 128, 129, 143, 174–175, 176–178, 183, 193, 196, 204
  native 65, 174, 183, 224
  non-native 7, 64, 65, 68–69
Speechley, William 119–123
Spence, Joseph 126
spring wood management 3, 55–70
  decline and extinction 55–57, 63–67
  and fuel supplies 60–63
  revival of woodmanship 68–70
  and wood pasture tradition 57–60

standards, coppice with standards 8, 47–49, 60–62, 65, 67, *68–69*, 83
Stenton, Sir Frank 19
Steward of the Forest Courts 95–96
sustainability 5, 161, 162, 219
    and industrialization 6
    and over-utilization 3, 73–81, *80*
    and Schlagwaldwirtshaft 84, 88, 90–91

tanning, and bark 63, 64, 144
tenants, access to woodland 44, 136–153
Ternan, J.L. *et al.* 200–201
terracing 7, 167, 194, 197–201
thinnings, uses 67, 117, 122, 123, 128–129
Thomas, Keith 7, 223–224
Thoresby Park 101, 102, 110, 118, 123–126, *124*
timber
    in ecclesiastical buildings 39–42, 44–45, 49–50
    and Exchequer income 34, *35, 36*, 45
    imported 43, 50, 220
    sale 12, 34, *35, 36*, 44, 67, 85, 97, 119, 159–161, 175
    shortage 50, 75–78, 85–86, 128, 136, 218, 220
    vs. agriculture 7, 64, 75, 80, 85, 86–88, 91, 194, 196
    vs. protection 84–85, 192–194
    see also building; conservation; scaffolding; shipping
Torrington, Viscount 126–128
trees
    cataloguing 1–2, 8, 49, 94–101, 159–161, 182, 183, 189
    cultural meanings 1, 93–111, 116
    cutting *see* fines
    deciduous 2, 68, 116, 118, 123, 177–178, 225–226
    fee 95–96, 100
    planting 115–118, 119–122, 126–129, 137, 143
    see also bark; nurseries; species

underwood 45, 50, 62, 96, 117
urbanization, and estate management 117
utilization
    and conflicting interests 74–75, 77–78, 203–205
    and management strategies 75–80, 85, 162

Verderers, Sherwood Forest 95–96

Walpole, Horace 118
Warburg, Otto 166–167, 168–169, 171–172

Warwickshire, medieval forests 31–32
water conservation 4, 84, 172, 191–212
wavers 61–62
Weald, English 20, 28
wealth, land as 115
Weitz, J. 165, 170, 174, 175, 177
Welbeck Park 41, 101, 118, 119–123, *120*
Welldon, Finn, R. 23
Wentworth estate, Yorkshire 6, 62–67, 68–70, *68*
Wheatley, Francis 129–131
whitecoal 61, 63
Wickham, Chris 4, 5
'Wildbann' 11, 13, 15, *15*, 17
Wiltshire, and medieval forests *22*, 23, 25
Wolfson, David 167, 168
wood-names 3, 11, 13, 18
woodkeepers *see* foresters
woodland
    access to 44, 75, 90–91, 136–153
    ancient, and forest policy 4, 8, 215, 219, 222–228
    conversion to forest 64–7, 68, *68–69*
    definition 2, 4
    and landscape gardening 116
    primary/secondary 223
    regeneration 19, 25, 64–65, 97
    and royal forest 23
    see also forests; management; pasturing; *silvae*; spring wood management
woodmanship, decline and revival 63, 64, 68–70
woodwards 45–47, 64, 94–97
Worcester cathedral
    estates 27–8, 45–51, *46, 48, 49*
    timbers 42, 45, 50
Worcestershire, medieval forests 20, 27–28, *29*, 31–32
Worksop Manor 101, 118, 121
Wright, P. 222
Wuerttemberg
    forest boundaries 12–13, *12*
    place-name evidence *13*

Yisrael, Miqwe 166
Yorkshire
    and fuel supplies 60–63
    and revival of woodmanship 68–70
    and rise of high forest 63–67
    and royal forests 34, 37, 41
    and spring wood management 3, 55–70, *56*
    wood pasture tradition 57–60
Young, Arthur 122–123, 126, 128

Zionist Movement, and afforestation 3, 5, 165–178

Compiled by Meg Davies (Registered Indexer, Society of Indexers)